T0364126

Bilbao's Modern Business Elite

The Basque Series

EDUARDO JORGE GLAS

Bilbao's Modern Business Elite

University of Nevada Press Reno Las Vegas

The Basque Series
Series Editor: William A. Douglass
A list of books in the series follows the index.

University of Nevada Press,
Reno, Nevada 89557 USA

Manufactured in the United States of America

Library of Congress
Cataloging-in-Publication Data
Glas, Eduardo Jorge, 1957–
Bilbao's modern business elite / Eduardo Jorge Glas.
p. cm. — (The Basque series)
Revision of the author's thesis (Ph.D.—Columbia University, 1993)
Includes bibliographical references and index.
ISBN 0-87417-269-1 (alk. paper)
1. Bilbao (Spain)—Economic conditions. 2. Businessmen—Spain—Bilbao—History. 3. Iron mines and mining—Spain—Bilbao—History.
I. Title. II. Series.
HC388.B55G58 1996
338.946'63—dc20 96-32403
CIP

The paper used in this book meets the requirements of American National Standard for Information Sciences—Permanence of Paper for Printed Library Materials, ANSI Z39.48-1984. Binding materials were selected for strength and durability.

First Printing
05 04 03 02 01 00 99 98 97
5 4 3 2 1

This book was published with the support of The Program for Cultural Cooperation Between Spain's Ministry of Culture and United States' Universities.

To the memory of my parents, Isaac and Isabel,

and to my brothers Manolo and Ricardo

Contents

Tables

Acknowledgments

This book is the culmination of a long process that started in 1983, when I entered the Graduate School of Arts and Sciences at Columbia University to pursue a Ph.D. in history. I defended my thesis in November 1992, and finally got the coveted degree in 1993. The book is a revised version of my dissertation.

Many people and institutions helped me immensely during those years. The Columbia History Department honored me with several fellowships. I would like to thank specially Professor Edward Malefakis, my dissertation adviser, for his help in finding sources to support my research. I am also grateful to him and his wife for treating me to a frequent flyer ticket so I could go to Spain to tie a few loose ends in my research in April 1991. I did most of the field research for this book during two prolonged stays in Spain in 1987 and 1988. Those stays were generously supported through fellowships from the Social Science Research Council (through grants from the Andrew W. Mellon Foundation and the Ford Foundation) and the Program of Cooperation between the Spanish Ministry of Culture and American Universities.

I wrote more than half of my dissertation while working for the Legal Aid Society (Criminal Defense Division) as an interpreter in the Bronx office from April 1989 to September 1991. I am grateful to my then-supervisor, Norma Connolly, who accommodated my needs by letting me work on a part-time schedule during the summer of 1990. Thanks to a Whiting Fellowship, I was able to write full-time and finish the manuscript during the 1991–92 academic year.

During the research phase of this project in Spain, many people offered me their hospitality and friendship. In Madrid, Ambrosio Aznar and Pilar Ceballos opened the doors to their houses, made me feel at home, and entertained me with hard-fought Trivial Pursuit games and excursions outside the city. Alfonso Otazu also regaled me with his generosity and encyclopedic knowledge of Basque subjects. I am deeply indebted to him for his constant encouragement and help.

In Bilbao, I owe special thanks to Enrique Ybarra. The spelling may confuse some readers, but Enrique is a direct descendant of the Ibarras, who play a prominent role in this book.[1] Enrique allowed me to analyze papers in an old office that had been the family's business headquarters in the nineteenth century, and that he was trying to convert into a museum. In that office, I found important documents related to the mining activities of the family. In addition, Enrique provided me with wills and estate inventories of several of his ancestors. All these documents have greatly enriched this book. Ana María Rivera, my housemate during my last months in Bilbao in 1988, has been an inspiration, as she maintains a busy and active scholarly life in spite of frail health. She kindly helped me collect data from the local archives while I was in Bilbao and even after I came back to New York. Manu Montero shared much of his vast knowledge of the industrialization of the Basque region; and Manuel González Portilla generously let me photocopy his copy of the Actas del Comité de Madrid de Altos Hornos de Bilbao, which he had used for his *La siderurgia vasca.*

Many people read some of the chapters of the entire manuscript and offered valuable suggestions. I benefited in particular from Professors Malefakis's and Paxton's comments. My classmates at Columbia, loosely associated in the Social History Group, also provided incisive critiques. My dear friend Kathryn Parlan deserves special praise, since she witnessed the project from its inception to its final form. I greatly appreciate her advice and encouragement. Jim Berg patiently enlightened me on the subtleties of English that befuddle me as someone who grew up speaking Spanish.

On both sides of the Atlantic, numerous archivists and librarians eased the difficulties of research. María Teresa Tortella and her staff at the Archivo del Banco de España were extremely helpful, as were Pilar González of the Archivo Histórico Provincial de Vizcaya and Miguel Cruzado of the Archivo Histórico del Banco de Bilbao. In New York, the staff of Butler Library at Columbia University, especially the interlibrary loan office, tracked some very hard-to-find books and articles for me.

I would also like to thank Professor William A. Douglass and the University of Nevada Press for publishing my work and improving its content through a thorough editorial process.

Finally, I wish to thank my family. I regret that my parents died before this book was even a project; I suspect that they would have been happy to see the family name in print. As always, my elder brothers Manolo and Ricardo have been unflaggingly supportive. My sister-in-law Gloria and my niece Elena also participated in this work, tutoring me in the intricacies of the spreadsheet program and treating me to so many home-cooked meals that I am forever indebted to them. Although Sasha came into my life after the manuscript was completed, I am confident that I will share many future projects with her.

Introduction

While the nineteenth century was a period of economic modernization for Northern Europe, the southern regions of the continent did not share equally in this development. Most economists and historians have used the words *failure* and *stagnation* to describe the process of industrialization in the Mediterranean countries.[1] Nevertheless, there were important regional exceptions. In Spain, the Basque country during the second half of the nineteenth century is an excellent example of a rapid and solid regional development within the context of a stagnant national economy. This study analyzes this period of Basque economic growth, focusing on its main protagonist: the local entrepreneurs.

The rapid economic growth of the Bilbao region was tied to the massive export of iron ore to Great Britain, which began in earnest after 1875 and peaked in 1899. The prosperity created by the mining industry allowed the local businessmen to diversify their regional economy, establishing modern metallurgical, paper, chemical, shipping, banking, and hydroelectrical companies. By the early twentieth century, the Bilbao area concentrated the bulk of Spain's heavy industry, the largest merchant fleet in the country, and some of the most powerful financial institutions in the Peninsula.

While this work analyzes in detail the mining boom and the opportunities that it created for the Bilbao businessmen, it also tries to establish a broader chronological and analytical framework to explain the success of the local entrepreneurs. In fact, it is argued in this study that the modernization of Bilbao's industries preceded the mining boom, and can be chronologically located around 1850. It was during those mid-century years that the traditional iron manufacturing methods used in the region were replaced. In 1849, the Santa Ana de Bolueta company lit the first blast furnace ever built in the Bilbao area, and in the early 1850s, the Nuestra Señora del Carmen factory became the first ever in the province to use mineral coal in its furnaces to smelt iron. These attempts to modernize the Vizcayan metallurgical industry were not without difficulty, but they preserved an uninterrupted tradition of manufacturing activities in the region. The mining boom did not create an industrial base from scratch, but built upon those early experiences.

Although other historians such as Manuel González Portilla, Emiliano Fernández Pinedo, Luis María Bilbao, and Manuel Montero have published several admirable studies about the economic growth of the region, the approach and the archival sources used in the present work have complemented, enriched, and sometimes questioned their interpretations of the industrialization of the Bilbao area. For instance, notarial documents and the private records of the Ibarra family have clarified the process by which a small number of businessmen appropriated the most productive iron mines. Similarly, those sources have helped to shed new light over an on-going debate about the profits generated by the mining industry and its overall contribution to the general economic development of the region.

In addition to studying purely economic factors such as the unprecedented international demand for the local ore during the last quarter of the nineteenth century, this work marshals cultural, social, and political arguments to account for the reasons behind the economic growth of the region. The focus on the entrepreneurs allows me to address issues which have not been discussed by Vizcayan economic historians. The study argues that, as much as favorable economic circumstances, social networks, inheritance practices, a strong work ethic, and other cultural traits also encouraged the industrialization process. Indeed, it is within this context that one must try to understand one of the most intriguing factors in the development of the region: while foreigners took the lion's share of the profits generated by the exploitation of mineral resources in other Spanish provinces, this was not the case in Bilbao, where the mines remained under the control of a dynamic group of local businessmen.

This work also differs from the current historiography in its departure from the widely accepted notion that the Bilbao businessmen opposed the provincial laws (*fueros*) and the local governing institutions because they hindered economic growth. In most historical studies, those laws and institutions are equated with an Ancien Régime that had to be destroyed to allow for the rise of the bourgeoisie and the development of capitalism. On the contrary, this book tries to show that the businessmen were not opposed to the provincial laws and institutions from which they derived tangible benefits. They welcomed certain changes in the provincial laws imposed by the Spanish central government, but they did not feel any kind of antagonism against the ruling local elite. Nor should they have, since the provincial government actually promoted economic development, thus belying the idea that it stood against business interests.

Although this study encompasses many subjects in order to explain the economic growth of the Bilbao area, it cannot claim to have exhausted all the factors that influence such a process. In fact, a very glaring omission is the lack of any discussion about the relationship between businessmen and workers. This particular omission should not be interpreted as an indifferent or uncaring attitude toward labor problems. Rather, the reason for the absence of this important subject is the existence of two books that have already treated the topic in a competent manner: Ignacio Olabarri's *Relaciones laborales en Vizcaya (1890–1936)* (Durango, 1978), and J. P. Fusi's *Política obrera en el País Vasco 1880–1923* (Madrid, 1975). Labor issues during the period 1850–75 remain terra incognita. Unfortunately, the archival sources used in this study did not provide enough information to allow a serious discussion of the subject.

The book is divided into six chapters. The first one is a general background of Vizcayan history, stressing the importance of the *fueros* and the related issue of local autonomy which set the Basque country apart from all other Spanish provinces. In addition, a brief discussion about the agricultural poverty of the region, and the traditional Vizcayan reliance on commerce and manufacturing sets an adequate framework to understand the post-1850 rapid development of the Bilbao area.

The second chapter analyzes the mining industry, explaining how a small group of businessmen acquired and retained the vast majority of the deposits prior to the export boom. After a description of the mining industry and the technological changes in metallurgy that caused the strong international demand for the Vizcayan ore, there is a quantitative evaluation of the wealth created by this activity.

The third chapter outlines the growth of the other sectors of the region's economy, emphasizing the dynamic behavior of the local entrepreneurs in the creation of new companies. It includes a calculation of the total capital invested in new companies during the period 1880–1900, concluding that the mining profits alone could not account for the rapid growth experienced during those years.

The fourth chapter looks into the social origins of the local entrepreneurs, and maps the changes in the composition of the business elite during a period of fifty years. It stresses the importance of Bilbao's merchants in leading the way from the traditional economic activities of the region to the development of the modern industries. In addition, the chapter shows that the business elite was a relatively open social group, whose diversified investments helped to ease the potentially different economic interests of the various Vizcayan industries.

The fifth chapter discusses how certain sociocultural Basque traits shaped the entrepreneurial behavior of the business elite. In particular, it argues that religious attitudes, social networks, family structure, education, inheritance practices, and lifestyles contributed to promote economic activities in the region.

Finally, the last chapter probes the connection between politics and the industrialization of the Bilbao area. It studies the direct participation of the local businessmen in the regional and national political institutions prior to and after the abolition of the *fueros* in 1876. It shows that the most active involvement of the businessmen in national politics did not take place until the 1890s, when tariff protection and the promotion of industries through government contracts became a rallying cry for Bilbao's business elite. The chapter also argues that the high degree of autonomy and the fiscal advantages that existed in the region even after 1876 improved the position of the local businessmen vis-à-vis their competitors in other Spanish provinces. A final section links the increased involvement of the businessmen in national politics to lobbying campaigns demanding tariff protection and the promotion of industries through government contracts.

Although for reasons of clarity, political, economic, and sociocultural themes are treated separately in each chapter, the main point of the study is that, only by considering all these factors at the same time, can one begin to understand the industrialization of the Bilbao area.

Finally, throughout the book the terms *businessman* and *entrepreneur* have been used interchangeably. The study does not rely on Schumpeter's well-known definition of entrepreneur as an innovator or promoter of new technologies. Rather, the word has been defined more generally as a person who organizes, manages, and assumes the risks of a business enterprise.

Chapter One

Historical Background

Although this book focuses on the rapid economic growth in the Bilbao region during the second half of the nineteenth century, the process cannot be isolated from the previous history of the Basque country. A long tradition of mercantile and manufacturing activities characterized the Bilbao area for many centuries prior to its modern industrial development. In part, these activities were encouraged by the region's physical environment. But if geography provided a basic incentive, social and political reasons reinforced that tendency. Basque politics and society were shaped by regional laws, known as *fueros,* and institutions that set the region apart from the rest of Spain. Basque law, for instance, ensured that local inhabitants controlled their mineral resources, whereas royal ownership was the rule in other Spanish provinces. Similarly, tax exemptions and a special customs arrangement that survived well into the nineteenth century set the Basque region apart from the rest of the country. These local laws created friction with the central government, which sought to increase its authority by diminishing regional rights. Although the *fueros* were curtailed in 1844 and eventually abolished in 1877, the Basque region preserved a large degree of administrative autonomy, which was used effectively to promote the local economy. Thus, to explain the complex history of the Basque region, this chapter focuses on the local natural environment and the *fueros,* showing how they shaped the political, social, and economic life of the Bilbao area.

Vizcaya's Environment

The Spanish Basque country is a small region located in the north-central sector of the Iberian Peninsula. It includes the provinces of Alava, Guipúzcoa, and Vizcaya (see map 1). This last province is the western outpost of the region. Its area of roughly 850 square miles makes it the second smallest province in Spain after Guipúzcoa. Vizcaya is compressed between the Bay of Biscay and two mountain ranges, the Cantabrian Mountains and the Pyrenees. The Cantabrian peaks on the southern border separate the region from Old Castile. Although this range presented a major obstacle to trans-

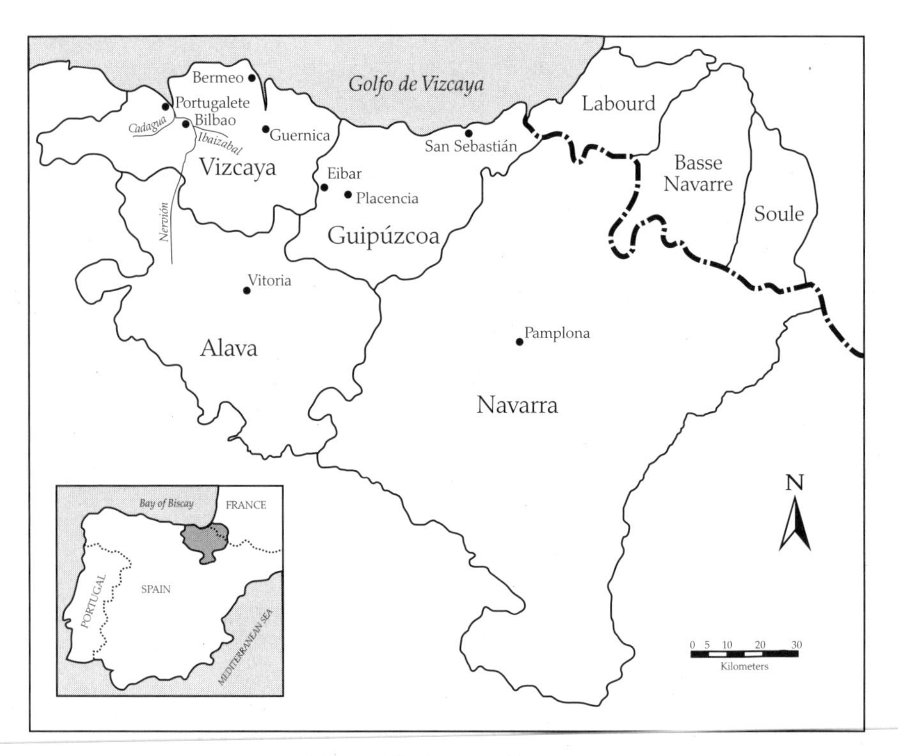

Map 1. The Basque Country

portation, some of the best routes to reach the interior of the peninsula were located in Vizcaya, since they were rarely blocked by snow in winter. The Pyrenees did not present as formidable an obstacle as did the Cantabrian Mountains. As the Pyrenees approach the sea, branching out into Guipúzcoa, Vizcaya, and neighboring Santander, they lose altitude.

As the etymology of the name suggests, Vizcaya is a mountainous region, with a median altitude of 1,500 feet.[1] Its terrain is extremely rugged, with narrow valleys and little arable land. Its rivers are short and, with the exception of the estuary of the Nervión, not navigable. Nevertheless, since the rivers meander on a mostly south-to-north course, their valleys served as the natural setting for the roads connecting the coast with the interior of the peninsula. In this sense, Bilbao's position was a privileged one because it was located at the convergence of the three most important rivers in Vizcaya: the Nervión, the Ibaizabal, and the Cadagua. Communication within the province, however, was difficult, since most valleys run parallel to one another. As a result, many of them were isolated from the main urban markets and relied on self-sufficient agricultural activities. This situation was at the root of a long conflict in Vizcaya's history that pitted modernizing urban centers like Bilbao against a very conservative countryside.

In spite of heavy rainfall, which creates a lush countryside and an illusion of fecundity, Vizcaya's soils are not fertile. Crops were not sufficient to feed the province's population; inevitably, wheat had to be imported to make up the region's deficit of cereals. During the eighteenth century, the widespread cultivation of corn, which adapted well to the temperate and humid climate, helped abate the grain deficit. By the mid-nineteenth century, the province was self-sufficient in corn, but still had to import about 50 percent of the wheat it consumed.

Although Vizcaya's soils yielded only meager wheat crops, they supported an abundance of chestnuts and fruit trees, especially apple trees. Yet, when agriculture was the main economic activity and wealth was measured in terms of cereal production, the abundance of fruit trees could not dispel the notion of Vizcaya as a poor province. This notion helped establish, in the first codification of local laws in 1452, the Vizcayans' right to import goods for their own consumption without paying any customs duties.[2] As a result, the Spanish customshouses were not located at the sea, but in the Ebro Valley.

Vizcaya did possess rich iron deposits and abundant forests, however. For centuries these two natural resources permitted the existence of a great number of ironworks, known as *ferrerías*. In fact, the founding of Bilbao in 1300 coincides with an increase in the use of iron in Europe.[3] The existence of the iron mines around Bilbao had been known since antiquity; references to their rich deposits appear in Pliny the Elder's *Natural History*. Little or nothing, however, is known about the exploitation of these mines or the production of the *ferrerías* until late in the Middle Ages.

The production of iron goods and their trade put Bilbao in contact with the main commercial centers of northern Europe. Some scholars estimate that the Basque share of Europe's annual iron production during the fifteenth century was as high as one-third of the total.[4] A large percentage of Basque production was exported to England, France, and the Low Countries. These exports usually took the form of pig iron, the basic material for other manufacturing processes.

Several reasons account for Vizcaya's large share of the European iron trade during the early modern age. Bilbao's mines were easy to exploit, since the ore was very near the surface. Transportation costs were low because the mines were close to the *ferrerías.* In addition, the purity of the mineral made Vizcaya's iron a much-sought-after commodity. A measure of its fame is given by Shakespeare's use of the word *Bilbo* (the Basque name for Bilbao) in several of his plays to indicate a special kind of sword. Similarly, in French the word *biscaien* became the name of an iron musket.[5]

The *ferrerías'* golden age seems to have spanned the early fifteenth to the late sixteenth centuries. The general decline in Spain in the seventeenth century hurt them badly. During the following century, they prospered once again. However, beginning in the 1790s and the early 1800s, they entered a severe slump from which they never recovered. The *ferrerías'* reliance on production methods that had changed little since the fifteenth century doomed the industry. Production became extremely expensive, and even their markets within Spain were seriously threatened by Swedish iron.[6] The *ferrerías* were replaced during the nineteenth century by modern factories, but their prolonged existence helped establish an uninterrupted manufacturing tradition in the region.

In addition to fueling the ironworks, Vizcaya's forests supported another industry that brought much fame to the region: shipbuilding. In general, the Basques undertook naval endeavors relatively late, remaining for centuries a landlocked people dedicated mostly to pastoral activities. According to Julio Caro Baroja, the impetus to develop naval enterprises came to the Spanish Basques through their French brethren, who had been influenced by some of the Middle Ages' greatest seafarers, the Normans.[7] Once the Vizcayans acquired the skill, their reputation as shipbuilders, sailors, and fishers became established all over Europe. They provided fleets and sailors for the expansion of Castile, first during the Reconquest and then during the discovery and colonization of America. As Fernand Braudel mentions, in the rough seas of the Bay of Biscay, European sailors honed the skills they used to conquer the world.[8] Among the Basques, the point is perhaps best illustrated by one of the greatest achievements in naval history: the first-ever circumnavigation of the globe by Sebastián Elcano, who assumed command of Magellan's expedition after the latter's death in the Philippines.

As shipbuilders, the Vizcayans benefited from their geographical position, a crossroads on the route linking the Mediterranean world with northern Europe. They adopted naval engineering practices from both regions, but they did not create any technological breakthroughs themselves.[9] In general, Vizcayan shipbuilding underwent cyclical fluctuations that paralleled the fortunes of the Spanish economy as a whole. Its glory days were during the fifteenth and sixteenth centuries. A profound crisis shocked it in the 1600s, as Spain lost its preeminence in world affairs. In the eighteenth century, there was a revival in construction until the wars of the French Revolution. After the political turmoil and the civil war that plagued Spain from 1796 until 1841, Basque shipyards enjoyed another wave of prosperity until the early 1860s. In the 1850s, Bilbao had the largest number of sailing ships in Spain, and its fleet's overall tonnage was second only to Barcelona's. This growth, however, represented the swan song of the wood-hull ship before the advent of the steamer and iron-hull vessels. Yet, as in the case of the *ferrerías,* the old shipyards established a tradition that continued well into the twentieth century.

Ships, iron, and sailors were three of the pillars on which Bilbao's merchant community built its wealth. Throughout the late Middle Ages, Vizcaya's merchants were actively involved with the main commercial centers of northern Europe. Just as merchants in other great commercial cities during that period did, Bilbao's merchants established permanent trade missions in Bruges, Bordeaux, and other harbors to secure a niche in international trade. In 1511, a royal grant permitted the formation of a merchants' guild, or *consulado,* in Bilbao, which regulated all commercial activities until its abolition in 1829. The power of this institution grew with the city's trade, and the guild participated in the construction of roads, improved the harbor, and built warehouses to enhance Bilbao's commercial capacity. The *consulado* proved so adept at managing all trade matters that its rules and regulations (*ordenanzas*) were copied all around the Spanish Empire and remained in effect until a national commercial code was enacted in 1829.

Between the fourteenth and the early nineteenth centuries, wool also accounted for a large part of Bilbao's commercial traffic. Unlike iron, wool was not an indigenous product, but a Castilian staple. Since the late Middle Ages, Castilian wool had fed the looms of the Low Countries, England, and France, where its quality sustained a steady demand until cotton replaced it as the main raw material for textiles. Early in its history, Bilbao entered the competition among northern Spanish ports to attract this profitable trade.

Geographically, Bilbao offered several advantages as an export point for the wool going to northern Europe. The city was located a few miles inland, on the banks of the navigable Nervión estuary, which afforded protection against the usually tempestuous Bay of Biscay. Its protected position was an advantage not only against the rough weather, but also against the pirates who prowled these coasts until the eighteenth century. In addition, Bilbao was located closer to Burgos (the Castilian center of the wool trade) than were other ports on the Vizcayan coast, a fact that, for instance, caused the decline of the more ancient port of Bermeo.[10] Bilbao's geographical advantage over Bermeo was reinforced in its charter, which ordered that commercial traffic between Bermeo and the Castilian plateau pass through Bilbao. In the same charter, Bilbao was also granted monopolistic trade privileges over the length of the Nervión River. Such privileges were highly resented and caused extensive litigation between Bilbao and other cities (especially Portugalete, located at the mouth of the Nervión). Yet, Bilbao skillfully defended itself and was able to stop the rise of any competitors within Vizcaya.

According to Luis Vicente García Merino, it was Bilbao's political skills more than its location that accounted for its preeminent position on the northern Spanish coast.[11] Although this argument is appealing, it, surprisingly, plays down geographic factors. True, Bilbao was not an ideal port. A sandbar at the entrance of the harbor could be passed only during certain tidal conditions. Big ships also had difficulty navigating along the winding Nervión toward Bilbao; cargo had to be unloaded and transferred to smaller boats that took it upstream. However, these disadvantages could not offset the overriding importance of the city's proximity to Burgos, since land transportation was much more expensive than that over waterways, even when ships had to wait for tides or transfer their loads to smaller boats. In addition, Bilbao's iron assured another cargo to shippers calling at this port. No other port on Vizcaya's coast could offer the variety of products and proximity to Burgos at the same time. Without these advantages, Bilbao could not even have begun to exert political influence.

Despite its political influence and geographic advantages, Bilbao was unable to stop the rise of its biggest competitor on the Cantabrian coast, Santander.[12] Santander managed to gain an important share of the wool traffic because of the way this trade was structured. For most of its history, until the seventeenth century, the wool trade to northern Europe was not dominated by the port cities, but was directed from Burgos, a gathering place for wool production from different parts of the country. While Burgos dominated the trade, it had the ability to play off the different Cantabrian ports against each other, directing traffic to those offering commercial concessions. Competition among the ports became so keen that, for instance, Bilbao's proposal in 1685

TABLE 1

Comparison of Bilbao's and Barcelona's Populations, 1700–1877

Year	Bilbao	Year	Barcelona
1700	5,946	1717	34,000
1797	10,943	1787	111,410
1842	10,243	1842	121,815
1856	17,923	1856	188,787
1877	32,734	1877	272,481

Sources: For eighteenth-century Bilbao, M. Mauleon Isla, *La población de Bilbao,* pp. 60, 74; for nineteenth-century Bilbao, M. Basas, *Economía y sociedad bilbaína,* p. 327; for Barcelona, G. McDonogh, *Good Families of Barcelona,* p. 21.

to build a cart road over the major mountains between the city and Burgos was delayed by the opposition of San Sebastián, Vitoria, and other cities for more than fifty years.[13] Thus Bilbao never dominated the wool trade. Like the rest of the ports, it lent its harbor facilities and its fleet, yet the largest profits went to Burgos's merchants.

As Burgos declined in the seventeenth century, foreign merchants gained the upper hand in the wool trade. In the eighteenth century, after long and debilitating wars that hampered Spanish trade, Bilbao's merchants overtook the foreign competition. Although the reasons for their success during the eighteenth century have not been well studied, it is clear that they stopped being mere shippers. Instead, they went to Castile to buy the wool and to trade it on their own account.[14] Yet, they never managed to monopolize this trade because the Spanish crown started to promote the export of wool from Santander for political and fiscal reasons.

In spite of its relatively developed industrial and commercial life, Vizcaya was not very urbanized. Its cities had populations that in other Spanish provinces were matched by large villages. Even Bilbao, the undisputed commercial center of Vizcaya, had a surprisingly small number of inhabitants until well into the nineteenth century. The difference with Barcelona, for instance, is remarkable (see table 1).

According to J. Caro Baroja, the lack of great urban centers adversely affected the industrial development of the region.[15] Yet it would seem more appropriate to view the problem the other way around. Since Vizcaya was traditionally a densely populated province and, like the rest of the Basque country, a land of emigrants, industrial and commercial activities could not absorb the population, which was forced to leave a countryside that was unable to feed them.

This Malthusian problem was not overcome until the nineteenth century, when new technological developments permitted an unprecedented rate of commercial and industrial growth. Before that time, the expansion of Castile provided great opportunities for alleviating population pressure in Vizcaya. Large numbers of Basque emigrants staffed the Castilian bureaucracy in the peninsula and in America. The image of the Vizcayan as a bureaucrat was so widespread in the sixteenth century that Cervantes made fun of it in *Don Quixote.*[16] In America, Basques occupied important positions in the colonial administration and economic life. For instance, in Chile during the sixteenth century an unusually high number of governors had Basque origins.[17] In 1634, the bishop of Santiago complained about a clique of Vizcayan merchants who allegedly dominated the Chilean trade.[18] In Mexico, Basques conquered the northern territory, "as Durango's colonial name of New Biscay bore witness." In the eighteenth century, their predominance over Mexican commerce was so strong that it received official recognition when the *consulado* created two parties, the Basque and the Montañés, from which guild authorities were selected.[19] Although perhaps not as dominant as their position in Mexican commerce, there was nevertheless a strong Basque presence among the membership of the *consulado* of Cádiz, the merchants' guild that monopolized colonial trade until late in the eighteenth century. In the 1730–1823 period, the three Basque provinces accounted for almost 14 percent of all the registered merchants in that *consulado,* although their share in the total Spanish population was between 2.5 and 3 percent.[20]

On both sides of the Atlantic, Basques associated in confraternities or religious groups that helped them preserve their cultural identity and maintain their links with their homeland.[21] In these confraternities, piety and business were intertwined in an interesting fashion. D. A. Brading points to a Calvinist-like ethic among Basque merchants in Mexico, who led an austere life and came to believe that their careers were justified by material success.[22]

Whatever the ideological motivations to succeed, it is evident that Vizcaya's agricultural poverty presented a Toynbee-type challenge to its inhabitants. For some, commerce and industry within the province became the response. Others had to emigrate. But whatever their activity and wherever they were, Vizcayans retained a strong collective identity. To a large extent, the *fueros* came to symbolize the essence of this identity. As such, they played a very important role in Vizcaya's history.

The *Fueros* and Local Institutions

The Vizcayans' involvement in Castilian enterprises contrasted noticeably with the loose constitutional links that tied the Basque provinces with Castile. These loose links were the result of the protracted war against the Moors and its influence on the process of state formation in Spain. The burden of the Reconquest first fell on the northern inhabitants of the peninsula, who stopped the Moors' advance at the Asturian Mountains. From the early stages of the war, a series of independent regions emerged: Galicia, Asturias, the Basque provinces, Navarre, Aragón, and Catalonia. These regions spoke different languages and developed different laws.[23]

Unlike the language of these other regions, the Basque tongue did not have a Latin root. In fact, the origins of the Basque language remain a mystery after two centuries of research. The only conclusion that all researchers accept is that Basque is not an Indo-European language.[24] As clouded as the origin of the language is the early history of the Basques. The three provinces never united to form an independent state. For centuries, the only common link among the inhabitants of the three Basque provinces was their shared language.

During the eleventh and twelfth centuries, the Basque provinces wavered in their allegiance to the surrounding kingdoms of León (later Castile) and Navarre, as these two battled for the hegemony of the Christian territories. In the thirteenth century, Vizcaya joined the kingdom of Castile, which by then had become the largest and most powerful Christian state. After a long process involving similar unions, the Castilian kings became the rulers of all the component regions and kingdoms that constituted Spain. Yet, far from becoming a unified state, Spain was a federation of kingdoms that preserved their local laws and institutions. Known as *fueros,* these local laws limited the power of the Castilian monarchs in such matters as taxation and military recruitment. The *fueros* were particularly strong in the Basque provinces, Navarre, and Aragón.[25]

The *fueros,* then, were not unique to the Basque region. In fact, they were first used in Castile. Castile's *fueros,* however, never applied to large regions; their jurisdiction was confined mostly to towns. As a result, they did not foster any strong sense of regional identity. According to Stanley Payne, it was after their widespread use in Castile that the *fueros* were applied to the Basque region.[26] Roger Collins, while acknowledging this fact, has speculated that Castile's *fueros* might have had a Basque origin. His case rests on two arguments. First, when the *fueros* appeared in Castile, the region was heavily populated with Basque settlers. His second argument is tied to the meaning of the word *fuero.* As defined by the medieval legal code of *Las siete partidas,* the *fueros* were customs and traditions that were later sanctioned as written laws. According to Collins, the Basques were the only people in the peninsula who maintained a legal system based on custom; the rest adopted the Roman example of codified laws. Although plausible, Collins's case is based on circumstantial evidence; there are no hard facts to support it.[27]

If in theory the *fueros* had their origin in custom, they could, in practice, also represent a royal grant given to encourage the population of newly conquered areas or the development of urban life and commerce. Whether the *fueros* symbolized a contract between the Castilian crown and the Basque people as two equal partners has been the center of a political debate between those in favor of regional autonomy (or even independence) and advocates of Castilian centralism. The first group stresses the customary origin of the *fueros,* with the implication that they preceded the union of the region with Castile, whose kings could rule Vizcaya only as long as they respected the local laws. The advocates of centralism point out the similarities of the Basque *fueros* and those granted earlier in Castile and emphasize the aspect of a royal concession instead of the idea of a contract between equal partners. It is not my goal to settle this debate, an endeavor that might prove futile in any case, since both sides seem to have strengths and weaknesses. What is important is that the debate has existed for at least three centuries and that it has shaped the political and economic life of Vizcaya.

Many scholars agree that not all aspects included in Vizcaya's *fueros* had their origin in custom. For instance, one of the most important privileges established in the *fueros* was the universal nobility of the region's native population. This privilege appeared for the first time in the second compilation, the so-called Fuero Nuevo, of 1526. It is highly doubtful that such an important matter could have been inadvertently omitted in the first compilation, or Fuero Viejo, of 1452. In addition, the Fuero Viejo refers to the existence of a social cleavage between nobles and commoners at the time of its compilation.[28]

The privilege of collective nobility had few parallels in Europe. It strengthened the Vizcayans' sense of collective identity and promoted ethnic pride.[29] Although the reasons for establishing collective nobility are murky, its appearance in the Fuero Nuevo coincides with the defeat of Basque feudal chieftains (*banderizos*) by an alliance forged between the towns of the region and the Castilian monarchy.[30] As a result, the balance of power within Vizcaya was altered. The crown asserted its authority over urban governments and helped redefine the provincial governing institutions. Vizcaya's ancient nobility lost a large measure of its social and political authority, while the rural population became one of the socially most free in Spain. According to Alfonso Otazu, it was during this period of change that the Basque concepts of universal nobility and egalitarianism were forged to justify the rise to power of the social classes that had challenged the old nobility.[31]

In addition to that ideological justification, the concept of universal nobility had other important socioeconomic consequences within and outside Vizcaya. Implicit in the concept was the notion of "purity of blood" (*pureza de sangre*), that is, the idea that a person had no Moorish or Jewish ancestors. The proof of purity of blood was a sine qua non for those who aspired to noble status. In regions that had been under Moorish domination, such proofs required expensive genealogical investigations. Since Vizcaya was never conquered by the Moors, Vizcayans used their history to obtain recognition that they all had *pureza de sangre*.[32]

To preserve the claim and avoid the settlement of non-nobles and non-Christian people, Vizcaya's Fuero Nuevo established stringent rules for aliens who wished to reside in the province.[33] In general, these residency restrictions were applied to curb economic competition with the native Vizcayans. For instance, in eighteenth-century Bilbao, the rules were used to prevent large numbers of foreign merchants from settling down and taking business away from the locals.[34]

Furthermore, the egalitarian tendencies implicit in the idea of universal nobility did not foster the aristocratic prejudices against commerce and industry that were so common in Castile during the Ancien Régime.[35] To the Basques, these activities were vital, since their lands alone could not support an aristocratic way of life. Not surprisingly, Vizcaya's social elite participated in the major economic enterprises of the region.

Outside Vizcaya, universal nobility facilitated the careers of emigrants in Castile or America by permitting them entrance to administrative offices or army and navy positions reserved for the aristocracy. Even those engaged in more plebeian occupations could enjoy privileges like the right to be judged by a special tribunal for crimes committed outside the province.[36]

In addition to universal nobility, the *fueros* established the Vizcayans' exemption from direct royal taxation, from import duties within their province, and from military service outside Vizcaya. Their exemption from Castilian taxes did not necessarily mean that they did not contribute at all to the royal treasury, but their contribution took the form of a donation, not an obligation. After a formal request from the king, Vizcaya's governing bodies negotiated the payment of a lump sum, which they were in charge of collecting. For the central government, this system had many drawbacks. In negotiating the payments, the Vizcayans could delay matters and demand concessions from the central government. As a result, taxes within the province remained low, especially when compared with those of other Spanish regions. A further limitation to the crown's authority established in the *fueros* was the so-called *pase foral* (*foral* is the adjectival form of *fuero*), which granted Vizcayans the right to reject royal legislation that opposed *foral* regulations and laws.

The *fueros* also provided the Vizcayans a number of civil rights, such as protection against arbitrary arrest and security of the home from unwarranted searches. Moreover, they included many dispositions regulating property rights and inheritance matters, which had an origin in Vizcayan customs.[37]

In addition, the *fueros* set regulations regarding the main natural resource of the region: the iron mines. For centuries, they reserved the exploitation of these mines exclusively to Vizcayans. This regulation had important consequences in the later development of the region. Unlike what happened in other Spanish regions, where foreigners exploited natural resources, the huge profits made from the iron mines remained mostly in Vizcaya.[38]

The *fueros* thus emphasized privileges and traditions that separated Vizcaya from Castile while establishing links between the two regions. The last section of the Fuero Nuevo, for instance, explicitly recognized Castilian laws as the norm to follow in all matters left unregulated by *foral* dispositions.

The institutions with which the Vizcayans governed themselves were not described in the *fueros*. Nevertheless, they, too, were the result of traditional practices whose historical origin is difficult to determine. The highest institution was the provincial general assembly, or Junta General, which had convened at Guernica since the fifteenth century. This assembly gathered the representatives of the different territorial units into which Vizcaya was divided. The basic division separated the so-called *tierra llana* from the cities and towns.[39] The *tierra llana* was in turn divided in *anteiglesias*, or rural villages. Unlike these villages, the cities and towns were not completely ruled by the provincial *fueros*, but by special *fueros* granted at the time of their founding. This created a dual system of legislation within the province, since most of the cities came under the jurisdiction of Castilian civil law. This duality and some privileges granted in the urban *fueros* gave rise to numerous lawsuits involving jurisdictional authority between the cities and the *tierra llana*. For a long period between the late fifteenth and the mid-sixteenth centuries, the cities were barred from participating in the Junta General, and not until 1630 did they obtain full representation with voting rights in this assembly. Yet, even after their full incorporation into the assembly, cities like Bilbao were severely underrepresented, with the same number of votes as any rural village of the *tierra llana*.

The Junta has been hailed as the embodiment of Basque egalitarianism and democracy. Part of this argument rests on the idea that universal nobility precluded the formation of a society based on the three traditional European orders—the nobility, the clergy, and the third estate. The Junta indeed eschewed the orders as its constituting principle. Yet, it was not a democratic institution in a twentieth-century sense. Aside from the underrepresentation of city dwellers, the Junta adopted regulations that disenfranchised large sectors of Vizcaya's population. Although there is no good sociological study of the members of the assemblies, a general consensus among historians is that a landed oligarchy came to dominate this institution. Progressively, by restricting the eligibility of candidates through property qualifications, this oligarchy gained control of the Junta, and of the municipal government, which were in charge of selecting the representatives for the assembly. Nevertheless, to install such a political system in the seventeenth century was a precocious achievement for Vizcaya, if one remembers that in Spain, a similar system did not become prevalent until the 1830s, and even in France such a system was only established by the Revolution of 1789.

In the fifteenth century, the Junta decided to delegate power to administer the region to a smaller body, which eventually came to be known as the Diputación General. As a result of this delegation of power, the assemblies were convened less frequently, usually once every two years. Even then, they assembled only for short periods, during which they selected the members of the new Diputación, reviewed the business of the preceding administration, and dealt with matters pertaining to the economic and social life of the province.

The Diputación was headed by two general deputies who had to surmount an even higher barrier of landed wealth than did the members of the Junta. Given the rivalry between Bilbao and the rest of the province, it is strange that the permanent seat of the Diputación was established in that city. The Junta, however, took some precautions to avoid the city's domination of the Diputación, by forbidding the simultaneous holding of municipal and regional offices.

The Diputación's functions included the collection of taxes, the organization of provincial defense, the maintenance of public order, the construction of roads and other public works, the preservation of the *fueros* against any sort of breaches, and the convocation of the Junta General every two years. In addition to reviews by the Junta of the work of the Diputación, the deputies' actions were checked by an official, the *síndico,* who had the power to abrogate acts that were deemed to be contrary to the *fueros.*

Both the Juntas and the Diputación were under the supervision of an official of the central government. Until the nineteenth century, this official was known as the *corregidor;* after 1841, he became the civil governor. During the fifteenth century, this official was instrumental in the pacification of the region, which had been devastated by feudal wars, the so-called *guerras de linajes.* In addition, the early *corregidores* had a strong influence in shaping provincial laws and political institutions.[40] As the representative of the central government, this official assumed political, administrative, and judicial functions. Politically, he presided over the provincial assembly and had to work closely with the Diputación, making sure that the rights and laws of the government in Madrid were not violated. Judicially, his office could act as a first-instance tribunal and as an appellate court. Administratively, his most important function was repressing contraband, a common activity in the region.

The relationship between the *corregidor* and the Diputación was sometimes tense. Despite his supervisory powers, local governing bodies could leave the *corregidor* in the dark on many issues. Indeed, many complained that their attempts to gather statistical information on economic activity were always resisted by the Diputación. In general, Vizcayans were averse to cooperating in such matters because they feared that an exact accounting of the region's resources might lead to higher taxes.[41]

As mentioned earlier, Vizcaya's municipalities elected their representatives to the *Juntas Generales.* Unfortunately, very little else is known about the municipal institutions. Even in Bilbao's case, despite Teófilo Guiard's four-volume history of the city, municipal politics remains a mystery. An elected mayor assisted by several officials called *regidores* ruled Bilbao. Until the *foral* system was altered in 1841, persons engaged in retail commerce or as artisans could not hold municipal office. It is unclear, however, whether those forbidden from holding office could participate in the election of officials. Given the importance of commerce in Bilbao, it would be a fair assumption to think that big merchants played a key role in the city's government. Yet, Bilbao was not a Venice, ruled exclusively by merchants. The power of big urban and rural proprietors who resided in the city rivaled that of the merchants.[42] The interests of the proprietors and the big merchants usually coincided, however. Moreover, the line between these two social categories was not distinctly drawn. Throughout the eighteenth and early nineteenth centuries, the merchants' guild and city government worked together to improve the roads connecting Bilbao with Castile's interior and to upgrade the port facilities.

In general, contemporary observers tended to praise the Basque system of government. While the rest of the country suffered the consequences of rule by a ponderous central bureaucracy, Vizcaya's institutions acted promptly to solve regional problems. Even enemies of Vizcayan privilege saw local institutions as a model that could be applied to the rest of Spain.[43]

Vizcaya's Society and Castile's Centralizing Policies

During the seventeenth century, the loose constitutional ties linking the different regions of the peninsula started to tighten. The expensive foreign policies of the Habsburg dynasty squandered colonial treasure from the Americas. Castile, the Habsburgs' other important source of revenue, collapsed under the double burden of heavy taxes and inflation. The Habsburgs thus decided to make all the peninsular regions share the cost of their empire. Not surprisingly, the regions with *foral* laws resisted strongly. Catalonia seceded from Castile from 1640 to 1652. The Vizcayans also revolted in the 1630s to protest the centralizing policy, but they were handled more adroitly than were the Catalans and remained loyal to Castile.

The Habsburgs' attempt to centralize was a great failure. In the eighteenth century, their successors, the Bourbons, were much more successful. Catalonia, Aragón, and Valencia lost their *fueros* as punishment for supporting the Bourbons' rival during the Spanish War of Succession. The Basque provinces and Navarre, which supported the Bourbons in that war, preserved theirs. As Spain entered the eighteenth century, these two regions were the only reminders of the confederation character of the kingdom.

Although the Bourbons at first respected the Basque *fueros,* subsequent policy reversed that position. A major point of contention between the Vizcayans and the crown throughout the eighteenth century and the early decades of the nineteenth was the location of the customshouses. In 1717, the crown attempted to transfer customshouses from the frontier with Castile to the coast, but strong opposition in the provinces led to a reversal of this policy. For the crown, the issue was not just to assert authority for political reasons, but also to curb an always flowering illegal trade that deprived the royal treasury of revenue. The fiscal privileges of the Basque region made it an ideal large-scale warehouse for goods later smuggled into Castile.[44]

The royal government adopted stern measures to stop the illegal trade. As a result, Basque commerce, which had been free from supervision, came under the increasing scrutiny of the crown. This policy created friction between the royal administrators and the merchants, who perceived royal interference as a violation of their freedom to trade as guaranteed by the *fueros.*

In the mid-eighteenth century, claiming that the Basques had unfair trade advantages, northern Castilian ports like Santander bitterly complained against Basque fiscal privileges. The crown, sympathetic to these complaints, sought to divert trade away from Bilbao by promoting the development of Santander. The government in Madrid financed the construction of a new road between Santander and the Castilian plateau and granted tax incentives for the export of wool from that city. Santander was also permitted to trade with the Americas during the late eighteenth century. In contrast, Bilbao, which had requested permission to enter the American trade, was denied the franchise. To make matters worse, the central government imposed a tariff on Basque iron, flour, and leather as Basque goods entered Castile. Faced with this duty, Vizcaya's tanneries and flour mills closed down. The effects of the duty on the iron industry were also severe, although less crippling than in the other two cases because it did not depend on imported raw materials.[45]

Napoleon's invasion of Spain devastated Vizcaya's economy and its political autonomy. Although the French never officially abolished the *fueros,* Vizcayan liberties in essence ceased to exist. Military conscription was imposed; commercial exemptions were terminated; local institutions like the *Diputación* became puppets of the French military authorities; and heavy taxes crippled Vizcaya's business.[46] In this context, it is not surprising that the Basques violently resisted the French occupation.

The advent of liberalism and constitutional government in Spain also proved inimical to Basque privilege. In an attempt to establish a uniform legal system for the whole country, the Cortes of Cádiz abolished the *fueros.* The fight against liberalism in Vizcaya became associated with the desire to preserve the region's autonomy. In 1814, the return of Ferdinand VII ended the liberal government and restored the *fueros.* Soon, however, Ferdinand met the same problems that his ancestors had faced when he tried to levy taxes on and conscript men from the Basque provinces.[47]

Toward the end of the eighteenth century, as the *fueros* and local institutions came under increasing pressure from the central government, Vizcayan society split into two basic groups, the traditionalists and the reformists. In the historiography of the region, the conventional "villains" in the slow and progressive demise of the *fueros* are the wealthy provincial landowners and Bilbao's mercantile bourgeoisie. Like Esau, who gave up his primogeniture for a dish of lentils, these two groups allegedly sold out their "birthright" to advance their economic interests. In the first half of the nineteenth century, they supported reforms sponsored by the central government that threatened to end *foral* privileges, even though the great majority of Vizcayans favored preserving the *foral* system without any changes.

In the 1830s, the traditionalists' cause became associated with that of Don Carlos, the pretender to the Spanish throne in the seven-year civil war that followed the death of Ferdinand VII. The spark that started this war was a legal dispute over the succession rights of Don Carlos, a younger brother of the king, and those of Ferdinand VII's infant daughter, Isabel. The issue, however, was not just who would rule, but what kind of kingdom Spain would be. Carlism represented an extreme reaction against any political or social reform; it meant the preservation of absolutism. In contrast, to ensure the succession of his daughter, Ferdinand VII adopted a conciliatory policy toward reformers in the last years of his rule. After his death in 1833, his wife, the queen regent Cristina, continued this policy and, as the threat of civil war turned into reality, called on the moderate liberals to create a constitutional monarchy.

The *fueros* did not represent the central focus of Carlist ideology; religion and the absolute power of kings were its two main tenets. However, in the Basque provinces, Carlism became associated with the *fueros* almost by default. Bourbon kings had had problems with the *fueros,* but had never abolished them. In contrast, Spanish Liberals had shown a strong bias toward centralization and abolished the *fueros* during their two brief stints in power, between 1812 and 1814 and 1820 and 1823. Thus, Don Carlos's promise to uphold tradition was perceived in Vizcaya as the best hope to maintain the *fueros.*

As appealing and elegant as the explanation of the split in Vizcayan society may be, it does not answer many questions. The study of Vizcaya's social history is in its infancy. Further research is needed to clarify the political choices of the different social classes. Until this task is completed, one must bear in mind that many explanations are based on educated guesses. Because we lack the necessary empirical evidence, we must note the problems in the accepted explanations.

Until the mid-nineteenth century, the wealthiest landed proprietors were at the top of Vizcaya's social pyramid. They constituted a rather open elite, which had earned its fortune through commerce or the holding of official positions. Since the sixteenth century, this elite had progressively merged with or replaced the old medieval aristocracy. In the 1630s, one of King Philip IV's administrators pointed out this process in a letter to the count-duke of Olivares: "[In Bilbao, the wealth of the old lineages] has been transferred to less prestigious families, who have created their estates and fame by trading and marrying in the Indies."[48]

Once wealth was achieved, these nouveau riches acquired land and founded *mayorazgos* (entailed estates). The size of their properties was modest when compared with the large estates of Andalucía. Since land in Vizcaya was not very productive, many of these proprietors continued to be or became involved in industrial and commercial activities.

Vizcaya's landed elite had strong ties to the Castilian monarchy. Some held high noble titles from Castile; others were knights in prestigious Castilian military orders. These noble Vizcayans gained their titles by serving the crown in different functions in the administration, the army, or the navy. Emiliano Fernández Pinedo has noted that, as a result of the connection and because of pressure from the crown, this group willingly compromised regional prerogatives. The elite feared that if it refused such compromises, it would jeopardize the careers in the royal administration of second sons, who needed those positions because of the practice of primogeniture.[49] Interestingly, in the 1820s, a Castilian minister toyed with the idea of excluding the Basques from the central administration unless they agreed to pay higher taxes to the crown.[50] But if such considerations shaped the conduct of the elite within the province, Vizcayans in the Castilian bureaucracy seem to have acted differently. Some historians have argued that many Basque officials in the administration in Madrid helped advance their region's interests by constituting an informal lobby within the central government.[51]

Some members of the provincial landed elite favored political and economic reform. Many of them owned houses in the major Basque cities, where they spent long periods and kept abreast of the latest European trends. The French Enlightenment influenced the Basque provinces probably more than any other region in Spain. In the second half of the eighteenth century, members of the landed elite and some merchants from the major Basque cities founded the Real Sociedad Bascongada de Amigos del País. Like the French academies on which it was modeled, the Sociedad Bascongada sought to promote education and new techniques for the development of industry, commerce, and agriculture. It would be inappropriate, however, to perceive the elite as a thoroughly modernizing class. Despite their dabblings with French Enlightenment thinkers, they remained loyal Catholic believers and, regardless of their adoption of new agricultural techniques, they continued the traditional practice of entailing their lands well into the nineteenth century.

What *foral* reforms was this class seeking? As owners of *ferrerías,* many of the landed proprietors were interested in the abolition of the duties the central government had imposed on Basque iron that entered the Castilian provinces. The ruling project of the liberals involved the creation of a national market without barriers to internal trade, which meant the disappearance of customs like those separating Castile from the Basque provinces. Many owners of *ferrerías* favored this solution, even though a modification of the *foral* laws also meant that the Basques would lose their privileges to import duty-free goods once the customs frontiers had been established at the seacoast.

It has been suggested that another reform that attracted the wealthy provincial elite toward the liberals' position was the massive sale of municipal and clerical lands.[52] Throughout Spain, large landowners and merchants took advantage of the disentailment program devised by the liberals to buy property cheaply. Such sales, however, were not incompatible with the *fueros.* Indeed, the Basque municipalities' sale of communal land preceded the triumph of liberalism. It started in the eighteenth century and accelerated during the War of Independence. Nevertheless, it is possible that those who benefited from these early sales supported the liberal cause as a way to consolidate and further expand their holdings.

Thus, the economic interests of some wealthy landowners coincided in some cases with liberal reforms sponsored by the central government. But willingness to accept reform must not be confused with rejection of the *fueros.* In fact, the provincial elite had much to gain from preserving as much autonomy as it could, since the local governing institutions seem to have been under its control.

If the Carlist War provides a test for liberal or reactionary sympathies, the jury is still out regarding Vizcaya's landed elite. Contemporary observers of the war noted that the "immense majority" of landowners sided with the liberal faction;[53] however, more recent studies have adopted a more cautious view.[54] The presence of some very wealthy landowners among the Carlist leadership seems to contradict the remarks of contemporary observers. One historian suggests that those landowners who supported Carlism were more likely to live on the land or in small towns and were engaged primarily in agriculture.[55] Although not proven, this hypothesis would fit the general framework of rural-urban antagonism that has marked Vizcaya's history.

Below the group of wealthy proprietors, there was a class of middle and small landowners. Historians agree that the region was not a rural arcadia worked only by small proprietors, as Basque propagandists assert.[56] Still, land was more equitably distributed than in other Spanish regions, and small proprietors constituted a sizable number of the population. In general, little is known about this group. Scholars seem to agree that they provided much of the leadership of the Carlist movement. Some observers have speculated that their support of Carlism stemmed from the perception that they were being squeezed out of the countryside by the larger proprietors and city-based merchants.[57] Property became concentrated in the region, but it remains to be proved that it was at the expense of these smaller landowners.[58] Nevertheless, their economic condition probably worsened in the first decades of the nineteenth century. Highly dependent on the income from their land, the deflation of agricultural prices that followed the Napoleonic wars must have lowered their standard of living.

To these lesser landowners, liberal reforms could not have been an appealing prospect. They must have seriously opposed the transfer of customs-houses to the coast. Since they were heavily dependent on imports for their consumer goods, such a move would be a severe blow to their shaky finances; and since their property assured them participation in the region's political life and administration, they had a high stake in preserving local autonomy. Nor could they take much advantage of the sale of communal and clerical land because they probably did not have enough capital to enlarge their holdings by such means.

If the class of middle and small landed proprietors provided the leadership for the Carlist movement, rural tenants and wage laborers constituted its rank and file. Tenants outnumbered day laborers by a large margin. Many of them held long-term leases that amounted to virtual ownership of the land. Unlike in southern Spain, where the rural population lived in large agro-towns, Vizcaya's peasantry was scattered throughout the region in hamlets or isolated farmsteads. The typical farm was a small holding of six to ten hectares, usually worked by a single family. To avoid subdivision of the land, local laws and customs required that the lease or the property be passed on to only one child in the family. Despite this practice, the size of many farms decreased after the mid-eighteenth century because of considerable population growth. Property rights continued to be inherited by one child, but farms were split into smaller operating units.[59]

The vast majority of the peasantry was engaged in subsistence agriculture. Their earnings had to be supplemented by income from communal lands or through artisanal labor. Thus it was not uncommon for many peasants to work as miners, woodcutters, charcoal burners, muleteers, or iron forgers during the agricultural off-season.

The plight of the peasantry worsened during the second half of the eighteenth century. Rents increased sharply because of the combined effect of population growth and a limited supply of arable land. The Napoleonic wars left behind an impoverished countryside. Supplemental income from communal lands eroded as these properties were sold by the villages to pay off debts incurred during the wars.

Life in the Basque countryside was imbued with a set of values directly opposed to the individualism characteristic of liberalism. Despite the geographic isolation of the Basque farms, strong community ties and neighborly customs governed many rural tasks. Some of these customs were completely informal; others found a legal and institutional expression in rural confraternities that regulated the division of work and the resources of the communal land.[60]

Rural confraternities, with their patron saints, reflected a key characteristic of the Basque countryside: its fervent Catholicism. Religion was present in practically every aspect of the peasants' lives. The rural priest had a strong ascendancy in the countryside. Usually he held an unchallenged ideological monopoly in the isolated Vizcayan villages, where he was the only educated person. The church could not sustain such a privileged position in cities, where it had to compete with the secular ideas of educated people.[61] Furthermore, it perhaps cannot be overemphasized that a language barrier accentuated the cultural isolation of the peasantry. Basque was the language of the countryside, while Castilian was the tongue of urban, secular people. Rural priests, recruited from the same peasantry to whom they preached, had no trouble communicating the traditional message of the church. In Vizcaya, unlike in other regions, this traditional message was not undermined by the existence of an extremely wealthy clergy. The local church was neither a major landowner nor a great tithe collector.

Liberalism had nothing to offer the peasantry. Peasants had good reasons to resent the sale of communal lands, and in areas close to Bilbao, they resented the purchase of land by merchants and other urban dwellers. During the war, however, the peasantry made no attempt to occupy land or destroy property records, as the French had done in 1789. Undoubtedly, the worsening economic conditions of the countryside created tensions, but they do not by themselves explain the peasantry's support of the Carlists. Don Carlos's slogan of "God, Country, and King" appealed to their traditional way of thinking.[62]

There is no doubt that many Carlists believed that they were fighting a crusade against the heretical Liberals. Many Liberals were themselves religious persons, but they wanted to reduce the role of the church in the economy and politics. For the peasants, such ideas were a direct attack on their way of life. Since such heresies originated in the cities, country priests found it easy to exacerbate traditional urban-rural antagonism. In addition, Carlist emphasis on the divine right of kings mirrored the peasants' "view of the world as inherently sacred and directly governed by divine providence."[63]

Without achieving the status of a religious icon, the *fueros,* too, were a key element in the traditional way of life of the Basque peasantry. They gave them a sense of identity and pride in their unique privileges. The *fueros* also granted them material advantages, like the exemption from the military draft and a better standard of living because of the low taxes. For the Basque peasant, the Carlist message of support for Spain's constitution was associated with a defense of the *fueros.*

In the traditional historiographic view, Bilbao and other urban centers are supposed to have been bastions of liberalism. Many historians of the Carlist War have remarked on the rural versus urban character of the struggle. The territories held by each side during the conflict tend to confirm this view. At their peak, the Carlists overran the countryside without opposition, while the Liberal strongholds were the cities.

Without challenging completely this view, more recent studies have shown a more diverse political outlook within Vizcaya's towns. Barahona asserts that support for the Carlist cause among urban artisans was overwhelming. In his view, most ironworkers "would eventually find themselves squarely in the Carlist camp."[64] First, he claims that there existed a class antagonism that led the artisans to join the side opposite the "liberal" merchants and landowners who controlled the iron trade. Further discontent stemmed from the fact that many of these artisans might have been recently displaced peasants. Third, Barahona claims that artisans opposed the Liberals' attack on the traditional guild system.

The last reason, however, does not seem very relevant to the Vizcayan case, since the guild system was virtually nonexistent in the province. Moreover, other historians have perceived a Liberal sympathy among ironworkers based on their need to protect the industry by transferring customs to the coast.[65] Yet, as logical as this explanation seems, it is still not the answer. Ironworkers fought on both sides, as the examples of the Guipúzcoan towns of Eibar and Placencia seem to suggest.[66] Any strong conclusion about the political attitude of the artisans remains elusive until further research is done.

Similarly confusing is the situation of clerks, lawyers, and notaries. Although not numerous, this group was very influential in public life. According to some historians, these professional groups sided with Don Carlos. The reasons why they might have supported Carlism are simple in the case of the local bureaucrats. Their job was at stake if centralization was carried out thoroughly. In addition, lawyers and notaries considered themselves "legal guardians and interpreters of the Fueros and of tradition."[67] As such, they opposed the Liberal proposal to homogenize the Spanish legal system. There is some empirical evidence that supports this view. For instance, in Bilbao, the Liberals dismissed numerous employees of the Diputación because of their Carlist inclinations.[68] A contemporary observer noted that the lawyers and notaries sided predominantly with the Carlists.[69] Citing a report by Carlist agents in charge of surveillance of suspected Liberals in several Vizcayan towns, John Coverdale finds that "the number of scribes, clerks, and lawyers whom the authorities considered disaffected from the Carlist cause is sufficiently large in comparison to their weight in the population to suggest that this group inclined disproportionately toward liberalism."[70] Once again, the Carlist litmus test fails to explain the political choices of a social group.

The urban group that supposedly became the strongest supporter of liberalism in Vizcaya was Bilbao's mercantile community. Yet, even this "safe" assumption has been undermined lately. A contemporary liberal merchant commented in a letter in 1834 that "the Carlists had 'put down roots' in the [Bilbao] junta of commerce."[71] With the exception of a few well-known names, which are quoted in most studies as the anointed ideologists of the bourgeoisie, it is surprising how little is known about Bilbao's merchants. There is obviously a problem with sources, because most merchants left no written record of their political views. As a result, historians have attributed their support of liberalism to their economic needs. Yet, these needs were complex and did not necessarily require jettisoning the *fueros,* as has sometimes been argued.[72]

After the 1750s, the central government adopted measures designed to erode Bilbao's trading advantages. The government hoped that a strategy of sticks and carrots would end the problem of having a region outside its fiscal control that was a sanctuary for smugglers engaged in highly profitable illegal trade with Castile. First came the sticks: the government tried to divert a large part of the wool trade to Santander; Bilbao was denied permission to trade directly with the American colonies; and some restrictions on the transfer of money from Castile to the Basque provinces created obstacles to Bilbao's commercial prosperity. If the customs privileges were given up, rewards would follow: the restrictive measures would be lifted, and Bilbao could obtain the much-coveted franchise to trade with the colonies.

Despite such measures, Bilbao prospered in the second half of the eighteenth century. Its municipal government, its *consulado,* and Vizcaya's Diputación responded to the challenge of Santander's ascension. In 1764, these institutions financed the construction of the Pancorbo Road, connecting Bilbao with the Castilian interior. This road seems to have at least doubled commercial traffic and may also have helped boost Bilbao's population growth, which between 1760 and 1790 increased nearly 40 percent.[73] By almost any measure, the Pancorbo Road was an effective response to the road constructed by the central government between Santander and Burgos.

In facing the challenge from Santander, Vizcaya's government, the municipality of Bilbao, and its *consulado* managed to put aside the traditional antagonisms that had pitted the city against the rest of the region. This harmony, however, did not last long. In 1777, when the *consulado* requested permission to trade directly with the Americas, the Vizcayan government did not back the petition because it was trying to reconcile the *fueros,* the merchants' desire to expand trade, and the fiscal requirements of the central government.[74] The Vizcayan government feared that, if accepted, the petition would lead to a taxation of imported goods, a direct violation of the *fueros.*

The situation remained unchanged, but a tug-of-war between Bilbao's merchants and the rest of the province continued. In the 1790s and early 1800s, in opposition to Bilbao's preeminent commercial position, some provincial notables sponsored projects to build ports that would be under the control of the Vizcayan government and outside the jurisdiction of the *consulado.* The most famous project was that of Simón Zamacola, who wanted to build a port in Abando, directly across from Bilbao, on the banks of the Nervión River. Interestingly, to increase the chances of the project's success, he sought the support of the king's prime minister, Manuel Godoy, who was willing to grant permission for it in return for promises regarding the contribution of Vizcayan soldiers to the Spanish army.

Whether these promises violated the *fueros* is not clear. Nevertheless, this ambiguity was cleverly exploited by the Bilbao notables, who, this time, championed the *foral* cause. They spread the rumor that Vizcaya's military privileges were about to be violated. The population responded by rising against the provincial government, which was discussing Zamacola's project. After Castilian troops quelled the uprising, the project fizzled out in the chaotic political situation at the turn of the century.

The position of Bilbao's merchants was threatened on two fronts. Within Vizcaya, a group of rural notables, influential in the regional government, resented the city's preeminent position in the province. Outside Vizcaya, the central government was seeking to throttle the region's commerce unless local privileges were given up. The merchants required a firm hand at the rudder to navigate between this Scylla and Charybdis. Despite pressure from the central government, the merchants held steadily to their *foral* privileges. One reason may have been that they feared the wrath of the great majority of Vizcayans, who wanted to keep the customs frontier inland.[75]

Yet, the *fueros* were not a complete liability for the merchants. In fact, they extracted tangible benefits from them. For instance, few government officials supervised their activities, so it was difficult to assess their income for tax purposes. *Foral* autonomy also benefited them. The *consulado* financed maintenance and improvement of Bilbao's port facilities by taxing the harbor's commercial traffic. In addition, since the local merchants did not pay import duties at the port, they had the advantage of requiring less capital for their commercial transactions than did their colleagues in the rest of Spain. Finally, despite the prohibition against reexporting certain products to Castile, the capacity of a duty-free zone to promote trade should not be underestimated.

Bearing in mind the advantages that the *fueros* offered the merchants, the *consulado*'s proclamations in the *fueros'* favor should be viewed as something more than Machiavellian endorsements to assuage the animus of the traditionalists. In 1717, when the newly established Bourbons ordered the transfer of the customs frontier to the sea, the head of the *consulado* wrote: "Supposing that the piety of the King allows us to choose between respect for the *fueros* or the transfer of commerce, or [between] the shift of the wool trade to Santander or the establishment of customs, it was less inconvenient to lose the profits that trade might generate than the honor involved in respecting the fueros."[76] In 1828, the central government offered to allow Bilbao to trade directly with the American colonies. Once more fearing that the offer implied a change in the customs situation, the *consulado* rejected it. It is possible that the conservative Diputación forced the hand of the *consulado.* However, the circumstances in 1828 were much different from those in 1777, when the *consulado* requested the trade franchise with the Americas. By the 1820s, the Spanish colonial empire in America was irremediably lost. True, Cuba and Puerto Rico remained attached to the metropolis, but despite their rich sugar plantations, what were these two markets compared to almost a continent? Furthermore, who could ensure in the 1820s that after the pan-American independence movement, the two Caribbean islands would not follow the same path? Thus, if the franchise meant a quid pro quo for giving up customs privileges, it is not surprising that the merchants in 1828 answered the king by saying that they, "like all other Vizcayans, your loyal vassals, appreciate their *foral* laws above all mercantile profits."[77] In this context, too, it is unlikely that the Diputación had to strain itself to convince the merchants.

Nevertheless, the bargaining position of the Bilbao merchants was not one of strength. During the first half of the nineteenth century, the city's trade declined from the peak in the second half of the 1700s. As in the rest of Spain, Bilbao's merchants suffered as a result of the Napoleonic wars and the independence of the American colonies. The postbellum situation did not favor their two main exports, wool and iron.

The decline of the wool trade has traditionally been attributed to the technological changes introduced by the Industrial Revolution. When the textile industry first mechanized, cotton replaced wool as the main raw material in the production of clothing. In addition, the wars greatly reduced the flocks in the peninsula. At the same time, competition from Saxony and other regions of northern Europe that had developed merino flocks of their own from sheep imported earlier from Spain ended the country's centuries-long preeminent position.

Bilbao's commerce was more deeply hurt by the slump in the sales of iron products. The loss of the American colonies was a devastating blow. Despite the prohibition on trading directly with the colonies, Vizcaya's iron was exported to the Americas through other Spanish ports. In the Americas, this product enjoyed a privileged position, since royal decrees in 1702 and 1770 had granted Vizcaya a monopoly.[78] Toward the middle of the eighteenth century, as Vizcaya's iron became less competitive in northern Europe, it came to rely more on the colonial and peninsular markets. Table 2 shows the evolution of this trade.

The situation was made even worse because many Basque merchants resorted to importing cheaper iron products from other countries. This trade provoked the wrath of the central government and the local iron manufacturers. For the latter, it was a disgrace that, while they were trying to gain pro-

TABLE 2
Bilbao's Iron Exports, 1773–1825

Year	Tons	Year	Tons	Year	Tons
1773	3,600	1805	1,020	1815	—
1774	4,080	1806	550	1816	950
1775	4,120	1807	530	1817	920
—	—	1808	450	1818	610
1780	3,520	1809	1,910	1819	740
—	—	1810	1,550	1820	620
1790	3,140	1811	1,270	1821	1,350
—	—	1812	990	1822	620
1800	1,400	1813	250	1823	510
—	—	1814	280	1824	960
				1825	300

Sources: Fernández Pinedo and Bilbao, "Auge y crisis," p. 216; and M. Nájera, "El comercio a través del puerto de Bilbao 1800–1825," p. 210.

tection from the central government, the merchants at home were torpedoing their efforts. For the central government, it was one more proof of the need to abolish the inland customs. Disguising the foreign production as indigenous, some Bilbao merchants illegally reexported the iron to Castile, taking advantage of the tariff differential between foreign and Vizcayan goods.

Despite its high profitability, this illegal trade only emphasizes the weakened position of Bilbao vis-à-vis the central government. Madrid was not going to suffer gladly these abuses for too long; and the Basque merchants could not afford to defy the central government constantly. After losing their colonial and northern European markets, they needed their Castilian customers. Therefore, they had to compromise. In the process, however, they tried to preserve as many of their *foral* rights as they could. Of course, not all merchants shared this position. Those involved in local wholesale or retail did not have anything to gain by obtaining access to the Castilian market. As a result, they probably adopted a less compromising position on the customs issue.

In summary, the historiographical tradition of associating the *foral* question with the ultraconservative Carlists has tended to overlook the broad support that the *fueros* enjoyed among *all* Basques, regardless of occupation or social status. This support was akin to the way E. P. Thompson describes the Englishman's pride in his civil liberties. In England, these liberties were upheld as a defense against arbitrary rule; in Vizcaya, the *fueros* were conceived as a barrier to the central government's encroachment on traditional Basque institutions. In both cases, these liberties helped shape a communal identity. Paradoxically, in both England and Vizcaya, these ideas could serve to mobilize a crowd either for "Church and King," or for a reformist cause.[79] Only within this framework one can explain how the Carlist motto of "King, Religion, and Fueros" could stand side by side with the assertion of a Republican sympathizer that "only democracy is compatible with the *fueros.*"[80]

The Carlist War seemingly contradicts the notion of a widespread consensus on the *fueros.* But among Vizcayans, the struggle did not exactly confront those who favored or opposed them. The problem was that, in the minds of the local traditionalists, Vizcayan reformers were equated with Castilian liberal centralizers. However, these last two groups did not share the same outlook on the *foral* issue. The local reformers protested strongly the central government's repeated violations of the *fueros* during the war. Yet, their protests did not ring true in their adversaries' ears. At the same time, the Carlists tried to respect the *fueros* as much as they could, given the belligerent conditions under which they acted. In the end, the central government's acceptance of the "peace and *fueros*" formula vindicated the position of the Vizcayan reformers. This compromise of 1839 finished the first war after seven years of fighting. The *fueros* were preserved, but remained a source of tension between the provincial and the national governments. In 1844, the central authorities imposed two major modifications: the transfer of the customs frontier to the seacoast and the abolition of the *pase foral.* Having lost the right to stop Castilian legislation, Vizcaya became more exposed to centralization, a process that was intensified as the judicial administration was adjusted to conform to general Spanish rules. Also important for the later economic development of the region was the application of national laws in areas like mining, which had previously been entirely regulated by the provincial authorities. Yet, the local institutions retained part of their old autonomy. The Junta and the Diputación continued their administrative functions, and the civil regulations of the *fueros* remained in effect. Although Vizcaya came under the Spanish fiscal umbrella, its contributions were limited to special quotas, at rates lower than in the rest of the country. The collection and distribution of all taxes within the province remained a prerogative of the regional government.

The new *foral* arrangement proved beneficial to Bilbao's economy. The city started growing again after years of war and political turmoil. Under the protection of Spanish tariffs, the port area underwent a reindustrialization process with the founding of modern iron factories and flour mills. Pent-up demand from the war years also gave a boost to shipbuilding. And most important, as mining was freed from regulations that had restricted the export of iron ore, the legal stage was set for the intensive exploitation of one of the richest deposits in Europe.

Chapter Two

The Development of the Mining Industry

In that part of the Cantabrian coast
which is washed by the ocean, there
rises a high and steep mountain,
which marvelous to relate, is
composed entirely of iron.
Pliny, *Natural History*

Since Roman times, geologists and naturalists have praised the richness of Vizcaya's iron deposits. Pliny's description probably refers to Mount Triano, in the heart of Vizcaya's mining district. The Roman geographer was wrong; the mountain is not entirely made of iron. However, the approximately one hundred million tons of ore extracted in the second half of the nineteenth century came close, at least metaphorically, to being a veritable mountain of iron. Thanks to this intensive exploitation, the Vizcayans achieved the alchemists' goal of transmuting iron into gold, since the export of ore brought vast fortunes. The "iron fever" allowed Bilbao to transcend its regional importance and become one of Europe's busiest harbors. When the profits of the ore trade began to be invested, modern steel factories replaced the traditional *ferrerías,* and other industries were developed. As a result of its economic strength, Bilbao started playing a major role in Spanish politics. The city came to resemble an American frontier town, attracting immigrants from other regions. All of these unprecedented developments were rooted in the Vizcayan subsoil.

Mine Location and Types of Ore

The mine district stretched from the left bank of the Nervión River to the neighboring province of Santander, and from the seacoast to San Miguel de Basauri, located two and a half miles southeast of Bilbao. The district can be divided in three main groups of mines, according to their location (see map 2): (1) those of Abando, Basauri, Ollargan, and Begoña were closest to Bilbao; (2) those of Somorrostro and Triano were the richest deposits; and

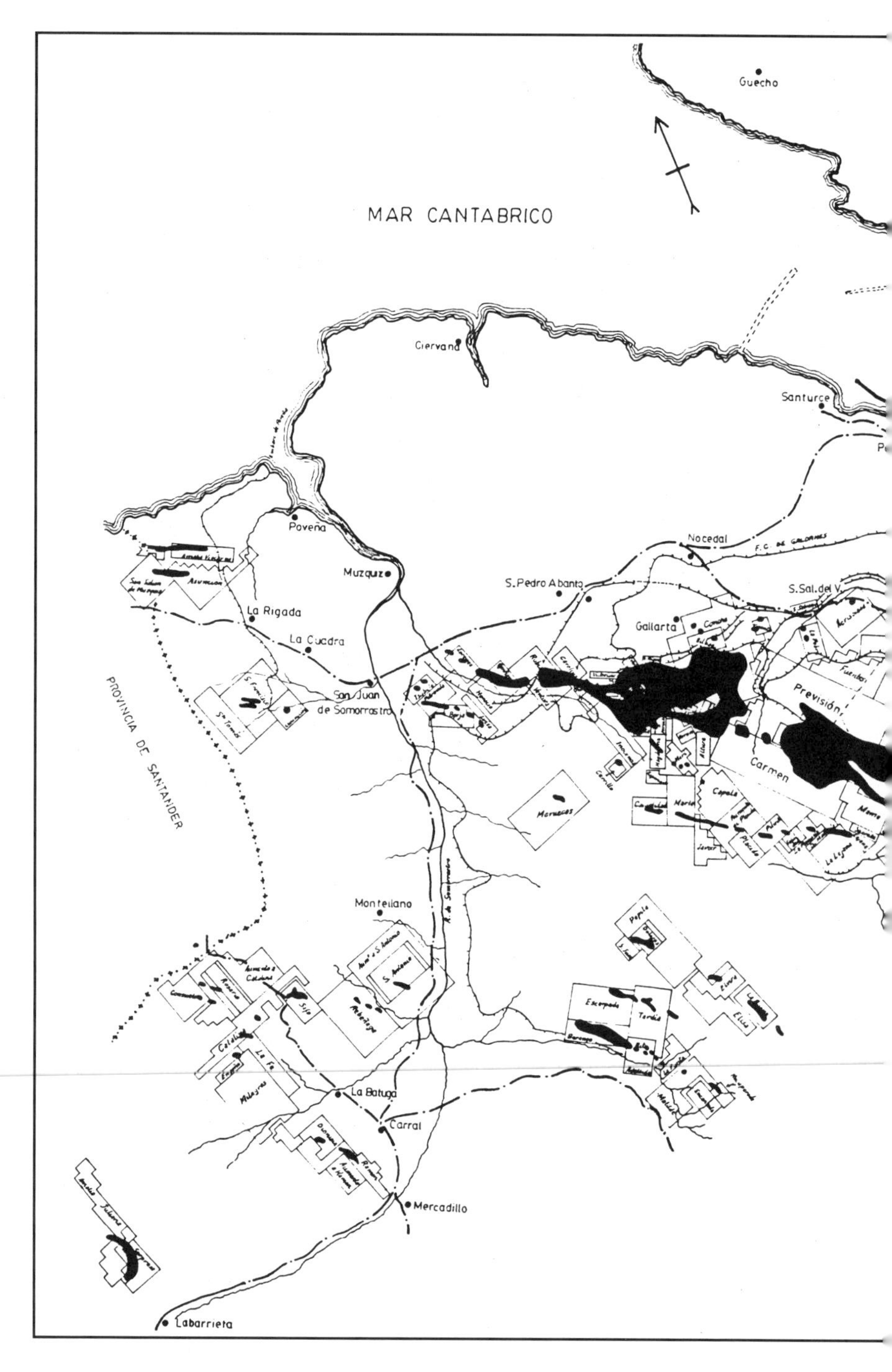

Map 2. Viscaya's Mining Districts

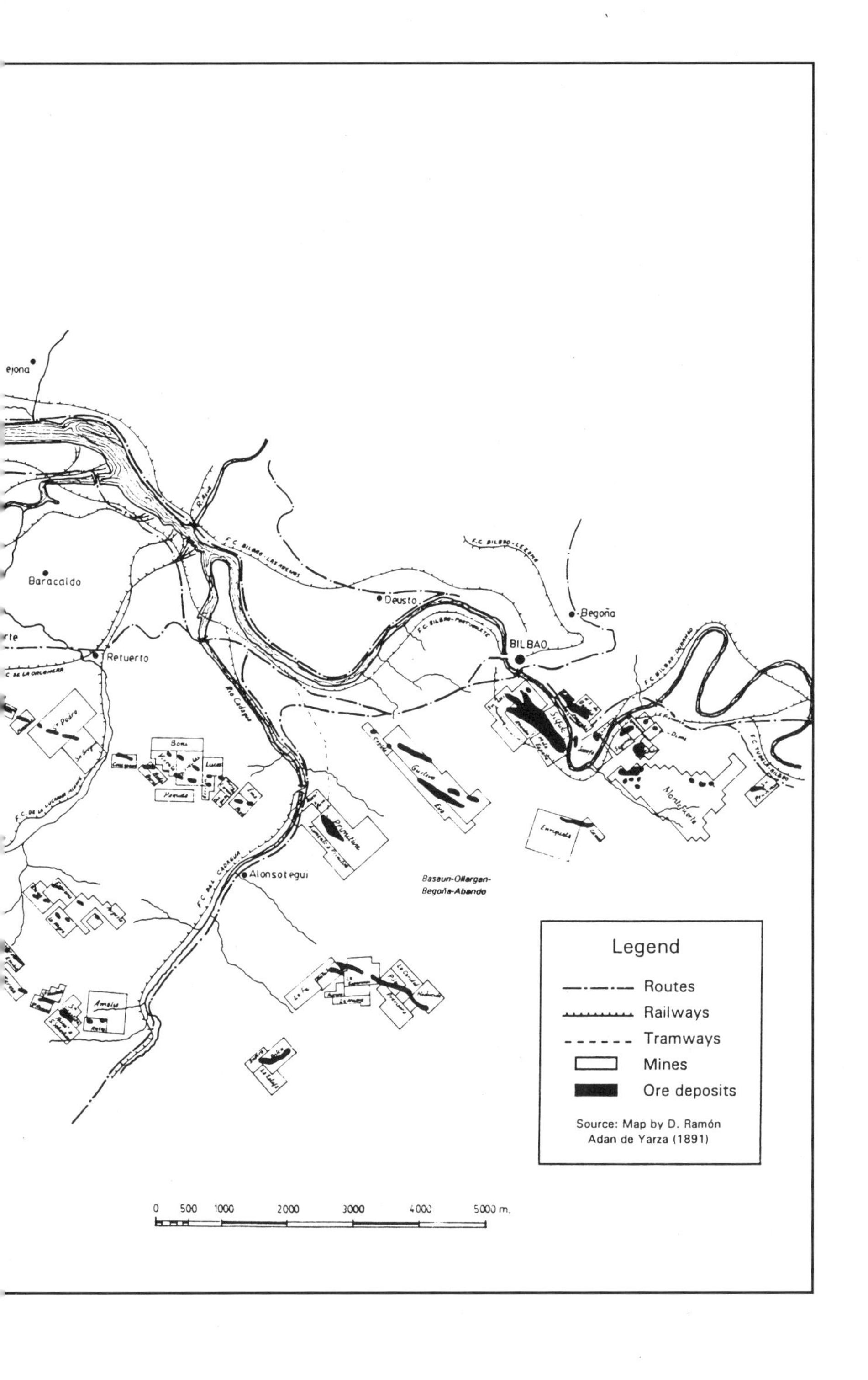

Baracaldo
Retuerto
Deusto
Begoña
BILBAO
Alonsotegui
Basaun-Ollargan-
Begoña-Abando
Legend
Routes
Railways
Tramways
Mines
Ore deposits
Source: Map by D. Ramón
Adan de Yarza (1891)
0
500
1000
2000
3000
4000
5000 m.

(3) those of Galdames and Sopuerta, near the border with Santander, were not fully exploited until later, since they were farther inland than the other areas.

Four classes of iron ore were commonly found in the region: (1) *vena* (a purple-colored hematite), (2) *campanil* (a red hematite), (3) *rubio* (a brown hematite, also called limonite), and (4) *carbonato* (a siderite or spathic ore). *Vena* was the purest, yielding 60 percent of metallic iron with the use of modern smelting techniques. For several centuries, the *ferrerías* consumed this kind of ore exclusively. Thus by the second half of the nineteenth century, the *vena* was practically exhausted. *Campanil* was the most-sought-after ore at the beginning of the mining boom. Found mainly in the Triano-Somorrostro region, the *campanil*'s yield was also very high, around 55 percent. By the 1880s, this variety too was almost exhausted. *Rubio* abounded in the three mining districts and accounted for the bulk of Vizcaya's production during the 1880–1900 period. Its yield was between 50 and 55 percent. During the 1870s and the 1880s, the *carbonato* was thrown away with the refuse of the mines, since, unlike the other classes of ore, this kind of mineral required some refining before it was sent to the iron factories. Yet, as the other types showed signs of exhaustion, and experiments showed that the *carbonato* could also yield a high percentage of metal, kilns were built to refine it, a fact that prolonged the life of the mines at the turn of the century. In addition to their high metallic content, these four types of ore were practically devoid of sulfur. This combination of qualities explains the high demand for Vizcaya's ores after the 1860s, with the invention of a new method for the production of steel, the Bessemer converter.

Early Exploitation and Regulations

Despite many centuries of exploitation, little is known about the property system and the methods of extraction used in the Bilbao mines prior to the export boom of the nineteenth century. The subject is not merely an erudite concern; the evolution of the property system and mining regulations helps us understand the appropriation of the iron deposits by a few families who eventually reaped handsome profits from this activity. Unfortunately, sources for studying the early history of the mines are scarce. The few extant documents do not deal primarily with the issues of property and methods of exploitation. Laws, which usually provide basic information on these matters, were not enacted in Vizcaya until the early nineteenth century. Before that, custom ruled the mines. Yet, like the ore extracted from the mines, those unwritten rules left traces after they were uprooted. It is through these traces that the early history of the mines can be re-created.

Vizcaya's mines, unlike other Spanish mineral deposits, did not belong to the crown. In other provinces, where Castilian laws were fully enforced, mines could be exploited only by private individuals through royal concessions. In Vizcaya, however, the *fuero* reserved the right for the region's inhabitants to exploit minerals, without the need of royal permits. This situation persisted until 1841, when, because of the weakening of the *foral* system after the First Carlist War, the mines came under the jurisdiction of national laws.[1]

The earliest indication of mine ownership seems to date back to the fifteenth century. At that time, a family of feudal chieftains, the Salazars, apparently derived their control of the Triano-Somorrostro mines from their ownership of vast extensions of land in that region.[2] It is unclear how the mines were exploited during this period, but by the sixteenth century a lawsuit against the Salazars demonstrates the direct involvement of other persons in the extraction and transport of ore.[3]

It is not known exactly when the Salazars lost control of their mines, but according to local tradition, they sold the mountains of Triano to the municipalities of the Somorrostro district for 14,000 ducados.[4] Whether the story is accurate or not, the municipal governments did preserve the ownership of those lands until at least the early nineteenth century. Despite what was probably a long period of communal ownership, the municipalities do not seem to have obtained any direct benefits from the mines. They apparently did not collect any taxes or rents derived from the extraction of ore. Nevertheless, they restricted access to the mines exclusively to the neighbors of the Somorrostro Valley.[5]

Communal ownership of the surface land did not imply a collective exploitation of the mines. Local customs granted the discoverer of a mine the exclusive right to exploit it for an indefinite period. However, the person could forfeit this right if the mine remained idle. According to local custom in northern Navarre, it was enough to work a mine one day per year to preserve the rights to it. Yet, since this led to abuses, by the middle of the eighteenth century, the minimum period of work was extended to a month per year.[6] Similar practices were enforced in Vizcaya, as Wilhelm von Humboldt remarked at the beginning of the nineteenth century.[7]

A more detailed discussion of the traditional practices regulating exploitation rights was given by Fausto Elhuyar, a professor of mineralogy who visited the Somorrostro mines in 1782. Elhuyar was struck by the facility with which any neighbor of the Somorrostro Valley could obtain rights to the local mines. He stressed the fact that only the native population of the valley could enjoy use of the mines, "without any distinction among them, all being free to extract the ore wherever and however they pleased."[8] Although such comment seems to deny the existence of any sort of private right to exploit the mines, Elhuyar went on to distinguish three kinds of exploitation: one in which the proprietors themselves worked the mine they owned; another in which owners and day laborers worked side by side; and a third in which the mine was partly or totally rented out by its owners.[9] Clearly, there was some form of private appropriation of the subsoil, and the rights involved in such appropriation were broad enough to permit the rental of a mine. However, since it is not known how the mines were demarcated, it is difficult to determine how the owners' rights were enforced. According to Ignacio Goenaga, custom regulated that the entrance to any new mine should be dug at least five *brazas* (8.36 meters) away from adjoining mines. Since it was not clear what procedures were followed to measure that distance, Goenaga speculated that each mine must have formed a circle with an 8.36-meter radius and a surface of approximately 220 square meters. The borders of the mine, however, were not projected vertically toward the interior of the mountain. Curiously, the miners had the right to continue their galleries in any direction and even beyond the surface boundaries.[10] If during the course of the excavations two mines met, the miners had to backtrack 17 meters and renew their digging in a perpendicular direction to the old gallery. Similarly, if the gallery of one mine was dug on top of another mine, the owner of the top one had to back away and continue work in another direction.[11] According to Humboldt, these traditional rules created uncertainty over mining rights and prevented investment in this activity.[12]

In the 1780s, there were approximately 120 active mines in the Somorrostro district. Elhuyar does not mention which of the three kinds of exploitation was the most common; however, since he emphasizes the ease with which a local inhabitant could obtain a mine, it is hard to imagine that the number of day laborers or renters was significant. Four or five persons worked in each mine.[13] In general, these miners were peasants who complemented the meager income from their fields by extracting ore in the agricultural off-season. Mining was usually carried on during the spring and summer, but even then, it was not a full-time occupation. In fact, one of Elhuyar's suggestions to improve what he considered subpar mining production was to double the working hours. Yet, the problem of Vizcaya's mines at that time was not the short workday, but the weak demand for ore.

Only the purest ore, the *vena,* was extracted to supply the *ferrerías* of Vizcaya and neighboring provinces. In order to protect the local industry, a *fuero* strictly forbade the export of ore to foreign countries. According to Fernández Pinedo and Bilbao, in the 1780s, Vizcaya's *ferrerías* produced annually up to seven thousand tons of iron. The ratio of ore needed to produce a ton of metal was approximately three to one. Hence Vizcaya's iron producers needed about twenty-one thousand tons of ore each year. In addition, the neighboring provinces required approximately another twenty thousand tons for their *ferrerías*.[14] Therefore, the total annual demand for ore hovered around forty-one thousand tons.

Since 4 or 5 persons worked each of the 120 active mines of Somorrostro, the total work force ranged between 480 and 600. According to Elhuyar, the daily production of each of those miners averaged 15 *quintales* (1 *quintal* equals approximately 0.046 tons).[15] Thus total daily production oscillated between 331 and 414 tons. Since the mines were worked only from mid-May to the end of September, they would have been open for a maximum of 135 days. Assuming that only 110 of those 135 days were workdays, the miners produced somewhere between 36,400 and 45,500 tons per year.

Therefore, even working short hours, the miners could produce enough to satisfy the demand of their sole client.[16] It cannot be argued that if the *ferrerías* had had more ore at their disposal, they could have produced more iron. In fact, during the 1780s, the output of the *ferrerías* reached a level never again achieved; the production levels of that decade were almost halved during the first third of the nineteenth century.[17]

Nevertheless, even if the pull of the demand for ore had been stronger, there existed important barriers to major increases in mining production. Transportation, for instance, was a serious bottleneck. The roads became useless in winter because of the heavy rainfall, and the limited agricultural resources of the region were not enough to feed the large number of draft animals needed to convey the ore from the mines to the docks and the *ferrerías*. Furthermore, until the early nineteenth century, local officials, believing that mining subtracted human and animal resources from agriculture, limited the number of animals that each miner could have and restricted their use in the mines to the spring and summer months.[18]

In addition, all observers agreed that the mines were exploited in a chaotic way. Humboldt's remarks in this sense were devastating: "Nowhere is mining practiced with less skill than here. Peasants, who don't have the slightest clue of this activity, . . . and who cannot be called miners . . . dig holes haphazardly, strike the mineral at hand with their picks, and after they have worked a certain time and the hole has an uncomfortable depth, or water has become a problem, they abandon the site, and dig a new hole, as haphazardly as before."[19] Ironically, the lack of methodical exploitation was in part due to the richness of Vizcaya's deposits, which almost guaranteed that no matter where the miners dug, they would strike iron. As a result of the poor extracting methods, the face of the Triano Mountains looked like a battlefield, crisscrossed by trenches and dotted with craters.

Elhuyar's and Humboldt's accounts indicate that a good part of the ore was extracted from underground galleries. In contrast, open-shaft mines prevailed during the intensive exploitation of the latter half of the nineteenth century. Why the early miners worked underground is unclear. Perhaps their interest in the *vena* type of ore forced them to do so; or perhaps, given the small scale of exploitation, it was easier to concentrate resources by digging in and following the ore vein in whatever direction it went rather than trying the systematic, and more labor-intensive, earth removal involved in the exploitation of open-shaft mines.

Although mining had always been one of the most important economic activities of Vizcaya, profit margins prior to the export boom were meager. Elhuyar remarks that "there is little or no distinction at all between the mine owners and the day laborers as far as the earnings that they obtain from these works."[20] Indeed, the Bilbao merchants involved in the iron trade reaped the highest profits from this activity. Their predominance was based in control of all the markets available for Vizcaya's ore. According to Manuel González Portilla, during the early eighteenth century, the merchants took control of the *ferrerías'* production by means of the putting-out system.[21] Actually, González Portilla's description does not correspond to that of the classic putting-out system, in which merchants control manufacturing processes by buying raw materials and farming them out to artisans, who produce goods for the merchants at a fixed rate. In Vizcaya, merchants gained control of production by acting as the *ferrerías'* bankers. According to the *fuero,* the merchants could not buy ore and resell it to the *ferrerías.* This provision was included in the provincial code precisely to avoid the iron manufacturers' dependence on intermediaries or speculators.[22] Nevertheless, the merchants eluded the prohibition by giving the iron manufacturers credit to buy the ore themselves and, instead of being repaid in cash, they were given iron goods at lower market prices. A handful of merchants came to dominate these practices, so there was little competition to lower the interest rates on the loans. In addition to tightly controlling the internal market, the merchants in effect monopolized the export of ore to other provinces, since Vizcayan authorities imposed restrictions on alien traders.[23]

The merchants' control of the iron trade did not require direct investment in either the mines or the *ferrerías.* Only when mine property rights were tightened through legislation and restrictions on the export of ore to foreign countries were lifted did the mines become a more attractive investment for the merchants. Vizcaya's first mining law dates to 1818. It tried to formalize some of the customs that had regulated the sector for centuries. Its primary concern was to introduce some order into the chaos. Yet, its strong reliance on tradition apparently defeated its purpose.[24]

The 1818 law continued to grant Vizcayans the exclusive right to exploit the mines. It increased the size of each mine from the traditional limits (cited by Goenaga). The law was not explicit about the mine boundaries, but again one may speculate that they formed a circle with the center at the entrance of the gallery, and with a radius of 33.5 meters. The right to dig in any direction beyond the superficial boundaries of the mine was maintained. Provisions to separate galleries from adjoining mines also seemed to have been based on custom, since they resembled the practices described earlier. To avoid accidents, the law allowed local officials to survey operations and gave them the power to close down those mines where personnel were endangered. For safety reasons, the new legislation also banned night work and forbade the extraction of ore during the winter months, establishing a mining season that went from mid-May until the end of September. An important innovation of the law was the prohibition against holding more than two mines at a time. Finally, the rights to a mine were guaranteed as long as it was worked at least two months every year.

The 1818 law was ambiguous on two key points. First, it was unclear whether it applied to proprietors whose operations antedated the enactment of the law.[25] The issue of retroactivity seems to have been especially important in relation to the limit of mines per owner. Second, there was no indication of how one had to proceed to obtain a mining concession, nor was there any mention of a property register. Consequently, the enforcement of property rights and the two-mine rule must have been plagued with difficulties.

The two-mine rule was controversial. The reason for its inclusion in the legislation was straightforward: to spread ownership among a large number of people. Immediately after its enactment, the official in charge of the supervision of the mines launched a two-pronged attack on the rule. First, he argued that the cap would jeopardize prospecting activities, since it would create complacency among two-mine owners, who, satisfied with their lot, would be unwilling to run the risk of opening a new mine that might not be as good as the ones they already possessed. Second, there was no need for the limitation because, historically the mines had never been concentrated in the hands of a small group of owners.[26]

Although the complacency argument had logic behind it, the historical reason was rather weak. The fact that the provision was included in the law may indicate that the traditional situation regarding mine ownership was changing. In the long run, had the provision been effectively and continuously enforced, it would have prevented the concentration of ownership that did take place in the 1840s. Yet, in the 1820s such concerns seemed premature, since the slump in Vizcaya's iron industry and export restrictions precluded, perhaps more effectively than the law itself, the concentration of ownership.

The 1818 law was an important first step in trying to regulate the mining sector. Yet, it was immediately apparent that the legislation had many shortcomings; realizing this, provincial authorities appointed a commission in 1824 to draft a new version. As the commission started its proceedings, the Spanish crown enacted a mining code. In 1825, the royal code reaffirmed the principle of the crown's ownership of the country's mineral resources. Since this principle was contrary to the *fuero,* Vizcaya's government vetoed its application within the province. Despite the veto, Vizcaya's new mining law of 1827 drew heavily on the royal legislation, adopting most of its provisions except the principle of royal ownership.[27]

The 1827 law was more precise than its 1818 predecessor. It included detailed instructions on how to obtain the rights to a mine. Claims had to be registered before the secretary of the Diputación. After officials surveyed the field and the claim was approved, the registration served as a property title. The law established that each mine would form a rectangular figure of twenty thousand square *varas* (approximately fourteen thousand square meters). Unlike its predecessor, the 1827 law required that all excavations be carried out within that rectangle, called a *pertenencia.* The old mines maintained their size as long as they did not exceed twenty thousand square *varas,* but had to be demarcated anew to establish their boundaries. The two-mine rule was preserved; however, exceptions were allowed, the most important being that no limit was applied to the number of mines that could be acquired through purchase, donation, or inheritance. Purchases at this time seem to have been rare; most mines were acquired by claiming new concessions. The intention of the lawmakers thus appears to have been to prevent a few individuals from staking a claim over the whole district, leaving the majority of those involved in the extraction of ore without the possibility of owning mines. Nevertheless, the exceptions had the potential of rendering the two-mine rule completely useless in the long run.

Under the 1827 legislation, mining concessions were granted for an indefinite period of time, as long as the owners fulfilled the obligations established by the law. The main obligation was to keep the mine active every year. If the mine was left idle for four continuous months or for a total of eight nonconsecutive months during one year, the owner lost the rights to it. Unlike the old legislation, the 1827 law permitted the exploitation of the mines during the whole year without interruption.

Finally, several provisions to protect the local iron industry were included. For instance, the ban on the export of ore to foreign countries was preserved, and the *ferrerías* were granted the status of preferred buyer (i.e., they had priority over other purchasers to any shipment of ore).[28]

Two of the 1827 law's most important innovations were the creation of something akin to a register, which should have permitted easier enforcement of property rights, and the idea of mines whose boundaries could not be trespassed through underground digging. These regulations set the legal framework for more orderly mining practices. Yet, centuries-old customs die hard; consequently, it is not surprising that, although the law was enacted, it took a long time before it could be enforced.

When the law of 1827 was enacted, the situation in the mines was similar to the conditions Elhuyar had described fifty years earlier. According to a report prepared by Gregorio González Azaola for the Vizcayan government, neither the number of workers (four hundred men) nor the annual production (approximately forty thousand tons) had changed by 1827.[29]

Despite stagnant production levels, a subtle change was taking place during those fifty years. González Azaola records the more direct involvement of the iron merchants in the exploitation of the mines. Singling out Ibarra, Mier, Chavarri, and Llano as the biggest traders, González Azaola mentions them, too, as the owners of the only systematic operations in Somorrostro that deserved to be called mines. A few months after the report was written, these four traders formed a partnership under the name Ibarra Mier and Co.[30] The creation of this company constituted a landmark in the history of Vizcayan mining. During its early years, its domination of the ore trade was so overwhelming that it came close to being a monopoly.[31]

From the late 1820s to the 1840s, Ibarra Mier and Co. functioned primarily as a commercial company and specialized in the trade of ore and iron manufactures. An 1849 balance sheet lists total assets of 2,582,500 pesetas, but of that figure, only 9,325 pesetas were invested in mines and tools for the extraction of ore.[32] This amount is deceptive, reflecting as it does not the mines' true worth but their still-low valuation. Table 3 shows that the company already owned a number of mines and shares in other mines in the Somorrostro-Triano region in 1849.

Most of the mines had been acquired only recently. In 1840, Ibarra Mier and Co. declared to the Vizcayan government that it fully owned only three mines, only one of which, Orconera (acquired by Mier in 1813), appeared on the 1849 balance sheet. In addition, the company also stated in 1840 that it owned shares of two other mines in the Triano area.[33] This shows that the firm was complying with the ownership regulations required by Vizcayan law. By 1849, the number of mines that the company owned had increased by a factor of three. Very few of those mines were bought; the majority were discovered and registered directly by the company. Moreover, Ibarra Mier and Co. had perhaps underreported its holdings in 1840 to comply with the law. On the other hand, many of the mines might have been registered during the 1840s, as the Spanish government took over the jurisdiction of Vizcaya's mines. Most likely, a combination of both possibilities took place.

Contemporary observers commented that the national authorities applied

TABLE 3
Ibarra Mier and Co.'s Mine Holdings, 1849

Mine Name/Location	% Ownership	Valuation (pesetas)
Barga	100	1,500
Blanca	100	300
Bomba del Cadegal	100	375
Bomba Vieja	100	375
Cayuela in la Cerrada	100	3,000
Orconera	100	375
Pasadiza	100	190
Plazuela	100	250
Primavera y Otoño in Tarresuela	100	250
San Benito in Los Castaños	100	375
TOTAL		6,990
La Cerrada	50	150
El Cadegal	50	100
La Bomba	40	375
Fuente Fría	37.5	500
La Plazuela	33	425
La Cerrada	25	250
Los Cobachos	20	375
TOTAL		2,175

Source: I.F.A., Gabriel María Ibarra's unpublished paper on José Antonio Ibarra's estate.

the mining legislation more zealously than did provincial officials.[34] In principle, the most important innovation implied in the change of jurisdiction was that the Vizcayans could not continue to enforce their monopoly on ownership. Yet, in practice, such privilege was not seriously challenged during the 1840s and the 1850s, and the mines remained under the control of local owners.

Despite the national officials' interest in limiting the size of the mines to the twenty thousand square *varas* prescribed by law, this process was slow. The mines were so close to each other that to divide the space into neat rectangular boxes was impossible without merging and redistributing exploitation rights. To complicate matters even further, at least during the early years, the

TABLE 4
Vizcayan Ore Production, 1840s

Year	Production (tons)
1840	22,275
1841	21,640
1842	24,860
1843	24,850
1844	24,310
1845	24,090
1846	24,635
1847	24,470
1848	24,150

Source: Aldana, "Descripción," pp. 365–368.

change of jurisdiction itself created new problems. Some individuals claimed property rights to mines before the national authorities, even though those mines had been previously registered with the Vizcayan government. Such was the case of José Ignacio Ustara and Ramón Maruri. The former had registered the San Agustín mine with the provincial authorities, whereas the latter, years later, did the same with Spain's central government.[35]

Due perhaps to these uncertain conditions and the protracted decay of the *ferrerías,* mining production never even approached the level it reached at the end of the eighteenth century, when it hovered around the 37,000-ton level. The anemic state of the trade in the 1840s is clearly shown in table 4.

The Concentration of Mine Ownership in the Triano-Somorrostro Region

In 1849, two legal developments shook the Vizcayan mines out of their doldrums. First, a new mining code was enacted, permitting larger units of exploitation. Second, the prohibition on the export of ore was abolished.

The new mining legislation and its zealous enforcement created the most important changes in the property structure of the mining region. In the 1850s, a large section of the Somorrostro-Triano district was demarcated according to the new laws and ended up in the hands of a few individuals. Such development was not devoid of difficulties. Lawsuits and complaints were

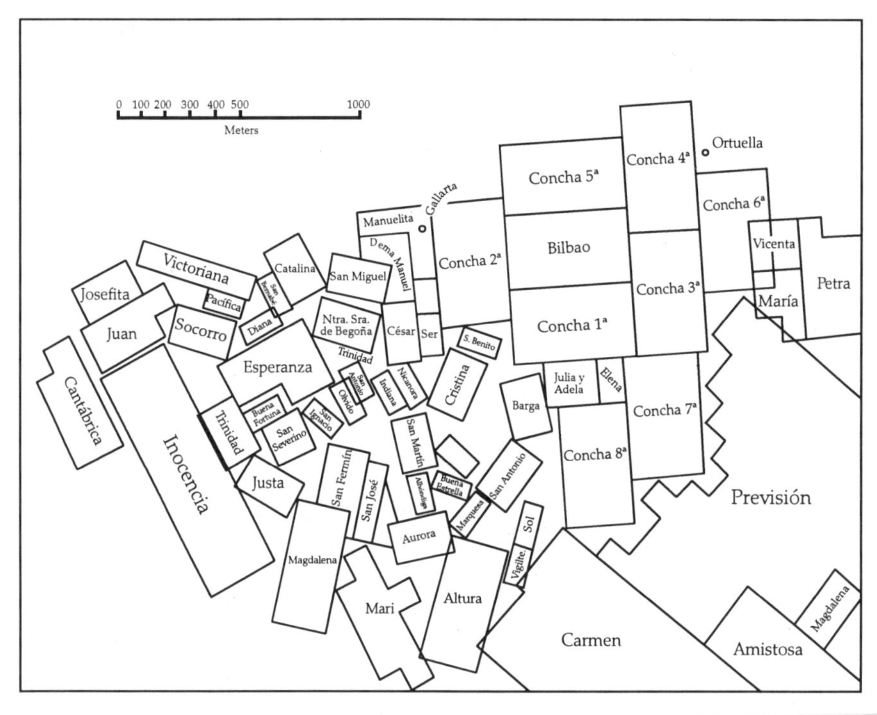

Map 3. Principal Triano Mines, 1896
Source: William Gill, "The Present Position of the Iron Ore Industries of Biscay and Santander."

part and parcel of a process that was akin to the enclosure of lands in England in the eighteenth century. In many cases, mines came to resemble Chinese boxes in that earlier claims came to be included in later ones.[36] Although it dates from a period when most of the suits had been settled, map 3 still shows the proximity of the mines to each other, the pell-mell orientation of the *pertenencias,* and the difficulty involved in demarcating them.

Among the complaints that have surfaced in Vizcaya's archives, two by residents of the Somorrostro Valley show the other side of the process of ownership concentration: the proletarianization of the peasant-miner. Dated April 30, 1850, one of the complaints notes: "During other not so distant times the neighbors of the Seven Councils were less unhappy because they obtained a livelihood from the mines in Mount Triano, but today the majority is deprived of this livelihood because a few mines, registered according to the special royal mining legislation, have occupied the whole mountain." The document ends with a petition to the Vizcayan authorities to annul the registration of these mines. The second complaint, dated July 1859, sheds some light on the means used to gain control of the mines: "under the shadow of stipulations that regulated the exploitation of the mines, monopoly was developed and nurtured, since a few with knowledge of the bureaucratic processes, which for the people in general are incomprehensible, took over the very best of the immense iron wealth hidden in the Triano Mountains, Somorrostro, etc."[37]

In addition to the mining laws, which figured as the main reason for the woes of the peasant-miners in both complaints, the renewed interest in the mines during the 1850s was due to the liberalization of the iron trade in 1849. The opening of new markets must have created a rush to acquire mines and to take advantage of an expected export boom. By 1850, approximately 30 percent of Ibarra Mier and Co.'s shipments of ore went to French ports.[38] In fact, France became the major foreign customer of Vizcaya's mines during the 1850s. Unfortunately, the only statistics available for the period do not indicate volume, but only the value of this trade. Although it is perilous to infer production increases from statistics that record value, it is likely that the abrogation of the ban on exports boosted Vizcaya's mining industry. However, as the figures in table 5 show, the level of exports quickly reached a plateau. Even at its peak in 1854, the ore trade represented only a tiny fraction of total exports from the Bilbao harbor. In fact, during the mid-1850s, the prosperity of the port was based primarily on the export of flour and grains, which increased notably because of the Crimean War and bad harvests in other European countries. Although the situation was improving for the mine owners, the years of abundance were yet to come.

The concentration of ownership in the Triano-Somorrostro district con-

TABLE 5
Vizcaya's Ore Exports, 1849–1860

Year	Value (pesetas)	% of Bilbao's Total Exports
1849	110,000	1.10
1850	205,000	1.80
1851	360,000	2.70
1852	322,000	1.60
1853	340,000	1.50
1854	410,000	1.70
1855	368,000	1.00
1858	285,000	1.20
1859	353,000	1.40
1860	385,000	1.40

Source: F.C.R., vol. 7 (1851–1860).

tinued relentlessly. An 1861 list (appendix table I-1) reveals that there were seventy active mines in the region, owned by forty-five different companies or persons.[39] Actually, the number of owners was smaller, but this fact was masked by the use of front men to register titles. In many cases, the title holder was merely an employee of the real owner(s); in others, the name of only one of several owners was used to register. A more accurate picture of the distribution of mining concessions emerges after studying notarial documents. Eventually, to clarify property rights, the person who had registered a mine for others would say so in a written statement before a notary.

Why was the use of front men necessary? The mining laws of 1849 and 1859 still set limits on the area that one company could register within a given mineral deposit. Not until the law of 1868 were those limits done away with. Yet, before this law was enacted, the large mine proprietors could easily circumvent the spirit of the law by using front men. For example, although Ibarra Mier and Co. appeared to possess only four mines in the 1861 list, the company actually owned twenty (slightly more than 28 percent of the total). Its other mines were registered under the following names: Ibarra Hermanos, José Gorostiza, Sebastián San Martín, Ramón Castaño, Leonardo Zuazo, Matías Salcedo, and Francisco Recalde.[40] Not all of those twenty mines were fully owned; some were shared with other proprietors. Eventually, Ibarra Mier and Co. acquired a larger number of mines in the district through purchases and boundary deals with owners of adjoining mines. By

1880, Ibarra Mier and Co. had property rights to the mines listed in table 6.

Sixteen of the 31 mines listed in table 6 were registered or claimed before 1861.[41] Interestingly, only five appear to have been registered prior to 1849. I have shown, however, that the company did own many mines before that date. Yet, since their official demarcation probably did not occur until the late 1840s and the 1850s, their dates of registration reflect this fact. As the successive laws permitted the ownership of larger mines, Ibarra Mier and Co. expanded many of its holdings. Thus a mine like the Orconera, which the company owned practically from the beginning of the nineteenth century, came to occupy an area of more than one million square meters.[42]

Ibarra Mier and Co. was by far the largest mine owner of the region, but the 1861 list included other proprietors who also had been active in the iron trade since the early nineteenth century: Juan Durañona, Pedro de la Bodega, José Arana, José Chavarri, and José Antonio Ustara.[43] It is hard to determine exactly how many mines these five persons actually owned in 1861. Once again, the registration procedures cloud the issue. For instance, the Julia mine, listed only under Arana's name, had been co-owned since 1855 with Bodega.[44] Similarly, José Chavarri, who had only two mines listed under his name, also owned half of the Justa mine, and one-fourth of the Diana.[45] Nevertheless, the mines listed under the names of those five owners and the twenty mines that belonged to Ibarra Mier accounted for half of the active mines in 1861.

The 1861 list features a fair number of small owners who managed to maintain their rights to mines that they probably worked themselves. This group, however, dwindled until it had almost disappeared by the 1870s. Unfortunately, since many of these small owners sold their mines prior to the export boom, they missed the opportunity to obtain higher prices for their property. The appreciation in the value of the mines between the 1860s and the 1880s can be illustrated with the example of the San Fermín mine. Sold in 1861 to Alejandro de la Sota for 670 pesetas, it was purchased in 1867 by Cirilo María Ustara for 2,000 pesetas. In 1880, Ustara leased the mine to a British company for a lump sum of 6,750 pounds (around 168,750 pesetas) and a royalty of 0.875 pesetas per ton of ore extracted.[46]

Not just the small owners sold their mines prior to the export boom. José Chavarri, for example, bought all of Pedro de la Bodega's concessions during the 1860s.[47] All these sales concentrated mine ownership even further. Table 7 shows the distribution of property rights circa 1880 of some of the most important mines in the Triano-Somorrostro area.

Between 1880 and 1900, the mines that were actually worked among those listed in table 7 represented close to 70 percent of the active concessions in the Triano-Somorrostro district. In addition, as table 8 shows, they accounted

TABLE 6

Mines Owned by Ibarra Mier and Co., ca. 1880

Mines[1]	Date Registered or Claimed[2]	% Ownership	Area (square meters)
Altura	1864	100.00	130,095
Alhóndiga	1849	100.00	13,974
Barga	1849	100.00	41,924
Bilbao	1869	50.00	150,000
Buena Fortuna	?	30.00	13,974
Nuestra Señora del Carmen	1864	100.00	1,200,000
Catalina	?	12.50	41,924
César	1860	50.00	37,312
Concha	1848	100.00	1,200,000
Cristina	1853	50.00	41,924
Despreciada	1848	100.00	18,591
Esperanza	?	40.00	159,784
Indiana	?	36.00	13,974
Magdalena	1852	100.00	53,317
Mingolea	1872	100.00	60,000
Nuestra Señora de Begoña	1860	50.00	58,049
Olvido	1865	100.00	13,974
Orconera	1849	100.00	1,050,000
Perseguida	?	100.00	?
Presentación	1871	50.00	?
Previsión	1864	100.00	2,290,000
Rubia	1864	100.00	300,000
San Antonio	1845	50.00	40,456
San Benito	1848	100.00	13,974
San Bernabé	1849	100.00	13,974
San Ignacio	1849	50.00	40,799
San Martín	1854	100.00	34,516
San Miguel	?	16.66	51,924
Ser	1847	50.00	33,732
Socorro	1852	10.00	41,924
Trinidad	?	20.86	27,652

TABLE 6
Continued

Sources: A.H.P.V., S. Urquijo, L. 6256, no. 425; and F. Uribarri, protocols for 1885, nos. 389 and 390; I.F.A., reports prepared by Pedro Celis in 1895 and 1898.
Notes:
[1] In Celis's 1898 report, the Ibarras appear as owners of the Parcocha and Unión mines. Their share of the Parcocha appears to have been contested, and it is doubtful that they obtained any gains from this mine. As for the Unión, no other document I consulted listed their ownership of this mine. It is absent, for instance, from the inventory of the estate of Juan María Ibarra, which lists all the other mines.
[2] The time between the claim and the registration of a mine was sometimes long. When available, the claim date was used to indicate the earliest date of control of the mine.

for an extremely high percentage of Vizcaya's total production. During the 1880s, this group of mines represented about half of all active concessions in the province, but their percentage of total production was close to 80. Their decline in the share of total Vizcayan production during the 1890s was due to two factors: their gradual depletion, and the increased production of the other Vizcayan districts. In fact, the 36 active mines in 1898 represented only 25 percent of all the active mines in Vizcaya.[48]

Unlike the proprietors listed on table 7, not all the owners in the Triano-Somorrostro district had been involved in Vizcaya's iron trade since the early nineteenth century. Yet, the predominance of the group with the long presence in the district forced latecomers during the 1870s to register or purchase mines in the periphery, which was not as productive as the region's heart. There were, however, some exceptions. A British company, the Somorrostro Iron Ore Co., in 1871 bought two of the best mines, the Unión and the Amistosa, whose output almost rivaled that of the Ibarra mines.[49] In 1876, the mercantile partnership of Vitoria Maruri and Suñol, which had not been involved in the ore trade until then, bought half of a very productive mine, the Parcocha, from Manuel Moya for 15,000 pesetas. When Vitoria Maruri and Suñol finally bought the other half of the mine in 1880, they paid 90,000 pesetas, reflecting a hefty increase in the mine's value.[50] In 1881, Matthew and Richard Curtis, from Manchester, paid a large sum for mines that belonged to the Arana family.[51] Yet, despite these examples, sales were rare in the Triano-Somorrostro district, and, although the owners could command high prices for their mining properties, they preferred to exploit them themselves or to lease them.

Among the companies that gained ownership by claiming and register-

TABLE 7

Distribution of Property Rights, Somorrostro-Triano, ca. 1880 (%)

	Property Owner							
Mine	Ibarra	Chavarri	Ustara	Arana	Amezaga	Bellido	Allende	Total
Adela		50.00						50.00
Altura	100.00							100.00
Alhóndiga	100.00							100.00
Asunción	50.00							50.00
Aurora		100.00						100.00
Barga	100.00							100.00
Bilbao	50.00							50.00
Buena Estrella		100.00						100.00
Buena Fortuna	30.00	8.30	15.00	10.00				63.30
Nuestra Señora del Carmen	100.00							100.00
Catalina		12.50		35.50	27.00			75.00
César	50.00					50.00		100.00
Concha	100.00							100.00
Cristina	50.00		50.00					100.00
Despreciada	100.00							100.00
Diana		40.00						40.00
Esperanza	40.00		33.00	15.00			10.00	98.00
Indiana	36.00	32.00		32.00				100.00
Julia		50.00			50.00			100.00
Josefita		33.30						33.30
Justa	50.00						50.00	100.00
Magdalena	100.00							100.00
Mingolea	100.00							100.00
Nicanora			25.00				50.00	75.00
Nuestra Señora de Begoña	32.00	32.00		27.00				91.00
Olvido	100.00							100.00
Orconera	100.00							100.00
Pacífica		70.00		30.00				100.00

TABLE 7
Continued

	Property Owner							
Mine	Ibarra	Cha-varri	Ustara	Arana	Ame-zaga	Bellido	Allende	Total
Rubia	100.00							100.00
San Antonio	50.00		25.00		25.00			100.00
San Benito	100.00							100.00
San Bernabé	100.00							100.00
San Ignacio	50.00		25.00					75.00
San Felipe		32.00						32.00
San Fermín			100.00					100.00
San Miguel	16.60	32.00		8.30	24.60			81.50
San Severino			25.00			50.00		75.00
Ser	50.00							50.00
Socorro	10.00	10.00		10.00			2.80	32.80
Sol			25.00		12.50	62.50		100.00
Trinidad	20.80							20.80
Victoriana		18.40						18.40

Source: A.H.P.V.

ing new mines was the Luchana Mining Co. This company possessed a large number of concessions registered during the early 1870s by its agent J. B. Davies. Its mines stretched from San Juan de Somorrostro to Retuerto, forming an arch along the outer limit of the district. Although numerous, these mines were not among the richest of the region.[52] Yet, perhaps carried away by the infectious enthusiasm that pervaded the region during the mining boom, the company invested heavily to exploit the mines. In a sense, its example illustrates the risks involved in the mining business. Not all the investors who flocked to Vizcaya dreaming of quick fortunes achieved their goal.

Clearly, a small group of Vizcayan owners took advantage of their early acquaintance with the iron trade by controlling the most productive mines. Their control preceded the export boom and was never relinquished. First they were helped by Vizcayan legislation, which prevented outside investors from acquiring mines. When national legislation ended this privilege, they still had an insider's advantage and a long tradition of managing the region's

TABLE 8
Percentage of Total Production, Vizcaya, 1880–1898

Year	No. of Active Mines	% of Total Production
1880	29	79.5
1885	36	78.5
1890	37	65.0
1895	34	45.0
1898	36	43.5

Source: B.O.V.

iron trade. In a province with limited resources, this group of owners must easily have understood the key position held by the mines in the local economy, even before the international demand for ore exploded.

Property in the Other Mining Districts

Unlike the Triano-Somorrostro area, the Galdames district does not seem to have an extensive history of exploitation. In fact, most of its mines were not demarcated and registered until the 1870s. According to Goenaga, the area was known to have ore deposits long before then; however, since the type of ore was predominantly *rubio,* which was not used in the *ferrerías,* the region remained unexploited until the export boom. During the early 1870s, companies rushed to stake their claims in Galdames, expecting to discover mines as rich as those in Somorrostro. Such enthusiasm, however, did not correspond to reality, and the district never yielded more than 8 percent of total Vizcayan production during the 1880–1900 period.

Although many mines were claimed in the district, few actually became highly productive. The Bilbao firm of Uriguen Vildosola registered the main concessions in 1870.[53] This company also obtained the license to build a railroad connecting the Galdames district with the Nervión River.[54] In 1871, Uriguen Vildósola reached an agreement with a British company, the Bilbao River and Cantabrian Railway Co. (later the Bilbao Iron Ore Co.), for the leasing of the mines and the sale of the railroad concession.[55] The Berango, Escarpada mines and Tardia were included in the lease and accounted for almost the whole production of the district during the 1880s. In the 1890s, as other mines came into production, their share hovered between 25 and 30

TABLE 9
Galdames Mine Share of Vizcayan Production, 1880–1898

Year	No. of Active Mines	No. of Owners	% of Total Production
1880	2	2	6
1885	2	2	2
1890	12	9	7
1895	13	8	6
1898	18	12	8

Source: B.O.V.

percent. Table 9 shows the number of active mines in the district and their percentage of total Vizcayan production.

The third mining district, Basauri-Ollargan-Begoña-Abando, had perhaps a longer history of exploitation than the Galdames region. Yet, since the deposit consisted mainly of the *rubio* type of ore, it is doubtful that the district was widely exploited prior to the nineteenth century. The first demarcations seem to have taken place right after the First Carlist War. Around 1850, these mines were producing between 7,300 and 8,700 tons per year.[56] The *rubio* from this region was especially apt for the charcoal blast furnaces that started to replace the *ferrerías* in the 1840s. Not surprisingly, one of the main mine owners of the district was the Santa Ana de Bolueta iron factory. In 1886, the company's balance sheet listed six mines in Ollargan, valued at 680,000 pesetas.[57] Ibarra Mier and Co., which had founded an iron factory with blast furnaces in the early 1860s, also owned some mines in the district. Yet, in 1872, the company leased them for sixty years to a British firm, Somorrostro Iron Ore Co., for the lump sum of 250,000 pesetas.[58] Unlike the Triano mines, the Ollargan concessions changed hands several times. In 1883, the Somorrostro Iron Ore Co. sold its lease to those mines and the Malaespera mine (in Abando) to José Martínez Rivas for 100,000 pesetas.[59] Only two years later, Martínez Rivas sold the leases and the Malaespera mine to Juan Aburto y Azaola for 230,000 pesetas.[60]

Another important mine owner in the district was Manuel Lezama Leguizamón, who belonged to a landowning Vizcayan family involved in the mining business and the *ferrerías* from very early on. He died in 1884, and his estate included eight mines in the Basauri-Ollargan-Abando-Begoña area, valued at 620,000 pesetas.[61]

Julio Levinson, a foreigner, acquired mines in the Abando and Begoña

TABLE 10
Basauri-Ollargan-Abando-Begoña Mine Share of Vizcayan Production, 1880–1898

Year	No. of Active Mines	Tons	% of Vizcayan Production
1880	18	73,000	3.8
1885	14	40,000	1.6
1890	14	282,000	6.6
1895	15	471,000	10.7
1898	15	626,000	11.9

Source: B.O.V.

areas during the late 1860s and the early 1870s and became one of the major owners in the district.[62] Another foreign owner in the district was a British company, Landore Siemens. Although it did not buy a large number of concessions, its Primitiva mine, bought in 1872, was among the most productive of the region.[63]

Overall, these mines accounted for more than 90 percent of the district's productive concessions. Similarly, they represented the bulk of the district's production (see table 10).

The number of active mines in the Basauri-Ollargan-Abando-Begoña district was much smaller than the number in the Triano-Somorrostro district, but the most productive deposits were concentrated in both regions among a handful of owners. Also as in Triano-Somorrostro, many of the proprietors of the Basauri-Ollargan-Abando-Begoña district were involved in the iron trade prior to the export boom. Except for the Ibarras, few of the district's proprietors owned concessions in the other mining areas. In fact, there was very little overlap among the owners in the three principal districts. All seemed to concentrate their resources in one territory, where they expanded by taking advantage of the successive mining laws, which permitted the ownership of larger and larger concessions. By the time the export boom started, it was difficult to find unregistered surfaces within the mining districts. Thus, by the 1870s, almost the only way to obtain a concession was to lease or buy one.

Production and Markets during the Export Boom

In 1856, searching for a strong kind of cast iron needed for his new artillery shell design, Henry Bessemer discovered, by luck and genius, a way to produce steel cheaply and in vast quantities. The advantage of steel over wrought and pig iron had been known for some time, yet, the pre-Bessemer methods used to produce steel from pig iron were expensive, since they consumed vast quantities of fuel and were extremely slow. As a result, steel was used sparingly. Bessemer discovered that by blowing air through molten iron, he could speed up the chemical reaction needed to produce steel, at a considerable savings of fuel. What had taken a full day with old production methods his converter could achieve in a matter of minutes. This amazingly simple solution opened the door to what some historians consider the second Industrial Revolution.[64]

Although Bessemer first presented the results of his experiments in 1856, the converter was not fully functional until the late 1860s. A series of technical problems plagued the method, delaying its adoption by the iron factories. The converter proved to be effective with only one kind of ore, a hematite with practically no phosphorous. Bessemer had been lucky because he had inadvertently used this kind of ore in his experiments. It was by chance, too, that he lined the converter with a material that helped produce the desired result; this material, however, was seldom used by the iron manufacturers. Unaware of these requirements, the first factories that rushed to adopt the converter failed to produce good-quality steel. As the chemistry behind Bessemer's method came to be understood better, and after much trial and error, some of the technical problems were overcome; however, the dependence on nonphosphoric ore remained.

Hematites were rarer and therefore costlier than ordinary ironstone. Great Britain, the world leader in iron manufacturing until the end of the nineteenth century, possessed hematite deposits in Cumberland and Lancashire. When in the late 1860s and the 1870s, the widespread adoption of the Bessemer converter spurred the demand for this type of ore, the production from those deposits increased from around a half million tons in 1855 to slightly over two million in 1870.[65] At the same time, prices skyrocketed from an average of 13s. 11d. (16.70 pesetas) in the 1867–1870 period to 33s. 6d. (40.20 pesetas) per ton in 1873.[66]

The increased production of those mines, however, did not keep up with the demand of the British factories; consequently, the need to import arose. The Bilbao mining district proved to be the ideal source to supply the Bessemer converter's ever-growing appetite for hematites. There were multiple reasons for this preference. First, Bilbao's ore was rich in metallic content (above 50 percent on the average against a 30–33 percent yield from the British ore).[67] Second, the relative proximity of Vizcaya's mines to the coast facilitated the export of the ore to the northern European manufacturing centers. Third, the ore could be extracted very cheaply from the open-shaft

TABLE 11
Vizcaya's Iron Ore Production, 1861–1900

Year	Production (tons)	Year	Production (tons)
1861	55,000	1881	2,621,000
1862	70,000	1882	3,855,000
1863	71,000	1883	3,628,000
1864	120,000	1884	3,216,000
1865	102,000	1885	3,311,000
1866	90,000	1886	3,301,000
1867	136,000	1887	4,400,000
1868	154,000	1888	3,960,000
1869	165,000	1889	4,180,000
1870	250,000	1890	4,740,000
1871	403,000	1891	3,840,000
1872	402,000	1892	4,210,000
1873	365,000	1893	4,310,000
1874	11,000	1894	4,566,000
1875	34,000	1895	4,575,000
1876	350,000	1896	5,250,000
1877	1,040,000	1897	5,254,000
1878	1,306,000	1898	5,073,000
1879	1,263,000	1899	6,496,000
1880	2,684,000	1900	5,382,000

Source: A. Pourcel, "Mines de fer de Bilbao," p. 73, for years 1861–1875; for the period 1861–1875, García Merino, *La formación de una ciudad industrial,* p. 504.

mines. No hematite deposit in Europe could offer such advantages.[68] Only the Swedish mines could match the abundance of Bilbao's deposits, but since they were located deep inside the country, they were hard to exploit. As a report presented to the British Iron and Steel Institute recognized, "there was little hope of getting the ores of Sweden to the English coast at a cheaper rate than about 20s. per ton . . . , while such ore could be delivered under similar circumstances from Spain for . . . sixteen shillings."[69]

Bilbao's advantages, then, paved the way for a tremendous increase in Vizcayan mining production. As table 11 shows, the real production boom did not begin until the 1870s, when it was briefly interrupted by the Second Carlist War. More than 90 percent of Vizcaya's production during the 1870–

1900 period was exported. The boom was coterminous with the widespread adoption of the Bessemer converter. Production peaked in 1899, then began a slow decline. By the first decade of the twentieth century, most of the best ore had been extracted, but the use of more sophisticated mining techniques prolonged the life of the mines.[70]

During the last quarter of the nineteenth century, Bilbao cemented a strong commercial relationship with Great Britain, which became the main consumer of Basque iron. More than two-thirds of the British imports of ore during those years came from Bilbao, and Great Britain absorbed between 65 and 75 percent of the annual ore exports of the Bilbao mines.[71]

Bilbao did not remain an ideal source of iron ore forever. The technological puzzle of how to make steel with the more abundant sulfuric ores was solved in 1879 with the development of the Thomas-Gilchrist oven. According to the French consul in Bilbao, the news of this development caused a panic among local mining companies.[72] However, the fear was premature. It took some time until the steel produced with the new method could compete in price with the output of the Bessemer converter. Yet, even when production by the Thomas-Gilchrist method became cost-effective, it was adopted primarily by the continental countries, which lacked any significant reserves of hematites. In contrast, British factories, Bilbao's main customers, continued to rely heavily on Bessemer's method.[73] Thus, although the Thomas-Gilchrist innovation did not decrease production from the Bilbao mines, it dampened their marketability, reducing the potentially higher profits of the local mining entrepreneurs.

Organization of Production during the Mining Boom

In order to satisfy the large international demand for its ore, the Bilbao mining district had to change. The concentration of mining concessions among a few owners facilitated the development of large-scale exploitation. Still, mine owners had to raise vast amounts of capital to extract what must have seemed in the early 1870s like mountains of iron. The infrastructure of the region had to be adapted to the needs of an intense and continuous movement of ore from the mines to the harbor. Also, the harbor itself had to be improved to permit the mooring of larger cargo ships, and drops had to be built on the Nervión's docks to allow the expeditious loading of those ships. These needs were financed through a combination of private (both national and foreign) and public funds.

Public funds were used to build the first railroad line connecting the mines to the harbor. The provincial government had planned to build such a line since the early nineteenth century;[74] however, bureaucratic problems delayed its construction and inauguration until 1865. Managed directly by the government of Vizcaya, the line remained the only mechanical means of transportation until the late 1870s, when three companies built railroads connecting their mines to the harbor.

The Triano Railroad, as the government-run line was called, was an unusual enterprise for a provincial government in nineteenth-century Spain. Such projects were traditionally managed from Madrid. In Vizcaya's case, however, a long history of autonomous local government, which had built a network of roads to increase commercial traffic, explains the enterprising attitude of the provincial authorities.[75] The line turned out to be highly profitable, becoming one of the government's major sources of revenue. Between 1876 and 1900, the Triano Railroad transported approximately 35 percent of the region's ore.[76] According to Manuel Montero, the prices the government set for its line were "political" and benefited the mineral exporters.[77] Yet, González Portilla has shown that the line enjoyed very high profit margins (around 50 percent of total revenues).[78] Moreover, the high profit margins explain why, during the 1870s, three mining companies built railways whose tracks paralleled the Triano line. In fact, the transport of ore from the mines to the harbor increased the f.o.b. cost of the ore by a factor of two.[79] It is unlikely therefore that the railway companies could have charged higher tariffs without affecting the competitive advantage of the whole district. Once all the lines were in service, competition among them kept prices at fairly stable, but still high, levels.[80]

The other major public expenditure that facilitated the tremendous increase in the export of ore was the construction of a modern harbor. For this purpose, in 1872 the national government created the Junta de Obras del Puerto de Bilbao. Under the direction of a Basque engineer, Evaristo Churruca, the Junta's main objectives were to eliminate the dangerous sandbar that dominated the entrance to the harbor and to improve the navigability of the Nervión. Since the national government granted the Junta the right to levy a tax on the traffic into and out of the port (and especially on the export of ore), Churruca had no problem financing the work. By the early 1880s, the initial and most important work was completed.[81]

Of course, public expenditures alone did not meet all the needs of the mine owners. In fact, the rapid growth of the district required massive infusions of capital to bring the dormant wealth of the subsoil to the surface. The case of the Chavarri family illustrates the point. Although one of the main mine owners of the Triano-Somorrostro district, the family was pressed for capital. In 1876, the Chavarris secured a loan for 100,000 pesetas from the rich merchant house of Sanginés Sobrino. The loan contract clearly stated the

Chavarris' needs: "[the Chavarris] owned some iron mines whose exploitation is difficult because they lack cash to pay for the costs."[82]

Another important mine owner, Cirilo María de Ustara, provides a further example. Early in 1878, he borrowed 125,000 pesetas from Felipe Abaitua, a merchant; after paying this debt in 1879, he obtained from the same lender a new loan, this time for 181,250 pesetas. Ustara's borrowing illustrates not just the need for working capital to extract ore, but also the need for investment funds to improve the infrastructure of the mines. By the time he secured the second loan, Ustara was planning the construction of an aerial ropeway to transport ore from his San Antonio mine to the railway terminus at Ortuella.[83] Such works were needed to end the use of animal transportation from the mines to the railway lines. The aerial transport systems, built mostly during the 1880s, were an ideal solution to the overcrowded mining space of the Triano-Somorrostro district, where many owners could not build a road out of their mines without disrupting adjoining operations.[84]

In these examples the mine owners secured loans from Bilbao merchant houses, a situation forced on them by the underdeveloped banking system in Spain.[85] The mine owners' need for capital might have caused the circumstances of the early nineteenth century to repeat, when merchants squeezed profits out of the miners. However, the scale of the trade, and the strong foreign competition, completely absent at the beginning of the century, averted this outcome. Foreign companies came to Bilbao not only as buyers of ore, but also as lenders and direct participants in the exploitation of the mines. Their investment in railroads, harbor drops, and other infrastructure was crucial in paving the way for the massive exports of the mineral.

The two most important foreign companies were multinational firms in which local entrepreneurs acted as partners. One of these, the Orconera Iron Ore Co., was formed in 1873 by the association of Bilbao's most important mine-owning family, the Ibarras, two British steelmakers that had pioneered the use of the Bessemer converter—Consett and Dowlais—and the famed German industrial firm Krupp. Each partner held a 25 percent stake in the company.

The Orconera Iron Ore Co. leased for ninety-nine years the following mines, which the Ibarras co-owned with the Mier family: Orconera, Concha no. 1, Carmen, Previsión, Magdalena, and the Ibarras' half of the César.[86] According to the terms of the lease, the Orconera Iron Ore Co. had to pay the Ibarras a royalty of 8 pennies for each ton of ore extracted from those mines, ensuring a minimum annual payment of 3,333 British pounds (for the first 100,000 tons of ore, which had to be paid for whether they were extracted or not). In addition, the transfer of the concession to build a railroad line from the mines to the Nervión River required the Orconera Iron Ore Co. to pay a royalty of 4 pennies per ton of ore extracted from the mines leased whether the mineral was transported or not. The mining company assured a minimum annual payment of 1,666 pounds. The transfer of the railroad rights also provided for the transport of ore from mines not worked by the Orconera; on this activity, which proved highly profitable for the mining company, the Ibarras did not obtain any royalties, since they forfeited them when the amount of ore transported from the Orconera's mines exceeded 350,000 tons annually—a figure that the company surpassed easily after its first years of production.[87]

The second major multinational company in which the Ibarra family became a founding partner was the Société Franco-Belge des Mines de Somorrostro. Formed in 1876, this mining company was an association of two French steel producers—Fonderies de Montetaire and Hauts Fourneaux et Forges de Denain et Anzin—and the Belgian manufacturing firm John Cockerill. The Ibarras owned only 10 percent of the stock in this partnership, while each of the other three partners possessed 30 percent each of the company's shares.[88] As in the Orconera case, the Franco-Belge leased for ninety-nine years the following mines belonging to the Ibarras: Concha nos. 2 to 8, Barga, San Benito, Despreciada, San Martín, Altura, Alhóndiga, San Bernabé, and Perseguida. The company agreed to pay the Ibarras a royalty of 1.25 francs (1 franc = 1 peseta) per ton, with a minimum annual guaranteed payment of 125,000 francs, whether the corresponding first 100,000 tons were extracted or not. Although the Franco-Belge also built its own railroad line, there was no transfer of rights involved in this matter, and therefore no royalty payment had to be arranged with the Ibarras. Still, the royalty paid by the Franco-Belge equaled the amount the Orconera paid the Ibarras for extraction *and* transport of the ore. The difference is perhaps explained by the fact that the Franco-Belge mines were richer in the *campanil* type of ore, which commanded higher prices than the *rubio* type, which abounded in the Orconera mines.

Why the Ibarras sought out foreign partners to exploit their mines is a difficult question to answer. At first sight, their financial position seems to have differed from that of other important local mine owners. Ibarra Mier and Co. was the biggest ore merchant and mine proprietor, and the partnership had been involved in the manufacture of iron since the 1840s. First they associated with the Vilallonga family, Andrés Gutiérrez de Cabiedes (a brother-in-law of the Ibarras), and Carlos Dupont in order to modernize and exploit an old *ferrería* in Guriezo (Santander). In 1860, a new partnership under the name Ibarra and Co. was formed to take over the Guriezo factory. It built a new facility on the banks of the Nervión. Ibarra Mier subscribed 60 percent of the capital of this new company, which was fixed at 1.5 million pesetas. The other partners, José and Mariano Vilallonga, J. J. Uribarren, and Cristobal Murrieta, each contributed an equal share of the remaining 40 percent of the capital.[89] The company eventually grew to become one of the largest in Spain. In addition, the Ibarras were involved in other business activities not just in Bilbao, but all around Spain. In general, Ibarra Mier's partners were fellow Vizcayans, and family bonds often cemented business associations. José Vilallonga, for instance, was married to a daughter of Gabriel María Ibarra, and Cristobal Murrieta's wife was a sister of the Ibarra brothers. The relationship with Murrieta and Uribarren was important because these two were the heads of Vizcayan banking houses located in London and Paris, respectively. These two financial firms were the fulcrum of Basque business affairs on both sides of the Atlantic, handling especially the flow of capital from Cuba to the peninsula and vice versa.

Thus it would seem as if Ibarra Mier had the necessary financial resources or connections to develop and fully exploit its mining properties. In fact, this idea must have crossed the minds of the partners, since in the early 1870s, before signing the contract with the Orconera Iron Ore Co., Ibarra Mier started the construction of a railroad line from the river to its mines.[90] Yet, perhaps the uncertainty created by the Second Carlist War kept the Ibarras from making large investments, and thus they welcomed the opportunity to bring partners into the business and to spread the risk.[91] Furthermore, considering that the Ibarras were familiar with the fluctuations of the ore trade, the deals with the Orconera and the Franco-Belge companies limited the Ibarras' risk by assuring them a fixed income no matter the ore price. And since the Orconera and the Franco-Belge had been formed primarily to supply ore to their constituent partners, market fluctuations had little impact on the production of the leased mines.[92] More important perhaps, although the Ibarras were certainly a rich family by Bilbao's standards, the large number of mines they owned required a proportional amount of capital to exploit. For example, by the end of 1877, the Orconera Iron Ore Co. had invested more than three hundred thousand pounds (approximately 7.5 million pesetas) to build the railroad and prepare the mines for working.[93] To give an idea of the magnitude of that sum, it was more than two times the capitalization of the Banco de Bilbao in 1873.[94] It would have been extremely difficult to raise that kind of capital locally for a private venture. Nevertheless, it might be argued that in the 1850s the Vizcayans had raised approximately 25 million pesetas to build a railroad connecting Bilbao with the Castilian interior. This project, however, was perceived as being in the public interest and required to maintain Bilbao's competitiveness as a port. Moreover, such construction, as was the case in Spain in general, received governmental subsidies and guarantees that attracted investors.[95] Mining, however, was never promoted in this fashion.

The Orconera and the Franco-Belge together with a company owned by a

TABLE 12

Production of Mining Companies, Vizcaya, 1879–1898

Mining Company	Production (tons)				
	1879–1880	1884–1885	1889–1890	1894–1895	1897–1898
Orconera	447,000	780,000	882,000	989,000	920,000
Martínez Rivas	111,000	399,000	688,500	844,000	751,000
Franco-Belge	71,000	216,000	384,000	367,000	538,000
Total	629,000	1,395,000	1,954,500	2,200,000	2,209,000
% of total production	33	56	46	51	42
Arana		41,000	124,000	197,000	188,000
Chavarri	182,000	160,000	394,000	207,000	195,000
Durañona-Gandarias		163,000		83,000	312,000
Echevarrieta-Larrínaga		38,000	134,000	137,500	170,000
Ibarra Hermanos	346,000	54,000	179,000	43,000	
Kreizner			26,000	105,000	154,000
J. MacLennan	143,000	142,000	276,000	144,000	130,000
Parcocha				91,000	165,000
J. B. Rochet	49,000	198,000	187,000	51,000	107,000
Sociedad Vizcaya		38,500	139,000	115,000	156,000
Total	720,000	834,500	1,459,000	1,173,500	1,577,000
% of total production	38	33	34	31	33

Source: B.O.V.

Note: Under the Chavarri name were combined the production figures that appeared listed with the names of Chavarri Hermanos and Víctor Chavarri. Similarly, the Ibarras' name in the table combines the production figures for Ibarra Hermanos, Herederos de Zubiría, Ibarra and Co., and José A. Ibarra.

local entrepreneur, José Martínez Rivas, were the largest mining concerns in Vizcaya. As table 12 illustrates, after 1880, they usually accounted for over 40 percent of total production.

Next in importance was a group of producers who worked more or less continuously during the 1880–1900 period and usually extracted more than one hundred thousand tons per year. Table 12 presents a rough estimate of the production levels of each of the companies or persons listed. Together with Orconera, Franco-Belge, and Martínez Rivas, the other companies on table 12 accounted for approximately 80 percent of total production. Fifteen of the remaining 20 percent was mined by companies that produced between ten thousand and one hundred thousand tons per year. Short-lived companies that operated for only one or two years and then disappeared, or firms that extracted fewer than ten thousand tons per year accounted for the final 5 percent.

Given the high concentration of mine ownership, it is not surprising that the extraction of ore was in the hands of a small number of companies.[96] Some of the largest mine owners, such as the Ibarras, the Chavarris, the Aranas, and the Durañonas, were also important producers. Yet, the three biggest producers do not seem to have owned the mines they worked. In the case of the Orconera and the Franco-Belge this is clear, since they signed long-term leases with the Ibarras. Martínez Rivas's situation, however, remains cloudy. The bulk of his production came from two mines, the Unión and the Amistosa, which were bought in the early 1870s by the Somorrostro Iron Ore Co. The ownership of those two mines is usually attributed to Martínez Rivas because his name, and not that of the British company, appears in the trimestral production reports of the B.O.V. A bond existed between Martínez Rivas and the Somorrostro Iron Ore Co., since the Vizcayan businessman had acted as the company's agent in the purchase of both mines in the early 1870s. In 1883, Martínez Rivas bought several leases from the British company in the Basauri-Ollargan-Abando-Begoña district. Unfortunately, I found no documents that showed that he also bought or leased the rights to the Unión and the Amistosa. Although the Somorrostro Iron Ore Co. left no traces in the B.O.V., the company did not cease to exist. A document in the archive of the Bank of Spain demonstrates the continuing operation of the firm until at least 1913, at which time the Martínez Rivas family was still the local agent in Bilbao.[97]

There seems to be no doubt that Martínez Rivas worked the Unión and the Amistosa mines; the question is whether he did so as an agent, a partner of the company, a lessee who had to pay royalties, or as an outright owner. Martínez Rivas was a wealthy entrepreneur in his own right. He made his first mark in the business world with his uncle, the first Marqués de Mudela, from whom he eventually inherited his title and acquired many of his companies. Thus it would be highly unlikely that he was simply an agent of the British company. Most probably, he leased the mines or was an active partner in the Somorrostro Iron Ore Co. in charge of ore extraction.

Among the other top producers, Echevarrieta and Larrínaga, Kreizner, MacLennan, Rochet, and the Sociedad Vizcaya also rented the mines they worked. The leases were similar, stipulating the duration of the contract, a royalty per ton extracted, and a minimum annual payment whether ore was extracted or not.

Clearly, foreign companies played an important role in the extraction of ore from Vizcaya's mines. According to calculations based on the trimestral reports of the B.O.V. during the 1880–1900 period, foreign companies extracted approximately 40 percent of the ore.[98] These calculations, however, do not take into consideration the Ibarras' shares in the Orconera and the Franco-Belge companies, and Martínez Rivas's production was regarded as completely local. In addition, only by including producers such as MacLennan and Rochet as part of the foreign mining companies does one arrive at the 40 percent share. Yet, in these two cases, historians seem to apply a parochial sense of nationality. Neither MacLennan nor Rochet was a Vizcayan (or a Spaniard, for that matter), yet if their companies operated in Bilbao and did not export their profits to other countries, there is no reason to consider them as foreigners. A French citizen, Rochet came to Bilbao during the early 1860s, prior to the mining boom. First, he was involved in trading and financial partnerships with another French emigrant, C. Jacquet.[99] Later, while engaged in mining activities, he continued to delve into other sectors of Bilbao's economy.[100] Rochet, then, appears to have been not a temporary, foreign resident, but a man with deep roots in Vizcaya.

MacLennan's case is slightly different. MacLennan, an Irish engineer, came to Bilbao in the 1870s to take advantage of the mining boom. And that he did. When he died in 1914, the documents of his estate show that he had amassed a vast fortune, with considerable investments in Spain (not just in mines, but in real estate, stocks, and bonds). And although he passed away in Paris, where he seems to have settled in retirement, he asked to be buried in Bilbao.[101] Bearing all this in mind, it becomes extremely difficult to assess the magnitude of foreign intervention in Vizcaya's mining industry, but it seems to have been smaller than is commonly assumed.

Today, when Europe's economic boundaries have collapsed, British, French, or German participation in Vizcaya's mining may seem irrelevant. It is, however, a controversial topic in Spanish economic history. Many historians have decried foreign "plundering" and "colonization" of the country's mineral resources.[102] Admittedly, Vizcaya's case has been treated more or less as an exception within a generally bleak situation.[103] Yet, this position, vigorously advocated by Manuel González Portilla and Luis García Merino among others, has recently been attacked in the work of Antonio Escudero. According to Escudero, Vizcaya's mining industry was also colonized by foreign companies. Escudero states that the profits captured by local business were in effect restricted to the royalties received by the Ibarras from the Orconera and the Franco-Belge companies, since "Bilbao Iron Ore, Luchana, Triano, Levison, MacLennan, Edwart [*sic*], and Parcocha worked their own mines."[104]

Escudero's assertions, however, are wrong for the most part. Many of the companies he mentions did not work their own mines. For instance, in 1872 Bilbao Iron Ore (also known as Bilbao River and Cantabrian Railway Co.) leased several mines in Galdames and a railway concession from the local firm of Uriguen Vildósola for which it agreed to pay three pennies for every ton extracted from those mines or transported on their railway (even if the mineral belonged to a third party). Whether ore was transported or not, the contract guaranteed an annual payment of 5,000 pounds.[105] That sum was estimated on the basis of a traffic of 400,000 tons of ore per year. Yet, not until the 1880s did the British company reach that level. Considering the important expenditures needed to build the railroad from Galdames to the Nervión (the longest of the district), quite some time must have elapsed before the British investors saw a good return on their money.

During the 1880–1900 period, the railway transported approximately 900,000 tons per year. Yet, the bulk came not from the mines that the company had leased but from third parties. In fact, the mines that the company had leased were sublet to a local steel manufacturer in 1883.[106] Hence, Bilbao Iron Ore was not actually a major mining company, but mostly a conveyor of ore through its railway. Undoubtedly, the company must have made considerable profits from this activity. Yet, the participation of Uriguen Vildósola was not negligible. Despite its lack of involvement in the capital outlay and management of the enterprise, the local firm collected, on the average, eleven thousand pounds annually in royalties during the 1880–1914 period.[107]

As far as the Luchana, Levison, and Edwards mining companies are concerned, Escudero is correct. The Luchana did own several mines in the periphery of the Triano-Somorrostro district in which it had invested 220,000 pounds (5.5 million pesetas) by 1876. Yet, it did not see any return on its investment until a decade later.[108] Indeed, until 1886–1887, its mines did not appear in the trimestral production reports published in the B.O.V. After 1887, production from these concessions hovered between 50,000 and 100,000 tons per year. Moreover, according to the B.O.V., the company did not always work the mines directly; in 1886–1887, 1889–1890, 1891–1892, and 1893–1894, it leased them to a local firm, Castaños and Co.

Edwards's case is similar to Luchana's, but on a much smaller scale. Edwards owned three mines in the periphery of the Somorrostro-Triano district; however, he rarely worked them himself, preferring to lease them to other companies.[109] In the same way, although Levison was one of the major owners in the Basauri-Ollargan-Abando-Begoña area, he only intermittently worked his mines. The highest production levels were not achieved when he worked them directly, but during the periods his lessee (Luis Núñez) worked the mines, during the 1890s.[110] Therefore, in these three cases, the implications of Escudero's argument would actually have to be reversed, since the foreigners did not make profits for considerable periods, and when they did, those profits came in large part from collecting royalties.

The case of Triano Iron Ore was quite different. Its most productive mines (Petronila, Lorenza, and Confianza) had been leased in 1880 by Francisco MacLennan for an indefinite period. A royalty of 7.25 pesetas per ton with a minimum annual extraction of 15,000 tons in the case of the Petronila and 30,000 tons for the Lorenza and the Confianza was agreed.[111] Yet, these mines were not listed in the trimestral production reports until the 1890s, when they were apparently worked by Otto Kreizner.[112]

The MacLennan cited by Escudero was probably José (or Joseph) MacLennan, who, beginning in the 1870s, appeared in the trimestral production reports as working the following mines: Socorro, Magdalena, Josefita, Justa, Rubia, Amalia Vizcaína, Diana, and San Francisco. Of all these, MacLennan owned only the Amalia Vizcaína; the rest were leased.[113] Again, the royalties paid for such leases were not negligible, hovering around the usual 1 or 1.25 pesetas per ton, and with guaranteed minimum payments whether mineral was extracted or not.

Finally, Parcocha Iron Ore did not gain control of its mines until the early 1890s. Previously, the deposit that gave the company its name was worked by local entrepreneurs and by the banking house of C. Murrieta.

If Escudero's argument dealt only with the issue of property, then the foregoing would be enough to refute it. However, he also argues that, for those who had to pay them, royalties were very low (0.5 pesetas per ton) until the 1890s.[114] Similarly, Fernández Pinedo claims that by virtue of their long-term contract with the Orconera and the Franco-Belge, the Ibarras actually received little rent from their mines and were not able to take advantage of the rise in royalties during the 1890s.[115]

Yet, the royalties contracted by the Ibarras in the 1870s were not so low, and in general, all these rents were certainly above the two reales mentioned by Escudero. In fact, the royalties that the Ibarras received appear to have had a fair market value, being comparable to those charged by mine owners in Great Britain. According to Lowthian Bell, during the 1880s, "The royalty dues paid in Cleveland for ironstone are about 6d. per ton. In Lincolnshire and Northamptonshire they are sometimes double this sum, because, instead of expensive shafts and machinery being required, the mineral lies near the surface."[116] The Ibarras's royalty was within that range. Nevertheless, the similar extracting methods used in Bilbao and Lincolnshire and the fact that British ore had on the average 20 percent less metallic content than the Vizcayan may lead one to think that the 8d. per ton that the Orconera paid the Ibarras was in fact a low royalty.[117]

Yet, such reasoning does not take into consideration the location of both districts. Since the Bilbao minerals had to be shipped to England, they had to offer certain advantages to make their extraction desirable. Furthermore, since the leasing contracts with the Orconera and the Franco-Belge set the royalties in foreign currencies (pounds and francs, respectively), the depreciation of the Spanish peseta in the 1890s did not affect the Ibarras adversely. On the contrary, their income in pesetas increased, as the local currency fell against the British pound. For instance, in 1876, the 8 pence equaled 0.83 pesetas reales, while in 1896 they represented 1 peseta; and at the Spanish currency's lowest level, in 1898, 8 pence became 1.3 pesetas.[118] Other leases signed for shorter periods in the Somorrostro-Triano area during the 1870s and the 1880s varied between 0.75 and 2 pesetas, depending on the location of the mine and the quality of its mineral. For instance, in 1877, Martínez Rivas leased from the Arana family its 32 percent stake in the San Miguel and Nuestra Señora de Begoña mines for ten years, agreeing to pay the owners a royalty of 1.25 pesetas.[119] Similarly, in 1884, Juan Aburto rented from Trinidad Ulacia the Inocencia mine for eight years. In this contract, the royalty was also set at 1.25 pesetas per ton.[120] Sometimes, the royalty was tied to the prevailing price of ore at the harbor, or that of the Bessemer-steel rails in England. For instance, in 1880, José MacLennan's lease with the Ibarras for the Rubia mine established a minimum royalty of 0.75 pesetas, and a minimum guaranteed payment corresponding to fifteen thousand tons. The royalty, however, increased in proportion to the average price of rails quoted in several British trade journals.[121]

Usually, short-term contracts lasting one, two, or three years with no option for renewal could command higher royalties. For example, in 1885, Sotera Mier leased to Víctor Chavarri and Pedro Gandarias her share of several of mines in Triano for three years. The royalty was set at 2 pesetas per ton.[122] In 1885, too, Manuel and Tomás Allende paid a comparable royalty, 1.87 pesetas to the Ibarras for the Cristina mine, according to the one-year contract they signed.[123]

During the mid-1890s, as many Triano mines were approaching depletion, those whose reserves were still plentiful could command higher royalties for their owners. In 1879, for example, the Socorro mine was leased to José MacLennan for a royalty of 1.25 pesetas per ton. When the contract was renewed in 1893, the royalty was doubled, to 2.5 pesetas per ton.[124] Hinting, however, at a decline in the mine's reserves, the annual minimum of twenty thousand tons in the first contract was reduced to fifteen thousand tons in the second.

Despite the stiff royalty increases during the 1890s, to argue that those who signed long-term contracts two decades earlier did not foresee this development is to abuse the historian's powers of hindsight. Such argument does not take into consideration the inherent security of a long-term contract, which must have induced the mine proprietor to forgo immediate short-term profits in exchange for a steady and prolonged flow of income. The royalty agreed to by the Ibarras with the Orconera retained a competitive market value for almost twenty years.[125] It would be hard to imagine that they expected it to maintain a high value for the whole contract. The best insurance the Ibarras had against the erosion of their royalties was their participation as stockholders. If royalties were low, the company enjoyed cost advantages and could generate higher dividends. As shown on table I-3 in Appendix 1, the Ibarras' income derived from royalties and dividends represented more than 50 percent of the Orconera's profits, a much higher proportion than their 25 percent control of the company's stock.[126]

Therefore, foreign participation in Vizcaya's mining industry was not deleterious to local interests. Without the foreigners' help, it would have been much more difficult to develop the region's mineral wealth. Of course, their participation was not a charitable activity, but it cannot be characterized as an example of economic colonialism. On the contrary, the situation proved to be a good symbiotic relationship in which local business captured the largest share of the profits generated by mining production.

Mining Profits

There have been several attempts to quantify the wealth created by Vizcaya's mining industry. The most recent have been those of González Portilla and García Merino. Several objections have been raised to their calculations. Scholars seem to agree on the production statistics and price series; however, production costs are contested. It has also been objected that these calculations do not take into consideration the fact that the Orconera and the Franco-Belge did not sell their mineral at market prices, but undersold it in order to assure their respective partners a supply of cheap raw materials. In addition, since production statistics did not always clarify the percentage that each type of ore represented of total output, and since there were important price differentials among the different types, it becomes difficult to quantify with certitude the total revenues generated through mining. It is thus worth analyzing early estimates in some detail, and then attempting a new, more refined calculation.

To quantify the wealth created by the mining industry, historians have mostly used the trimestral production reports of the B.O.V. Compiled for fiscal purposes, these statistics did not actually reflect the extraction cost of the mineral, but the price of the extracted ore at the mine site. González Portilla, who pioneered the use of this source, believes that this price equals the cost of extraction because the mineral was never sold at the mine site, but always at the harbor.[127] Yet, as García Merino demonstrates, González Portilla's assumption is wrong. Using documents from the archive of the *Círculo Minero* (an association of mining businessmen), García Merino shows that the price at the mine site was set after political negotiations between the businessmen and the fiscal authorities. The price thus reflected not just the extraction cost, but also royalties and a small surcharge that allowed the government to increase its revenue.[128] García Merino calculates that the price series was 40 percent higher than the actual extraction cost between 1877 and 1892, 30 percent higher for the 1893–1900 period.

Only the trimestral production reports give a complete cost series for the 1880–1900 period. Thus García Merino arrives at his percentages by using impressionistic data. Other historians have claimed that the f.o.b. costs were higher than García Merino's averages. For instance, Fernández Pinedo, basing his analysis on an 1896 article in an economic publication, breaks down the cost of a ton of ore in the following manner: extraction, 2.10 pesetas; transport to ships, 4.00; miscellaneous, 0.65; total cost, 6.75 pesetas.[129]

Fernández Pinedo's estimate is almost two pesetas higher than García Merino's average for 1896. Interestingly, Fernández Pinedo's extraction cost is lower than the 2.56 pesetas calculated by García Merino. The main difference between the two assessments lies in the transportation costs, which, according to García Merino, average only 2.5 pesetas for the 1877–1890 period, and 2 pesetas during the 1890s.

In fact, García Merino's averages are not completely off the mark, as documents uncovered after the publication of his work seem to demonstrate. The Orconera's, the Franco-Belge's, and García Merino's estimated f.o.b. costs for the *rubio* and *campanil* types are indicated in table 13.

During the 1880–1883 and 1897–1900 periods, the Orconera's costs were very close to García Merino's estimates; for the 1884–1888 quinquennium, if one averages the costs of the two companies, the result almost equals García Merino's calculations. Yet, the Orconera and the Franco-Belge had perhaps

TABLE 13
F.O.B. Costs, Orconera and Franco-Belge Mines

Year	F.O.B., Orconera (pesetas)	F.O.B., Franco-Belge (pesetas)	García Merino's Estimates
1880	3.65		3.59
1881	3.45		3.47
1882	3.31		3.55
1883	3.20		3.57
1884	3.09	4.15	3.58
1885	3.12	4.12	3.55
1886	3.30	4.04	3.58
1887	3.49	3.80	3.79
1888	3.40	3.72	4.39
1889		3.98	4.79
1890		4.23	5.42
1891		4.23	5.48
1897	4.61		4.81
1898	5.68		5.41
1900	4.81		5.64

Source: Orconera figures are based on the company's resident manager reports for 1883, 1885, 1886, 1888, and 1889. All these reports are in I.F.A., with the exception of the 1883 report, which I obtained from K.A., WA IV 835. Franco-Belge costs were obtained from company reports for the years 1886–1891 found in I.F.A.

the lowest production costs in all Vizcaya. Not only were the mines located in the richest area, but, by running their own railroad lines, these companies could deliver ore at the harbor at rock-bottom costs. It was their transport systems especially that made them more competitive than other producers.[130] The Franco-Belge could carry its mineral from the mines to the harbor at an average cost of about 0.85 pesetas per ton. The Orconera had to spend even less—between 0.5 and 0.75 pesetas per ton to convey its ore to the port.[131]

Other producers, who had to use the services of the several railroad lines, had much higher transport costs. For instance, those using the Triano line had to pay at least 2 pesetas to transport from the main ore depot at the Ortuella station to the harbor. In addition, they had to pay varying prices for the conveyance from the mine to the railroad depot. For instance, in 1884, Martínez Rivas signed a contract in effect until 1900 with Alonso Hnos. Uhagón and Co. to carry mineral in the latter's railroad from the San José and Perseguida mines to the Ortuella station. For this service, Martínez Rivas agreed to pay 1.68 pesetas per ton.[132] Thus Martínez Rivas's total transport cost from those mines to the port was around 3.68 pesetas. In contrast, the ore from Martínez Rivas's most productive mines, the Unión and the Amistosa, could be carried to the port for about 2 pesetas per ton.[133]

In 1880, the Ibarras also signed a contract with Alonso Hnos. Uhagón and Co. to carry mineral from the Esperanza mine. Their agreement stipulated a 1.56 pesetas per ton tariff from the mine to Ortuella, with the price to go down to 1.06 pesetas if the other owners of the Esperanza signed similar transport contracts.[134] Thus the Ibarras's total transport cost from the Esperanza mine hovered between 3 and 3.5 pesetas. Similarly, in 1879, Cirilo Ustara agreed with Carlos Hodgson, the builder of an aerial ropeway that connected several mines to the Ortuella station, to send ore from the San Antonio mine to that terminus using Hodgson's system. Actually, Hodgson was also in charge of the transfer of the ore to the Triano Railroad. For all these services, Ustara agreed to pay 3.75 pesetas.[135] Since the Triano line tariff from Ortuella was 2 pesetas, Hodgson was charging 1.75 pesetas for transport from the mine to the railway. However, if the contract was renewed for a second year, the price was supposed to go down 0.50 pesetas. In another example, the Chavarris agreed in 1882 with the Bilbao Iron Ore Co. to pay 1 peseta for the transport of ore by an inclined plane from the San Miguel, Begoña, and Diana mines to the Galdames Railroad, and 3 pesetas from the Bodovalle station to the river—4 pesetas from the mine to the harbor.[136]

All of these examples are of mines in the heart of the Triano-Somorrostro district. They were neither at a great distance from the railroad lines, nor far from the harbor. These mines, then, appear to have had transport costs that oscillated between 3.5 and 4 pesetas.[137] Consequently, approximately 35 percent of total production that came from those mines in Triano-Somorrostro not exploited by the Franco-Belge or the Orconera companies had transport costs that hovered around 3.75 pesetas.[138] This average might also be extended to the Galdames mines. Although these mines were farther from the river than those of Somorrostro, they did not require the use of different types of transport, which compensated for the longer distance to the port.[139]

Taking into consideration the different costs and the relative weight of the total production of Orconera, Franco-Belge, and Martínez Rivas, the average transport cost was around 2.60 pesetas per ton, a result that does not diverge much from García Merino's calculations. Nevertheless, his calculation of the

TABLE 14
Iron Trade Profits

Mining Company	Profits (thousands of pesetas)													
	1884	1885	1886	1887	1890	1891	1892	1893	1894	1895	1896	1897	1898	1899
Orconera[1]	902	1,007	1,064	1,205	1,243	1,268	1,388	1,449	1,405	1,384	1,547	1,547	1,853	1,695
Orconera[2]	1,043	1,257	1,487	999	1,698	1,640	1,284	1,982	2,155	1,882	2,022	4,290	3,958	3,143
Franco-Belge	880	1,080	922	1,046	1,122	1,384	941	1,381	1,589	1,160	1,735	2,082	2,273	2,125
Martínez Rivas[3]	1,421	1,360	1,691	2,177	2,851	2,083	1,650	1,688	1,909	2,415	4,140	6,427	8,053	5,400
Others	5,475	5,659	5,099	8,965	10,815	5,365	5,625	2,572	2,542	3,753	11,347	21,475	31,171	30,969
Total	9,721	10,363	10,263	14,392	17,729	11,740	10,888	9,072	9,600	10,594	20,791	35,821	47,308	43,332
García Merino's estimates		17,139	18,012	24,403	28,800	21,216	26,948	26,139	25,054	25,796	35,481	42,063	50,961	47,631

Source: I.F.A. and B.O.V.

Notes: [1]profits from sales to the company's partners; [2]the profits gained from sales to outside customers; [3]based only on the output of the Unión and Amistosa mines.

total profits generated by the mining industry are probably too high for two reasons. First, not all the companies sold the ore at market prices. In fact, the Franco-Belge sold ore only to its partners at a price that was fixed at 2.75 francs above the company's costs. The Orconera followed a similar strategy. Approximately six hundred thousand tons of its total production were sold to its partners at a set profit of 1 shilling 7 pence above the cost. The rest of the Orconera's output was sold at market prices to any buyer. These sales accounted for a good percentage of the company's profit, as can be seen in Table I-4 in Appendix I. However, the combined production of the Orconera and the Franco-Belge sold at below market price was sizable and must be taken into account when trying to estimate profits.

The second reason for García Merino's high estimates appears to be that his average sale price reflects more the price of the *campanil* type than that of the *rubio* variety of ore, which accounted for the bulk of the production.[140]

I recalculated the profits generated by the mining industry. Unfortunately, I could not complete the series for lack of data on the Orconera's, the Franco-Belge's, and Martínez Rivas's production for certain years. Nevertheless, my results for the years available are significantly below García Merino's estimates, as can be seen on table 14.

A rough calculation for 1888 and 1889 may give us a general idea of total profits. Given that the prices for 1888–1889 were equal to the average of the prices for 1887 and 1890, and that costs and production levels were roughly the same for both sets of years, the average profits must also be approximately equal. Hence, for 1888–1889, one may assume a gain of 32 million pesetas. A similar analysis based on the average price and production for the years 1877–1883 gives a gain of 53 million pesetas.[141] Hence for the period 1877–1899, profits equal approximately 293,553,000 pesetas. However, to this last sum one may add the royalties paid by the Orconera and the Franco-Belge to the Ibarras, which were around 30.5 million pesetas.[142] The total profits were, then, 324,053,000 pesetas, a much lower figure than the 529 million estimated by García Merino for the same period.[143]

The local entrepreneurs seem to have pocketed the largest portion of the 324 million. Around 30 million went to the Ibarras as royalties. Of the remaining 294 million, around 23 percent, or 67.6 million, was the Orconera's share. The Franco-Belge took around 9.6 percent (28.2 million). The Unión and Amistosa mines, worked by Martínez Rivas, accounted for 18 percent (53 million) of the gains. Thus these three producers accounted for 148.8 of the 294 million in profits. Given their respective share of production, 80 percent of the remaining profits, approximately 116 million, was retained by Vizcayan mining companies, while the other 20 percent (30 million) went into the coffers of foreign companies. However, since most of the latter paid royalties, their share should be reduced to 20 million.[144] And since the Ibarras owned 25 percent of the Orconera and 10 percent of the Franco-Belge, their portion of the total profits must be increased to 49.7 million. Bearing all this in mind, and without counting the production of the Unión and the Amistosa mines, the Vizcayans' share of total profits was 165.7 million. If one adds the share of those two mines, the local entrepreneurs earned 218.7 million pesetas. However, since Martínez Rivas's arrangement with the Somorrostro Iron Ore Co. remains unclear, it can be said only that Vizcayan business leaders obtained between 56.3 and 74 percent of the total profits.

Despite the reduction in overall profits, the 324 million pesetas represented a tremendous amount of capital. To put things in perspective, in the 1880s, the installation of a modern steel factory in Bilbao required around 10 million pesetas, which was about the same amount that the ore trade generated in just one year during the early years of that decade.

In addition to the 324 million pesetas generated through the extraction and sale of ore, services such as transport also produced high revenues. The Orconera, Franco-Belga, Galdames, and Triano lines accounted for approximately 90 percent of the mineral transported through railroads. During the 1877–1899 period, the Triano line, owned by the Vizcayan government, earned slightly more than 30 million pesetas in profits.[145] Unfortunately, there are no good statistics available for the other companies. However, table I-6 in Appendix I shows that the Orconera's profit on the transport of ore for other parties averaged around 2.30 pesetas per ton. Since the Orconera transported around 4 million tons for other companies, its profits from this traffic amounted to 9.2 million pesetas.[146] Assuming that the Franco-Belge had a similar profit margin, and estimating that about one-third (3.6 million tons) of its railroad load was occupied by noncompany ore, this traffic produced around 8.28 million pesetas.[147]

Finally, the Galdames line transported around 15 million tons during the years 1877–1899. Since this line serviced a longer route than the others, it probably had higher operational costs and, therefore, lower profit margins. Assuming that the latter were only 2 pesetas per ton, the line netted approximately 30 million pesetas, since it transported around 15 million tons. Thus the total profit of the railroad traffic amounted to 77.98 million pesetas.

Again, a good portion of the 77.98 million pesetas remained in Vizcayan hands. First, the 30 million earned by the Triano line stayed in Vizcaya; second, the Ibarras's 25 percent stake in the Orconera meant an additional 2.3 million, and their 10 percent share in the Franco-Belge produced around 0.83 million; third, the 3-penny royalty that Uriguen Vildósola received from the company running the Galdames line generated 187,000 pounds (approximately 5.2 million pesetas). All these sums add up to 38.33 million pesetas, thus indicating that the profits were almost evenly distributed between local and foreign companies.

The benefits that mining brought to Bilbao cannot be measured only by these profits. The close relationship that mine owners like the Ibarras established with the most important European steel producers facilitated the transfer of technology and business practices, which permitted the modernization of the industry in Vizcaya. Similarly, the high revenues that the provincial government collected from its Triano Railroad permitted Vizcayans to enjoy one of the lowest tax rates in Spain. Thanks to these conditions and the Vizcayans' entrepreneurial inclinations, the Nervión Basin was ripe for rapid economic development. Awash with cash, important mine owners like the Ibarras, the Chavarris, the Gandarias, and the Aranas boosted the local economy by investing in new factories, railroads, and financial institutions. To be sure, the origins of Vizcaya's industrialization preceded the mining boom. Yet, the boom instilled confidence in the local business community; this confidence together with the capital generated through the mines caused a quantum leap in Vizcaya's economy during the 1880s.

Chapter Three

General Economic Development

Don't even mention that fortune placed mountains of iron at our hands' reach. There are provinces in Spain where . . . more millions have been made than in ours, and yet they can't be compared in wealth. What have they done with their money? Bilbao's wealth comes from its people.

Miguel de Unamuno, "Del Bilbao mercantil al Industrial"
(all translations are mine)

Coming from one of Bilbao's most famous sons, Unamuno's reference may seem biased. Yet, it addresses the fact that other Spanish provinces, perhaps better or as endowed as Vizcaya, failed to prosper while the Basque province became the center of the country's heavy industry. In fact, the Vizcayan case must be analyzed as a regional exception to the overall poor performance of the Spanish economy during the nineteenth century. Bilbao's transformation from a small port city in the early 1850s into an industrial region in fewer than fifty years was a fascinating process that caused admiration not only among local observers like Unamuno but also among more detached commentators.[1] What caused this growth? Historians are still debating the questions Unamuno raised: How crucial were the existence of the mineral deposits and the ore exports to the region's industrial development? What was the role of the local businessmen in this process? This chapter describes Vizcaya's economic development by focusing on these questions.

Despite a long tradition of manufacturing activities, the start of Vizcaya's modern industrial development lagged considerably when compared not only with the earliest European industrialization but also with that of other regions in Spain. To a large extent, this delay is explained by the political turmoil and wars that ravaged the region between the 1790s and the late 1830s. In 1794 and 1808, Bilbao was looted by French troops; during the Carlist War of the 1830s, the city withstood prolonged sieges, which, despite the earlier capitulations to the French army, earned it the title of "unconquered." No other region in Spain suffered as much from war and civil strife. The contrast between Catalonia and Vizcaya during the 1830s illustrates this point: while the Mediterranean region was beginning to mechanize its textile industry, the Basque country became the site of a seven-year civil war that paralyzed its economy.

In 1840, the advent of peace paved the way for rebuilding the battered Vizcayan economy. While the local economy expanded almost constantly during the next six decades, the rate of expansion was not even. During the interregnum between the two Carlist Wars (1840–1872), the economy grew at a moderate pace; a new outbreak of civil war (1872–1875) brought new dislocations; but from 1876 on, growth was explosive.[2] Thus the end of the Second Carlist War seems to constitute a turning point in the economic history of the region. Since this period coincided with the peak years of the mineral export boom, it could also be argued that mining was the engine behind the economic growth of the region. These lines of reasoning neglect the fact that, during the interwar period, there were attempts to modernize Vizcaya's industrial sector after years of decay. The rapid industrialization of the last quarter of the nineteenth century was not a *creatio ex-nihilo;* it was built on these modernization attempts of the 1840s–1870s, which were another facet of the long manufacturing tradition of the region. In this sense, Vizcaya was not unlike other regions, where protoindustrial activities helped establish the conditions for modern economic growth.[3]

The Economy, 1840–1875

Seldom studied, the 1840–1875 period of Vizcaya's economic history has remained eclipsed by the more radiant years of the last quarter of the nineteenth century.[4] To a large degree, this neglect may be attributed to a lack of sources. Indeed, the biggest hindrance that economic historians face is not just to find reliable statistics, but to locate any statistics at all. Because of their fiscal exemptions, the Basque provinces were excluded from most nineteenth-century Spanish economic surveys. This omission would not have been so grave had the local government gathered information on the state of the economy. Unfortunately, this was not the case. Although the local authorities tried to compile statistics regarding the general wealth of the province, their attempts were always thwarted.[5]

Indirect ways of learning about the economy during this period are also difficult to devise. For instance, there are no tax rolls to use as a broad measure of regional development, since Vizcayans did not have to pay any direct taxes.[6] Similarly, important local sources such as the mercantile register, which has yielded interesting results in other regional studies, seem to have disappeared in Vizcaya.[7] Finally, the unfortunate closing of the Bil-

TABLE 15
Vizcaya's Population, 1704–1847

Year	No. of Inhabitants
1704	56,145
1787	116,042
1797	111,436
1822	112,802
1826	144,875
1826	132,004
1831	113,487
1832	113,598
1836	113,079
1837	111,436
1840	112,518
1842	96,755
1843	143,012
1844	92,115
1846	118,598
1847	120,626
1847	152,752

Source: Madoz, *Diccionario,* vol. XV, p. 402.

bao Municipal Archive because of severe flood damage in the late 1970s has aggravated the problem. The economic history of this period thus must be reconstructed from fragmentary sources.

By their very nature, fragmentary sources are uneven in quality; therefore, the historical contours drawn from them are not neatly delineated and become even more blurred when analyzed at close range. Yet, despite the loss of important details, the main features of Vizcaya's economic history can be recaptured. The population figures for the first half of the nineteenth century illustrate this point. Pascual Madoz's geographical and statistical dictionary records the estimates given in table 15.[8]

Taken separately, the figures Madoz cites appear to be inconsistent. For instance, the two figures given for 1826 do not agree, but both show an increase in the population after the Napoleonic wars. If such an increase took place, then the population numbers for 1831 and 1832 cannot be accurate.[9] Nothing happened during the intervening years that might have caused a 20 percent reduction of the population. The figures for 1836 and 1837, however, may well reflect a decline in the population due to the First Carlist War. Yet, the figures are more indicative than accurate, since no recount could have taken place during the war. The oscillations shown by the 1840s figures also cast severe doubts on their reliability. However, the 143,012 inhabitants recorded in 1843, and the almost 153,000 Vizcayans estimated in the second 1847 citation seem to be the most probable figures in light of the results of the first modern census, taken in 1860, which recorded 168,000 inhabitants.

Even if they are not completely accurate, these statistics do indicate at least a trend. After growing at a much faster pace than the national average during the prosperous eighteenth century, Vizcaya's population fell behind the national growth rate during the first forty years of the nineteenth century and remained stagnant or perhaps even decreased during those years. Following the First Carlist War (1833–1839), a noticeable increase took place, reflecting the overall recovery of the region after the troubles earlier in the century. Finally, between 1840 and 1860, Vizcaya's rate of population growth caught up with the national average.[10]

If it is hard to establish accurately Vizcaya's population prior to 1860, it is even more difficult to determine its distribution according to occupational classifications. The only information available is a report published in 1842 by Vizcaya's civil governor, Julián Luna. According to Madoz, this report was severely flawed, since, as a representative of the central government, Luna had to confront the outright hostility of the local population and authorities, who did not want to have their affairs probed by officials from Madrid.[11] Luna's occupational distribution of the population shows that 60 percent of the working inhabitants were engaged in agricultural activities, 15 percent were classified as artisans, and the remaining 25 percent were involved in the service sector (commerce, liberal professions, etc.).[12] The percentage of rural employment in Luna's figures is perhaps too low. Aguirreazkuenaga mentions that Luna did not classify 40.5 percent of the total population; among this group, the majority lived in rural areas.[13] Other sources also stress the fact that the majority of the population lived in rural areas. As late as 1857, Bilbao was the sole city in Vizcaya with more than ten thousand inhabitants. Only four other settlements had populations between two thousand and four thousand; the majority of the population lived in hamlets with fewer than two hundred inhabitants.[14] Although it was slightly urbanized, Vizcayan society did not differ much during those years from the rest of Spain, where rural activities occupied the great majority of the population.

The 1860 census confirms this conclusion. Even within the judicial district of Bilbao, the most urbanized and economically diversified in Vizcaya, agricultural workers constituted 60 percent of the labor force. The figure may be exaggerated, but undoubtedly their numbers were great.[15] Nevertheless, the percentage of agricultural workers was probably shrinking during the 1850s and the 1860s. In 1869, a census of the city of Bilbao recorded only 138 agricultural workers living within the city limits.[16] This small number cannot have been unique to the city, since, in fact, many of the manufacturing activities that prospered during this period were performed beyond the city limits.

Despite the large number of rural workers, Vizcaya's agricultural production did not satisfy the needs of the province.[17] Trade and manufacturing provided the local population with the resources to pay for the food that the soil did not yield. Unfortunately, there is no thorough study of the region's trade, but its contributions to the local economy were obvious. As the next sections show, rare was the project that was not financed or sponsored by local merchants. Their ubiquitousness in the economic life of the region is a testimony to the vitality and resilience of Bilbao's commerce.

Commerce

With fewer than twenty thousand inhabitants, Bilbao was not a large port in the mid-nineteenth century. In 1856, according to the value of its trade, Bilbao ranked tenth among the major Spanish ports, behind the most important Mediterranean harbors, and behind its old rival, Santander, on the Cantabrian seaboard. During the mid-1860s, however, Bilbao became the second most important port in Spain after Barcelona.[18] Thus its significance as a harbor in modern Spain's economic life preceded the mining boom.

Bilbao's ascent toward its privileged position on the Cantabrian seacoast was not an easy one. Its two traditional exports, iron products and wool, had greatly declined after the early years of the nineteenth century. The reduction in the iron trade in particular threatened Bilbao's competitive advantages vis-à-vis other harbors. Its traditional markets were shrinking fast, especially after the loss of the American colonies. Restricted to the peninsular market because of high production costs, this trade suffered another setback during the First Carlist War. Production difficulties during those years encouraged the development of iron companies in other regions. In Andalucía, for instance, new factories were established with production methods that were more advanced than those used in Vizcaya.[19] These developments successfully challenged Basque dominance of the peninsular market for iron products.

While the First Carlist War was an unmitigated disaster for Bilbao's trade, its end solved at least one of the most confrontational issues between Vizcaya's authorities and Spain's central government. The Basque privilege to

TABLE 16
Bilbao's International Trade, 1843–1866 (thousands of pesetas)

Year	Imports	Exports	Total
1843	12,200	1,500	13,700
1844	13,360	2,940	16,300
1845	11,890	2,180	14,070
1846	15,270	2,690	17,960
1847	13,630	3,040	16,670
1848	12,840	1,090	13,930
1855	21,000	14,500	35,500
1856	21,420	6,160	27,580
1857	24,430	4,360	28,790
1861	41,250	5,900	47,150
1862	37,750	3,650	41,400
1863	44,750	5,500	50,250
1864	84,250	6,600	90,850
1865	75,750	6,350	82,100
1866	25,250	8,150	33,400
Total	455,040	74,610	529,650

Source: Estadística de Comercio Exterior de España, and B.P.P.

import goods duty free was abolished. No longer did customhouses separate the Basque provinces from Castile, and Bilbao became fully integrated into the Spanish market. Although Bilbao's merchants lost the right to import duty-free products, they at least won easier access to the main peninsular markets, which in some instances had been barred to them. In addition, Bilbao's incorporation into the national market opened new trade opportunities with the remnants of the colonial empire—Cuba, Puerto Rico, and the Philippines. The local merchants involved in international trade adapted well to the new situation.[20] After the First Carlist War, the city's trade increased rapidly, as table 16 shows.

Imports grew consistently throughout the 1840s, the 1850s, and the 1860s, and were always much higher than exports. The trade deficit reflected the situation of the Spanish economy as a whole. The noticeable jump during the mid-1860s indicates the great quantity of railroad-building material imported during those years, when the line linking Bilbao to Tudela was about to be finished. While these increased imports help us understand Bilbao's jump from tenth to second place among the country's busiest harbors, the low 1866 figure reflects the crash in the railroad construction boom, which caused a severe depression in Spain.

The 1850s was the most prosperous decade for the export trade. The 14.5 million pesetas reached in 1855 was the peak of an active trade in grain and flour to England and France during the Crimean War. A Spanish saying cited by the British consul in Bilbao captured the bonanza caused by the war: " 'Agua y sol y guerra en Sebastopol.' Rain and sunshine and war in Sebastopol."[21] During the 1860s, exports never reached the levels of the early 1850s, but, averaging 6 million per year, they were slightly over 2.5 times higher than those for the period 1843–1847.

Great Britain, France, and Norway provided approximately three-fourths of the imports. From Britain came textiles, machinery (especially for the railroads), drugs, and chemical products. France exported luxury items such as silk and textiles. One of Bilbao's oldest and most lucrative commercial activities was the importation of codfish, a basic staple of the Spanish diet, which was bought mainly from Norway. A brisk trade in sugar and cacao imported from Cuba and Venezuela also took place during these years.[22]

In exchange for these products, Bilbao exported grain, flour, wine, and iron ore. Except for the years 1858 and 1863, which must have yielded scarce crops, grain and flour accounted for an average of 72 percent of the total exports from Bilbao during the years 1858–1866. In the mid-1850s, this percentage was possibly even higher.[23] During these decades ore exports, not legalized until 1849, paled when compared to the flour trade. Vizcaya's iron manufactures, although not part of international exchanges, retained their importance on the national level. According to French Consular Reports, the value of the trade of these products consistently surpassed that of all others except for the revenue generated by the sale of flour during the years 1853–1855.

Despite the large volume of flour and grain exported from Bilbao, the Basque port was actually a distant second to its main rival on the Cantabrian seaboard, Santander. However, this last port dominated the Cuban market, especially while Bilbao's trade was mostly directed toward England and France. These countries in turn channeled a good portion of their exports through the Basque port. For instance, in the 1860s, one-fourth of the total value of British imports to Spain passed through Bilbao.[24] Forging these commercial links with the most economically advanced countries in Europe helped the local businessmen realize the potential of the region's natural resources and made them receptive to the idea of seeking foreign partners to develop those resources.

To a large degree, the enterprising activities of the local merchants allowed Bilbao to acquire a preeminent position among Spanish ports. Unlike Santander's commercial community, Bilbao's merchants traded on their own account and not as agents of companies in other cities.[25] This direct involvement led them to create an infrastructure that encouraged the growth of the harbor's traffic. For instance, although during the 1850s Santander's overall trade (measured in pesetas) was larger than Bilbao's, the Vizcayan fleet was second only to Barcelona's, the busiest in Spain. In 1858, the ships registered at the Catalan port numbered 490, with a total tonnage of 69,200; while Bilbao's fleet was 739 vessels strong, its tonnage was assessed only at 68,200.[26] During the 1850s, around 55 percent of Bilbao's ships were small-sail cargoes weighing between 20 and 80 tons and assigned to the northern European or Mediterranean routes. The rest of the fleet, vessels between 80 and 400 tons, was dedicated to transoceanic trade.[27]

Practically all of these vessels were built in shipyards along the Nervión River. Between 1850 and the early 1860s, these yards enjoyed great prosperity.[28]

The feverish activity of the local shipyards explains the rapid increase of Bilbao's fleet from 30,000 tons in 1847 to 68,200 tons in 1858.[29] In the 1860s, construction of new ships declined considerably. According to the French consul, the decrease was due to the vast amount of capital invested in the construction of the railroad line joining Bilbao with the city of Tudela in La Rioja.[30] Yet, even after the line was finished, the shipbuilding industry failed to recover. The advent of the steamer and the iron-hulled vessel signaled the end of the era of the wooden sailing ships built in Bilbao's yards. The local shipyards could not compete against the powerful British companies, which virtually monopolized the construction of steamers during the second half of the nineteenth century.[31] For the most part, this industry disappeared from the Nervión's banks during the 1860s and the 1870s. In the 1880s, however, the tradition resurfaced in a valiant attempt to create a modern shipyard.

At the heyday of the local shipyards, a considerable amount of capital was invested in Bilbao's fleet. In 1846, Madoz calculated its total value at around 20 million pesetas.[32] Yet, his estimate seems overblown for two reasons. First, he exaggerated the total tonnage of the local fleet, which he estimated at six hundred vessels, to be between fifty and fifty-five thousand tons; however, documents from the navy's archive show a thirty-thousand-ton fleet in 1847. Second, he arrived at the 20-million figure by considering the value per ton of a ship at 400 pesetas, without distinguishing between newly constructed and used (therefore depreciated) vessels. A more accurate estimate seems to be the 9.87 million pesetas for the year 1853, which appears in the

Anuario Estadístico de España. In this estimate, the average value per ton was assessed at 187.5 pesetas. The same figure was used in documents I found in the navy's archive, which in 1858 assessed the total value of the fleet at 14.4 million pesetas.[33]

This smaller estimate of investment in the fleet represents a large sum for the period. For instance, the Banco de Bilbao was founded in 1857 with a paid-up capital of two million pesetas. In 1858, the investment in the three major iron factories of the region did not reach 5 million pesetas. With the exception of a railroad line, no other local business could match the level of capital invested in the fleet.

To protect the local fleet against the perils of maritime navigation, the local merchants formed insurance companies. For example, in 1850, forty of the wealthiest Bilbao merchants founded the Unión Bilbaína.[34] This company would eventually disappear, but others took its place. In 1861, seven insurance companies operated in Bilbao: Lloyd Vascongado, La Bilbaína, La Unión, Compañía General de Seguros Marítimos, Lloyd Español, Compañías Unidas, and Lloyd Barcelones. The first two were founded by local merchants; the rest were branches of insurance companies located in Madrid or Barcelona. Despite being outnumbered by outside companies, the two local firms underwrote slightly more than half of the policies issued during the early 1860s.[35] It is hard to determine what happened to these companies after the 1860s, but they seem to have ceased operation, due perhaps to the recession of the mid-1860s or to the Second Carlist War in the mid-1870s. The local maritime insurance business resurfaced, however, at the turn of the century, after a rapid expansion in the number of steamers in Bilbao's fleet.

Insurance and shipping taught the local businessmen the advantages of large partnerships to reduce risks. Prior to the formation of large shipping companies and to the specialization of economic functions in shipping and trading, a merchant preferred to share the ownership of various vessels with colleagues rather than to concentrate interests in fewer ships under sole possession. In this way, not only were risks reduced, but shipments could be scheduled at shorter intervals, since more ships were at the disposal of each merchant.[36] Thus shipping and insurance nurtured a spirit of economic association, which flourished in other businesses as well.

The establishment of the Banco de Bilbao and the construction of the railroad line from the harbor to Tudela also exemplify the local spirit of economic association. The timing for the creation of these two businesses was related to new laws permitting the formation of joint-stock companies to promote banking and railroad companies.[37] Founded in 1857, the Banco de Bilbao was a small institution, but it enjoyed the privilege, together with eight other

Spanish banks, of issuing paper money that served as legal tender.[38] During its first decades, this bank seems to have played a conservative role in the local economy, restricting its activities to the commercial field, acting as a clearing-house for mercantile bills of exchange, and altogether eschewing investment in industrial enterprises. Its only foray outside those limited areas seems to have been the rescue from bankruptcy of the local railroad company.[39]

The Bilbao-Tudela line was a much-needed link with the Castilian markets. After an unsuccessful attempt to make the line from Madrid to Irún (on the French frontier) go through Bilbao, local merchants sponsored the construction of a railroad to Tudela, where it would link with the lines servicing Madrid and Barcelona. Fear of being outperformed by Santander, which was also building a line to Castile's granary regions, accelerated the Vizcayan project. As the construction of both lines was proceeding, the British consul in Bilbao augured the Vizcayan railway a brighter future because of its potential links with the wine-producing region of La Rioja and the Catalan industrial areas.[40]

The Bilbao-Tudela line was an ambitious project that required around 25 million pesetas in investment capital.[41] Bilbao's inhabitants, from rich merchants to small investors, raised those funds.[42] The British consul marveled at the financial resources that a small town like the Vizcayan port could marshal.[43]

The line was finished in 1863, and like those of other railroads in Spain, its early years were extremely disappointing. In fact, the Bilbao-Tudela line went bankrupt in 1866, a year in which the whole Spanish economy took a tumble after a railroad construction boom failed to spur immediate economic growth. The bankruptcy of the line was a hard blow to the local population, which had invested heavily in this railroad. Only the intervention of the Banco de Bilbao averted a situation that might have had even more severe consequences.

While the line was a disappointment in the short run, its construction was necessary if Bilbao was to preserve its position as a port for the Castilian market. Ironically, the massive export of ore reduced Bilbao's dependence on the Spanish market after the 1870s. Nevertheless, the Bilbao-Tudela line retained its importance as a carrier of grain and wine for the region, and as an outlet for Vizcaya's industrial products.

Large projects such as the local railroad and the bank attracted many investors, but the promotion and management of the businesses rested on Bilbao's major merchants.[44] While trade and its related infrastructure remained the main preoccupation of the merchants during this period, they also tried to rejuvenate the manufacturing sector of the local economy.

Manufacturing

The full integration of Vizcaya into the Spanish customs system in 1841 gave a boost to the region's manufacturing activities. Prior to this, Basque iron products suffered because of the imposition of taxes as they entered other Spanish provinces. Similarly, the prohibition on transhipment of certain colonial products to the interior of the peninsula hampered the development of industries related to the colonial trade, such as sugar refining and tanning. Toward the end of the eighteenth century, the central government discouraged by fiscal means even the construction of flour mills in the Bilbao area. Not surprisingly, many of Bilbao's businessmen transferred their activities to neighboring Santander to avoid commercial and fiscal barriers.[45]

With the transfer of the customhouse to the seacoast, some of these industries revived and new ones emerged. Several tanneries, a sugar refinery, a glass factory, small textile workshops serving the shipyards' need for sails and rope, canneries, and paper mills were created during the 1840–1870 period. Yet, the growth of these industries remained limited, never achieving much importance beyond the provincial level.[46] Only the paper mills served as a springboard toward modern plants in the post-1875 period.

The two most important industries during the interwar years were the flour mills and the iron factories. Given the importance of flour among local exports, it is not surprising that the mills prospered after 1840. According to Madoz, there were 621 mills in Vizcaya in the late 1840s.[47] Citing statistics from 1861, another historian mentions that the three Basque provinces possessed 1,024 mills and 9 flour factories powered by steam engines or water. Among the other Spanish provinces, only the Castilian granary regions of León and Valladolid and Santander, their main port, seem to have produced more flour than the Basque country.

The 1,024 mills and 9 factories of the Basque provinces represented a total investment of 3,060,000 pesetas.[48] It is hard to determine how many of those mills and factories were located in Vizcaya. If the investment figure is correct, many of the mills must have been very small, used only for local needs. Mills dedicated to the export of flour required an investment between 125,000 and 375,000 pesetas during the 1860s. At least 5 or 6 mills of that size were located in or around Bilbao.[49] Madoz, for instance, describes a flour factory in Abando (across the river from Bilbao) in which more than 200,000 pesetas had been invested by the late 1840s.[50] Similarly, the factory built in Begoña by Máximo Aguirre, probably the largest in Vizcaya, was valued at 370,000 pesetas in 1863.[51]

The flour business seems to have been highly profitable. For instance, Juan Cruz y Oria (later Arteach and Co.) started in 1864 with capital of 62,500 pesetas to run a mill in Orozco. One year later, the firm expanded by renting a factory near Bilbao for 5,000 pesetas per year.[52] While the company's capital was set at 125,000 pesetas in 1865, three years later the firm declared a net capital of 390,500 pesetas before incorporating a new partner.[53] In just four years the partners multiplied their original investment by approximately a factor of six.

Thanks to those growth rates, flour producers such as Máximo Aguirre were among the richest merchants in Bilbao during this period.[54] Their activities were highly integrated and ran the gamut from production to commercialization and shipping. They were also involved in the formation of the railroad line, the bank, and the insurance companies. As mining became important, some of them invested in this sector.[55] During the 1875–1900 period, the flour industry retained its importance. Among other flour producers, Eduardo Coste and Arteach and Co. were still in business toward the end of the nineteenth century. Yet, as the local population increased rapidly after 1875, the mills' production shifted toward the local instead of the export market.

While perhaps not as profitable as the flour mills during the 1850s and the 1860s, Vizcaya's iron industry also progressed after the First Carlist War. The transfer of customs to the seacoast encouraged changes in the traditional ironworks by granting some tariff protection and assuring access to other Spanish provinces. Until the 1830s, merchants had dominated the iron trade by a means of a credit system that gave them control over the production of the *ferrerías* without requiring that they invest in them directly. By the 1840s, the situation had changed radically; by then the *ferrerías* were irremediably obsolete because of technological advances. New factories required large expenditures, and instead of acting only as bankers, several merchants founded companies that revitalized the local iron industry. Such changes in the merchants' economic activities mirrored the development in other European regions that made the transition from a long-standing manufacturing tradition into a modern industrial sector.[56]

In order to understand fully the transformation of Vizcaya's iron industry, some technological changes must be explained. Until the 1840s, the method used to produce wrought iron locally had remained almost unchanged since the Middle Ages. The *ferrerías* relied on small hearths fueled by charcoal and powered by water. The ore was transformed into wrought iron in a long, direct process that combined the smelting of the mineral and the refining of the metal. The quality of the iron was excellent, but it was very costly, as it consumed vast amounts of fuel and wasted great quantities of ore. Nevertheless, by concentrating on the production of high-quality iron, the *ferrerías* managed to retain their international competitiveness until the late eighteenth century.

In contrast, most of the main European metallurgical centers had long since abandoned the use of the small hearths in favor of blast furnaces. The use of these furnaces required a separation of the smelting and the refining processes. The blast furnace's advantages resided in its much larger output of an intermediate product, cast or pig iron, at lower costs than any *ferrería* could manage. In the early 1700s, the English began smelting the ore with mineral coal instead of charcoal, which reduced costs even further.

The product of the blast furnace was harder than wrought iron, but very brittle.[57] The brittleness was caused by a higher carbon content, an inevitable chemical result produced by the characteristics of the blast furnace. Thus in order to obtain wrought iron, the cast iron had to be refined with charcoal to reduce the carbon content. This created a bottleneck in production. Cast iron could be produced in rather large quantities, but its uses were limited unless it was resmelted.

The problem was solved by the introduction of the puddling system during the 1780s. Puddling permitted the complete substitution of coal for charcoal, since it resmelted the cast iron in a furnace that kept the metal and the fuel separated. If the blast furnace pushed Vizcaya's *ferrerías* out of the international market for low-quality iron, the puddling system forced the Basque ironworks to embrace technological changes or disappear.

Despite the severe threat posed by the new production methods, their adoption in the Basque region had to wait until the end of the First Carlist War. In addition to the unstable military and political conditions of the region, three reasons accounted for the delay. First, the most common type of ore from the Somorrostro region, the *vena,* did not yield good results in blast furnaces. Second, the country lacked mineral coal. Third, the protectionist policy of the Spanish monarchy tended to make the foreign products more expensive.[58]

While the external threat hung over Vizcaya's *ferrerías* like the sword of Damocles, it was the development of iron factories inside Spain that forced the Basque ironworks to modernize. As the First Carlist War was being waged in the 1830s, Manuel Heredia, an Andalucian businessman, built one of the first Spanish blast furnaces fueled by charcoal, in Marbella (Málaga).[59] In 1841, a group of Bilbao merchants took the first step to meet the new competition. Santa Ana de Bolueta, on the outskirts of Bilbao, was Vizcaya's first modern iron factory. Unlike Heredia's factory, Santa Ana did not start as a completely integrated plant, where the mineral was transformed from its natural state into industrial products. During its early years, Santa Ana possessed puddling furnaces and rolling mills, but no blast furnaces.[60]

In 1848, Santa Ana built the first blast furnace in Vizcaya and became a fully integrated mill. Morever, the company expanded to assure the provision of raw materials. It acquired the best ore deposits in the Ollargan district in the vicinity of the factory and forests that provided the charcoal to fuel the furnaces.[61] Lack of mineral coal in the region and high transportation costs made it impossible to use this fuel. In this regard, Santa Ana trod the path of other European countries, such as France, which used charcoal in their iron industry well into the nineteenth century.

Despite the reliance on vegetable fuel, the establishment of Santa Ana signaled a turning point in the industrial history of the region. During the 1840s, this factory alone had the same productive capacity as the one hundred or so existing *ferrerías*.[62]

The example of Santa Ana was followed in 1855 by a new factory, Nuestra Señora del Carmen, established on the banks of the Nervión, in Baracaldo, a short distance from Bilbao. The construction of this factory represented an expansion into manufacturing activities for the principal partners, the Ibarra family, the most important ore traders and mine owners in the region. Prior to the establishment of Nuestra Señora del Carmen, the Ibarras together with other partners had operated a small blast furnace in Guriezo (Santander) since 1846.[63]

Nuestra Señora del Carmen represented another turning point in the industrial history of the region. Whereas the Guriezo and Santa Ana furnaces were fueled by charcoal, Nuestra Señora del Carmen was designed to use mineral coal. The decision to build a new factory instead of expanding the old one was probably related to the fact that the coal had to be imported, and Bilbao's harbor offered greater advantages than Guriezo's location, which could be reached only by small vessels from the coast.

Lack of coal turned out to be a serious hindrance to the development of the factory. In the 1850s, for any given unit of production, the quantity of coal needed to smelt the ore and manufacture iron products was much greater than the combined weights of the ore consumed and the final product obtained.[64] As a result, it was more cost efficient to carry the ore to the coal fields than vice versa. Thus, factories located in the vicinity of the coal mines had a tremendous advantage over those near the ore deposits.

By the mid-nineteenth century, the only known coal deposits in Spain were located in Asturias. The development of these fields had been delayed by their difficult access. In the late 1850s, once railroads were built, the fields were opened, and modern iron factories were erected near them.[65] In the mid-1860s, Asturias surpassed Vizcaya and Andalucía in the production of wrought iron. As table 17 shows, production costs at the Felguera factory in Asturias were remarkably lower than those of Santa Ana or Nuestra Señora del Carmen. At the same time, the production costs at Vizcaya's factories were lower than those in Málaga. While Vizcaya's factories enjoyed an ad-

TABLE 17
Production Costs of Iron, 1865 (pesetas per ton)

		Vizcaya		
	Asturias La Felguera	Nuestra Señora del Carmen	Santa Ana	Málaga
Pig iron	85	106	125	158
Wrought iron	152	194	218	294

Source: Fernández Pinedo, "Nacimiento y consolidación," p. 13. The figure for Málaga was taken from Nadal, *El fracaso,* p. 173.

vantage over the Asturian companies in the cost of the ore (8.75 pesetas per ton versus 14.87 pesetas), the Asturians' edge in the cost of coke (19.75 pesetas per ton versus 54.25) was decisive.[66]

Given the importance of coal for their factory, it is not surprising that the Ibarras helped explore Vizcaya's neighboring provinces in search of this mineral. In 1864, they acquired, together with other partners, five mines in the province of La Rioja. Thus was born the Vasco-Riojana Co., with capital of 175,000 pesetas. Transportation from these mines to the Ibarras' factory would have been eased by the Bilbao-Tudela line. Yet, the lack of other information about the mines indicates that they were perhaps not productive enough to warrant exploitation.[67] Despite this disappointing result, the example illustrates the propensity of Vizcaya's businessmen to look for economic opportunities beyond the confines of their province. Such a tendency became stronger after the Second Carlist War, when, armed with greater financial resources, local businessmen created many new enterprises in other Spanish provinces.

The failure to find coal near the Bilbao region forced Ibarra and Co. to bring the mineral from Asturias or Great Britain. It also made the company experiment with other technologies. The Ibarras built a second blast furnace fueled by charcoal, but the product obtained through this method was even less competitive than the iron manufactured with coke. They also produced wrought iron by the Chenot system, a method that circumvented the use of blast furnaces by manufacturing the wrought iron directly from the ore. Yet, the wrought iron obtained was as expensive as that produced with charcoal.[68] Despite these serious problems, the factory grew to become the largest industrial concern in Vizcaya. In 1866, its investment in fixed capital amounted to

3 million pesetas. In contrast, Santa Ana's partners had invested only 1.4 million pesetas in fixed capital, even though this was a significant expansion for a company that had started with an initial investment of 200,000 pesetas in the 1840s.[69]

Nuestra Señora del Carmen and Santa Ana are the best known and the most successful factories established during this period; however, they were not the only ones. Other important Bilbao merchants turned their eyes toward iron manufacturing. For instance, Pedro Errazquin joined a group of investors from Madrid to create the Santa Agueda factory in Castrejana. Founded in 1859 with capital of 1,025,000 pesetas, this company appeared to be a serious competitor to Santa Ana and Nuestra Señora del Carmen. It was an integrated mill that smelted the ore in a blast furnace fueled by charcoal, and then transformed the pig iron into wrought iron by the puddling system. It also possessed rolling mills to give the wrought iron different shapes.[70] Yet, for reasons that are unclear, this company never achieved the prominence of its two local competitors. Lack of results, moreover, led to several changes in ownership.

There were unsuccessful attempts to revamp the *ferrerías* by relying on a system patented in France: the Tourangin furnace. Like the old method, the Tourangin furnace obtained the wrought iron directly from the ore. Although it used less fuel than the old hearths had, it was not efficient enough to compete with the metallurgical trilogy of the blast furnace, the puddlers, and the rolling mills.

While the preceding examples deal with companies whose final product was mostly iron as a raw material for further processing, there was at least one attempt to create a modern factory to manufacture beds, stoves, and other iron products. San Pedro de Deusto (later Agapito Zarraoa and Co.) was founded in 1862 by a group of local merchants. The main sponsor of the project, Agapito Zarraoa, already possessed a similar factory in Valladolid. Yet, as he explained to his partners, Bilbao's advantages over other potential locations like Asturias and Valladolid encouraged the establishment of the factory on the Nervión's banks. Unlike Asturias, Bilbao had a railroad line that facilitated access to the internal markets of the peninsula. Transportation costs also worked against Valladolid, where coal had to be taken overland, while this fuel reached Bilbao in cheaper sea carriers. Finally, the development in Bilbao of factories like Santa Ana and Nuestra Señora del Carmen assured the supply of iron bars at good prices.[71]

All these factories were built on a scale that surpassed the local needs of the province. In fact, they were created with the idea of supplying the needs of the whole Spanish market. Excessive production costs made exports unthinkable, but the supply of the national market was not a chimera. Tariff protection kept international competition at bay, except for a major loophole in the protectionist defenses the Spanish state erected: the import privileges of the railroad companies. In order to accelerate the construction of the lines, the Spanish government granted the railway companies the right to import duty free all the material necessary to build and operate them. Between 1855 and 1865, the length of the Spanish network grew from five hundred kilometers to five thousand kilometers. The bulk of this construction boom was financed by foreign companies. Among these firms, French investors played the most important role. Despite the predominance of French capital, most of the railroad material came from Great Britain. The enjoyment of the import privilege was not restricted to foreign companies only. The Bilbao-Tudela railroad, financed with local capital, also imported all its materials from Great Britain.

For some historians, the privileges given to the railway company represent a lost opportunity for the Spanish iron industry.[72] Antonio Gómez Mendoza has contested this thesis, however.[73] He argues that a protected national iron industry would have had to grow by a factor of twelve during the 1855–1865 decade to satisfy the demand of the railway companies. According to Gómez Mendoza, such an increase would have been impossible.

Analyzing the statistics in a different way, one may arrive at a different conclusion. The factor of twelve estimate seems exaggerated. Gómez Mendoza builds a counterfactual model in which he assumes that Spain's industry would be able to supply the demand of the railroad companies in any given year during the 1855–1865 period in the same way as foreign companies did. In 1855, the Spanish industry was producing approximately 13,000 tons of pig iron. Gómez Mendoza estimates that in 1865 the railroad companies used 155,000 tons. Thus he reasons that, in order to meet that need, the industry had to increase production twelve times from its 1855 level.

The choice of years exaggerates Gómez Mendoza's results. By 1865, the Spanish iron industry had actually increased its production to almost fifty thousand tons annually.[74] Thus, in order to meet the railroad companies' demand, Spanish iron producers needed to produce at least four times more, not twelve.[75] To increase production four times was still a difficult task, but not impossible. In 1865, the Basque factories were utilizing only a third of their capacity. Working at full capacity, Santa Ana and Nuestra Señora del Carmen were able to produce thirty-four thousand tons of pig iron annually.[76] Similarly, the Asturian factories, also underutilized during those years, may have reached similar production figures.[77] Overall, then, without any new investment, the production of the Spanish iron industry could easily have been close to eighty thousand tons of pig iron. While this still could not meet the requirements of the railroad companies, the investment needed to increase production to the desired level might not have been as impossible to achieve had output had to be multiplied by twelve. Moreover, 1865 was a peak year for demand. Over the whole 1855–1865 period, as Gómez Mendoza himself points out, the needs of the railroad companies would have required only a doubling of the production of pig iron in Spain.

Although nobody can deny the importance of railroad lines for a country with the orographic characteristics of Spain, the widespread bankruptcies the railroad companies suffered in 1866 suggests that construction might have proceeded at a slower pace without harming the economy as a whole.[78] After all, the network for all intents and purposes did not grow between 1865 and 1875. That extra decade could have given the iron industry additional time to catch up. The question is whether the foreign capital that financed most of the construction would still have been invested under different circumstances. No one can answer that. Yet, had foreign investment been available at the same level, supplying the railroad companies with Spanish rails would not have been impossible.

It was not just the loss of the railroad market, however, that caused the slow development of Spain's iron factories. In fact, in England, the United States, and France, the effect of railroad construction on the growth of the metallurgical industry was relatively minor.[79] In these countries, the agricultural sector provided a strong market that permitted the rapid expansion of the iron industry. In Spain, however, the agricultural sector grew very slowly, without much investment in machinery and tools.

Urban renewal projects, so common in most of Europe during the 1850s and the 1860s, also provided outlets for the iron industries of the most advanced nations. Bilbao, for instance, had to build water conduits to satisfy the needs of its inhabitants during the late 1860s. The iron tubes for the project were provided by Santa Ana and Zarraoa and Co. during the late 1860s.[80] Yet, these orders were not sufficient to warrant the rapid expansion of the industry, since their scale was limited by Spain's slow urbanization.

Facing a narrow national market, Bilbao's industrialists tried to sponsor other businesses to assure outlets for their factories. For instance, the Ibarras and their partners became associated with a Catalan metallurgical company, Herluison and Co., which was a buyer of iron products from the Guriezo factory in 1855.[81] In 1857, Ibarra Bros. and Co. also appeared as a partner in the Guipúzcoan iron factory of Nicolás Soraluce and Co.; and in 1862, the Ibarras participated in the formation of a weapons company in Placencia (Guipúzcoa), Zuarubircar Isla and Co.[82]

Despite the industrialists' efforts to increase the demand for their products, the inertia of the national market could not be overcome, and the production level of Bilbao's factories remained painfully low when compared to French or German levels. Given the technological and economic conditions of the period, the local iron industry had reached a plateau.

While the growth of the 1840–1875 period was moderate, the region at least rekindled its economic vitality. The resources and conditions created by the mining boom were not yet present; however, thanks to enterprising local merchants, the region recovered rapidly from the wars and political instability of the early nineteenth century. The city grew to become one of the most important ports in Spain. Its merchants built a railroad, founded a bank, created insurance companies, and possessed one of the largest commercial fleets in Spain. Flour mills, shipyards, and iron factories were the manufacturing pillars of the economy, although two of those pillars were on shaky grounds toward the end of the interwar period. The shipyards collapsed in the early 1860s, but during the 1840s and the 1850s, their work helped local businessmen regain a preeminent position in shipping.

Unlike the shipyards, which were phased out by new technology, the local iron industry was gradually breaking away from the obsolete methods of the *ferrerías.* It adopted production processes that were similar to those used in other European countries that lacked ample coal deposits. Its slow growth was mostly due to the lack of this fuel. Compounding this problem, scarce internal demand dampened the possibilities of rapid expansion. After the Second Carlist War, the lack of coal was partly overcome by new production techniques, and as the industry became competitive, it even found markets beyond Spain's frontiers and permitted the resurrection of that other traditional Vizcayan industry: shipbuilding. Thus, although Vizcaya's iron industry reached a plateau during the 1860s, factories like Nuestra Señora del Carmen were the foundation on which new companies would emerge during the next period.

The Economy after the Second Carlist War

The Second Carlist War (1873–1875) was shorter and less destructive than the first; as in the 1840s, the local economy recovered quickly with the added boost of ore exports. By almost any measure, growth during this period would warrant the use of catchphrases like a "Vizcayan economic miracle." While the population, for instance, had increased by only about 50 percent in the first half of the nineteenth century, it had almost doubled by 1900, from 160,000 inhabitants in 1857 to 311,000. The bulk of the increase was centered around the mining district and the Nervión Basin, where the population actually quadrupled between 1857 and 1900.[83]

During the 1858–1900 period, the median population rate growth in the Basque provinces was twice as high as that for Spain as a whole; no other Spanish region grew at such a rapid pace.[84] Moreover, the rate for the three Basque provinces especially reflects Vizcaya's rapid growth, since Alava actually lost population and Guipúzcoa grew by only 20 percent during this period.[85]

Much of the population growth during this period was due to immigrants from other Spanish regions. For the first time in modern history, the Basque region attracted population instead of exporting it. Lured by work in the mines during the 1870s and the 1880s, the immigrants eventually found jobs in other sectors of the economy. In 1887, 41.5 percent of the male population of the judicial districts of Valmaseda (the site of most of Vizcaya's mines) and 31.6 percent of Bilbao's male inhabitants had been born outside the province. In 1900, the proportion of immigrants increased to 47.7 percent in Valmaseda, and to 38.2 percent in Bilbao.[86] In certain places, such as Baracaldo (the site of the Ibarras' factory), 73 percent of the population growth during the 1857–1900 period can be attributed to immigration.[87] Although there are no comprehensive studies about the origins of these immigrants, the few available monographs tend to show that they came from northern Castile and from other Basque provinces.[88] This large immigratory wave caused changes in the occupational distribution of the Vizcayan population. By 1887, the industrial and the service sectors had nearly as many jobs as agriculture. By 1900, the proportion of the population engaged in primary sector employment (42 percent) and that involved in secondary and tertiary sector jobs (58 percent) was reversed from 1860 levels.[89] Manufacturing jobs accounted for two-thirds of secondary sector employment; mining occupied the other third.[90]

As late as 1900, the primary sector in Spain occupied almost two-thirds of the active population. In Vizcaya, however, the proportion was almost reversed. The only other region in Spain with this kind of occupational distribution was the province of Barcelona. In the Catalan region, only 39 percent of the inhabitants were occupied in the primary sector; 35 percent worked in industrial or mining activities, and the remaining 26 percent were engaged in the service sector.[91] Although Barcelona's and Vizcaya's work forces were proportionally comparable by the end of the nineteenth century, the Catalan region had experienced a more gradual growth than had the Basque province. In the 1860s, the population engaged in the primary sector in the province of Barcelona was already around 37 percent. In contrast, the majority of Bilbao's industrial and mining jobs were created between 1880 and 1900.

The changing distribution of Vizcaya's work force was made possible by an investment boom from 1886 to 1901. According to the Mercantile Register, 904 new companies were founded with a nominal capital of 1.05 billion pesetas. Approximately half of that figure was invested in 1900 and 1901. During 1901, the creation of companies in Vizcaya worth 483 million pesetas represented 55 percent of all new investments recorded by the Mercantile Register for all of Spain.[92] Not all these millions were actually paid-up capital. In fact, the Mercantile Register tends to exaggerate the amount of investment, since it usually recorded a company's nominal capital, which could be inflated when compared with the actual disbursement of its partners.[93] Nevertheless, the Mercantile Register gives a rough indication of the vibrant Vizcayan economy at the beginning of the twentieth century.

The Origins of Investment Capital

Where did the investment capital that permitted the creation of so many companies come from? This question has caused a heated debate among historians. On the one hand, Manuel González Portilla, Manuel Montero, and Vicente García Merino have emphasized the importance of the capital accumulated through mining and its subsequent investment in other industries. On the other hand, Emiliano Fernández Pinedo, Antonio Escudero, and Jesús Valdaliso Gago tend to underplay the direct contribution of mining to the development of the region.

Stressing the importance of mining does not necessarily preclude the possibility of other sources of investment funds. Still, some historians have adopted a narrow focus to determine the origin of those funds. For instance, Manuel Montero has argued that mining was the "only activity" that could have generated the capital behind Vizcaya's investment boom at the end of the nineteenth century.[94] While Montero has shown that the fluctuations in profits created by the export of ore and Vizcaya's investment cycles were synchronous, his evidence does not seem strong enough to buttress his argument. He calculates that during the 1878–1898 period the profits accrued by the mining industry amounted to approximately 405 million pesetas. This calculation, however, does not seem to take into account that not all the ore was sold at market price. Bearing this in mind, I calculate in chapter 2 that the profits for the 1877–1899 period amounted to approximately 324 million pesetas. Not all these funds belonged to Vizcayan businessmen; I estimate their share to be between 56 and 74 percent of the total profits. Even under the best circumstances, in which local entrepreneurs captured 240 million pesetas, this figure alone cannot account for all investment in the Vizcayan economy during the 1880–1900 period.

First, it is highly unlikely that all capital was utilized in productive activities. Second, even if it was, the 240 million pesetas still fall short of the funds invested. According to the British consul, between 1886 and 1899, over 14 million British pounds (around 420 million pesetas) were invested in new companies.[95] It is possible that the consul's figures reflect nominal capital instead of actual investment. On the other hand, his sums do not include the capital used in the creation of three of the most important Vizcayan steel factories, in the renewal of the local fleet, and in other companies founded during the early 1880s. These investments might have added at least another 100 million pesetas to the consul's estimate.[96] While actual investment figures are difficult to determine, it would not be farfetched to consider that the figure hovered around 400 million pesetas during the last quarter of the nineteenth century. If that was the case, the 240 million pesetas pocketed by the local entrepreneurs as profits from the mining industry could have accounted for only approximately 60 percent of total Vizcayan investment. This was a significant proportion, but it was not enough to brand mining as the only activity that could have financed the region's economic development.

Where did the rest of the money come from? There are only two possibilities: from other local savings, or from businesses outside Vizcaya. Montero downplays the first alternative and does not even consider the second. His analysis divides local businessmen into two groups: a "traditional bourgeoisie" more prone to direct its capital into government bonds or commercial activities; and an industrializing bourgeoisie connected to the mining industry and willing to finance new manufacturing ventures. By the 1890s, according to Montero, both groups participated in the creation of new industries, but those tied to the mining industry continued to play a more prominent role.

While businessmen tied to the mining industry did play a key role in the economic development of the region, Montero forgets to mention the presence of other Vizcayan and outside investors, which undermines his thesis. Fernández Pinedo and other historians trying to underplay the relevance of mining have called attention to those other investors.[97] Although my analysis has focused on just a few companies, it has shown that the mine owners or mine exploiters did not always contribute the majority of the local industries' capital. In the case of the Altos Hornos de Bilbao steel factory, investors from outside the province provided 57 percent of the capital. Similarly, the largest single stockholder of La Vizcaya, another steel factory, was the Olano Larrínaga shipping company, a partnership of Vizcayan businessmen based in Liverpool.[98]

While these examples show the important role played by investors from outside the province, there were also cases of businessmen from what Montero calls the "traditional bourgeoisie" who contributed to the formation of companies that transformed the local economy. Many local entrepreneurs actually defy classification, since they participated in more than one economic activity at a time.[99] This fact, which illustrates the entrepreneurial spirit of the local businessmen, has not been sufficiently stressed by historians on both sides of the debate. For instance, when Fernández Pinedo analyzes the list of investors in the Bilbaína de Navegación shipping company, he mentions that, among all the stockholders, only the Chavarri brothers had links to the mining industry. However, other founders, such as Benigno Salazar, Pedro Darío Arana, Ezequiel Uriguen, and Eduardo Coste, also had investments in mines.[100] Fernández Pinedo's overall point is valid: not just investors related to the mining industry were propelling the region's economy. However, by focusing so narrowly on the investors, he fails to see the overall strategic importance of the mining industry. It is doubtful that non-Vizcayan investors would have been attracted to the region had the mining boom and technological developments like the Bessemer converter not given rise to new economic opportunities.

Industrial Modernization in Vizcaya, 1880–1900

The modernization of the iron and steel industry toward the end of the 1870s and during the early 1880s demonstrates how local investors in partnership with outsiders took advantage of those opportunities. Given the long history of iron manufacturing in the region, Vizcaya's businessmen soon realized that the new technological developments that made the Basque mineral a coveted commodity in the international market also permitted the processing of the ore in situ. Between 1879 and 1882, three modern factories with powerful blast furnaces fueled by coke were erected on the left bank of the Nervión River: San Francisco de Mudela, Altos Hornos de Bilbao, and Sociedad Anónima de Metalurgia y Construcciones Vizcaya (commonly known as La Vizcaya).

The San Francisco had first been conceived as the project of a British company, the Bilbao River and Cantabrian Railroad. This company leased mines in the Galdames district during the early 1870s, and planned to manufacture iron bars in its local factory before sending them to England. In 1871, the firm bought land on the Nervión banks from a local businessman, Ciriaco Linares, who in the process became a partner in the enterprise.[101] The Second Carlist War and the slow development of the Galdames district made the company abandon the project. In 1879, it sold the factory, still under construction, to Francisco de las Rivas (Marqués de Mudela) for 0.75 million pesetas.[102] Rivas's investment in the factory was accompanied by the long-term leasing of mines in the Somorrostro district to assure a supply of ore.

Under the new owner, the factory prospered rapidly. By 1881, its four blast furnaces fueled by coke were producing almost thirty-six thousand tons of pig iron, approximately a third of total Spanish production that year.[103] Although a native Vizcayan, Rivas had made his fortune as a financier in Madrid. His case illustrates the point that, although it was not always the capital accumulated through the exploitation of the mines that was invested in the factories, the mines created business opportunities and attracted investment.

Altos Hornos and La Vizcaya soon followed the success of the San Francisco factory. Although a new corporation, Altos Hornos was actually the continuation of the two Ibarra and Co. factories in Baracaldo and Guriezo. The incorporation of new local and outside investors made possible the expansion of the company's capital to 12.5 million pesetas and the modernization of the Baracaldo facilities. The Ibarras' connections with the main steel producers in Europe helped them obtain the exclusive right to the Bessemer process in Spain. In addition, the Ibarras, in exchange for a royalty, transferred to the new company their rights to buy ore at a discount from the Orconera Iron Ore Co. and the Franco-Belge des Mines de Somorrostro.[104]

Also founded in 1882 with capital of 12.5 million pesetas, La Vizcaya was the creation of several prominent mine-owning families in association with the Olano Larrínaga shipping company. The promoter of the project and the first general manager of La Vizcaya was a young engineer, Víctor Chavarri. Thanks to its extensive mine holdings, the Chavarri family had been able to sign contracts in the early 1870s for the provision of ore with Belgian steel producer John Cockerill.[105] These contracts and the fact that Víctor Chavarri earned his engineering degree in Belgium must have contributed to the decision to adopt Cockerill's technology in the new factory. Interestingly, although many of La Vizcaya's stockholders were important mine owners, their mines did not supply the new factory. In order to obtain ore, La Vizcaya signed a long-term lease with a British company that controlled the exploitation of the richest mines in the Galdames district.[106]

Assured of their supply of mineral either by advantageous long-term leases or special contracts, as in the case of Altos Hornos, the three factories also enjoyed special arrangements that reduced the cost of transportation of the ore from the mines to the furnaces.[107] In addition, their location on the Nervión facilitated the provision of imported British coal, which came to Bilbao as a return freight for the ore exported to England. As a result of this two-way maritime traffic, the cost of British coal in Bilbao dropped dramatically during the 1880s.[108]

With the creation of these three factories, Vizcaya regained its hegemony in Spain in the manufacture of iron and steel. While Vizcaya's factories had produced 23 percent of all the iron bars manufactured in Spain from 1861 to 1879, they increased their share of total production to 66 percent during 1880–1913.[109]

The main center of heavy industry in Spain was thus transferred from the Asturian coal region to Bilbao. Such a displacement seems paradoxical because coal districts supposedly had locational advantages over iron areas regarding the construction of metallurgical factories. Yet, the savings in fuel introduced by the Bessemer converter and subsequent technological developments allowed regions with poor coal resources to develop factories to produce certain types of iron and steel that required relatively small quantities of coke. Bilbao's case was not unique in Europe. The Lorraine and Alsace regions, rich in iron and poor in coal, developed a symbiotic relationship with the coal-rich Ruhr Basin. Lorraine and Alsace specialized in the production of pig iron and cruder steel products, which were later sent to the Ruhr to be transformed into more refined types of steel that required intensive use of coal. Similarly, in England, the ore from the Cumberland and Lancashire districts was exported to Yorkshire and Wales until the adoption of the Bessemer converter permitted an efficient production in situ through the importation of coal from other districts.[110]

Therefore, it should not be surprising that pig iron and common steels could be produced at competitive prices in areas like Bilbao. However, competitiveness decreased as the manufactured products required intensive use of fuel. Historians like Gabriel Tortella are thus right in stressing the importance of coal in the location of the iron and steel industry. Yet, Tortella exaggerates when he argues that, since Spain could produce only iron ore at competitive prices, its iron and steel industries "should have been located outside Spain: in Cardiff, Newcastle, Essen or Pittsburgh; not in Bilbao, Avilés, Málaga or Sagunto."[111] In fact, during the 1880s, pig iron could be produced in Bilbao more cheaply than in some English metallurgical centers. As an article in the *Journal of the Iron and Steel Institute* pointed out:[112]

> blast furnaces in Great Britain are now largely employed in smelting Bilbao rather than British ores. This involves the carriage over sea of two tons of material at a present cost of about 10s. to 11s. Inasmuch as not more than half this weight of coke is required in making a ton of pig iron, and as the freight is about the same per ton, it would seem that ore can be smelted at Bilbao with coke imported from England at about 5s. per ton less for freight than the same ore can be smelted, say on the banks of the Tyne or Tees, whence the coke is exported.

During the late 1890s, the Bilbao factories could still produce Bessemer steel at one of the lowest prices in Europe, as table 18 shows.

Fernández Pinedo speculates that, in the late 1890s, the low costs of Bil-

TABLE 18

Production Costs of Bessemer Bars, 1890 (pesetas)

	Bilbao	Cleveland, Ohio	Liège	Loire	Pittsburgh, Pennsylvania	Westphalia
Ore	14.15	36.0	38.4	45.6	31.8	38.4
Coke	25.2	15.9	17.4	16.2	7.2	16.8
Calcium	1.8	2.4	1.8	1.8	1.8	1.8
Labor	4.5	3.3	4.2	3.9	3.0	3.6
Miscellaneous	1.8	1.8	1.8	1.8	1.8	2.1
Total	47.45	59.4	63.6	69.3	45.6	62.7

Source: Fernández Pinedo, "La industria siderúrgica," p. 157.

bao's factories were due to the peculiar circumstance of the stiff devaluation of the Spanish currency, which made local products highly competitive in international markets. The devaluation might have given the local factories a boost, but they were competitive before the depreciation of the peseta in the mid-1890s. Furthermore, the devaluation might actually be seen as a double-edged sword, since it increased the cost of imported coke for Vizcayan producers. However, the fact that the Vizcayan companies became more competitive during the late 1890s proves the logic behind the establishment of the factories in an iron-rich region. If the price of coke had been decisive, then the increase in its price caused by the depreciation of the peseta would have made Vizcaya's factories less competitive.

La Vizcaya, San Francisco, and, to a lesser degree, Altos Hornos put these cost advantages to good use by exporting a large portion of their production of iron bars. Exactly how much of the production was exported is contested. According to Fernández Pinedo, the export market accounted for approximately 30 percent of Vizcaya's production of pig iron.[113] González Portilla, on the other hand, believes that the proportion was even higher, reaching 60 percent. For La Vizcaya alone, González Portilla shows that, for the 1886–1891 period, exports amounted, on the average, to 62 percent of the region's total annual production. While Fernández Pinedo criticizes González Portilla's use of some statistical sources, the documents the latter utilized to obtain the 62 percent figure seem unimpeachable, since González Portilla extracted the data from the minutes of the stockholders' meetings in the company's archive.[114]

Between 1887 and 1890, Italy became the main international customer of

Vizcaya's factories, absorbing 65 percent of the iron bars exported. Germany bought around 9 percent, while France purchased approximately 7.25 percent of this product.[115] During the 1890s, these exports decreased noticeably, since the new metallurgical technology had created a market glut. In addition, protectionist winds swept over Europe, making it more difficult for Vizcaya's factories to enter foreign markets. Not immune to this trend, Bilbao's industrialists also lobbied hard to obtain tariff protection.[116] Once the Spanish government satisfied their demands in the mid-1890s, factory owners in Vizcaya concentrated their efforts on supplying national markets, where their profit margins were higher than in the competitive European arena.[117] These higher profit margins in the national market were the result of the formation of a cartel among the three factories in the late 1880s. The strategy was risky because, although profit margins were high, the Spanish market was not sufficiently developed to sustain a large production of iron and steel.[118]

Altos Hornos, La Vizcaya, and San Francisco were only part of an investment boom that lasted from the late 1870s until the mid-1880s. During those years, railroad lines, chemical factories, and shipping companies were also rapidly developed.

Between 1880 and 1885, railway lines parallel to both banks of the Nervión were built connecting Bilbao with the seacoast. In addition, lines to Guernica and Durango linked the main industrial region with Vizcaya's interior. During these years, too, the first urban tramway appeared in Bilbao. Without counting the lines dedicated exclusively to mining, the investment in railways and tramways during the 1880–1885 period reached almost eleven million pesetas.[119] All this railroad construction facilitated the movement of goods and of the increasing population along the Nervión Basin. It also contributed to an urban construction boom, which led to Bilbao's expansion into Abando.[120]

The effect of railroad construction on the growth of the region's manufacturing industries appears to have been small, at least during the 1880s. Valery Shaw attributes the construction of those lines to the sponsorship of prominent Vizcayan families with interests in the mines and the metallurgical factories. In her view, the railway was the last link in a vertical economic integration that led from the extraction of ore to the production of steel rails.[121] This relationship was not, however, so straightforward. True, in some cases, such as the Bilbao-Portugalete line, a company like Altos Hornos de Bilbao was a major stockholder and became one of the leading purveyors of rails. But for the other lines, the ties between the railroad companies and the steelmakers were weaker or nonexistent, while the provision of rails seems to have been far less than a monopoly of the local factories.[122] Their share of this market probably increased after the mid-1890s, when protectionist legislation gave a strong competitive advantage to the local mills.

Unlike the railroads, which developed feeble linkages with local manufacturing enterprises, the chemical industry received a boost from mining and the steel factories. The mining industry's increasing demand for explosives attracted companies like the Sociedad Española de Dinamita, which owned the rights to the Nobel patent to produce dynamite in Spain. The company was founded in 1872 by a group of French investors associated with a Bilbao merchant, Pedro T. Errazquin.[123] In 1884, Errazquin and his partners founded a second factory, La Cantabria, with the intention of producing the sulfuric acid needed in the production of dynamite and phosphates for use as artificial fertilizers.[124] The manufacture of phosphates created a link with the steel mills, which sold sulfur as a by-product of the chemical reaction produced by the coke batteries.[125]

In addition to these chemical factories, one of the earliest oil refineries in Spain was established in 1878 on the banks of the Nervión. The company was a joint venture of two French brothers, the Fourcades, and the local mercantile house of Gurtubay, whose members also participated in the foundation of Altos Hornos de Bilbao. Fourcade and Gurtubay's initial capital was set at one million pesetas, but two years later, that figure was doubled as the company expanded by opening another refinery in Alicante and a depot in Seville.[126]

The shipping industry has been poorly studied, but attracted vast amounts of capital during the 1880–1885 period. The overall tonnage of the local steamers rose from around six thousand in 1871 to approximately ninety thousand in 1885.[127] The capital that financed the growth of the local fleet amounted to approximately 30.3 million pesetas, a sum that surpassed the combined capital of La Vizcaya and Altos Hornos de Bilbao in 1882.[128]

The large amount of capital invested in shipping was related to the increased maritime traffic caused by the massive exports of mineral from Bilbao. Despite its remarkable growth, the local fleet's share of the ore traffic remained small, between 5 and 10 percent until the late 1890s. Before that period, the increased tonnage was not sufficient to deal with the enormous volume of mineral exports. While these exports tripled from 1 million to 3 million tons between 1878 and 1883, the local fleet's share also trebled, from 100,000 tons to 333,000 tons. Thus although the growth rate of the fleet kept pace with that of exports, its starting level was so low that it could not catch up with the demand for transport.

Metallurgy, shipping, and railroads were the industries that attracted the most capital during the early 1880s, if we exclude mining. It is hard to quantify the amount of this capital from the sources available. Yet, considering that between 25 million and 30 million pesetas were invested in Altos Hornos, La Vizcaya, and San Francisco, that another 30 million at least were used to finance the expansion of the fleet, that approximately 11 million were poured into railroad construction, and that 2.5 million were spent in the chemical industry, then approximately 73.5 million pesetas in investment capital can be well documented. If one takes into account the intensive activity of the housing construction industry, the modernization of the harbor, and the investment in the mining industry's infrastructure, 100 million pesetas would be a conservative estimate of the total investment in the region's economy during the 1875–1885 period.

After a brief respite during the mid-1880s, a second investment cycle commenced toward the end of the decade and lasted until the mid-1890s. In 1888, a competition to build three battleships for the Spanish navy rekindled the drive to form new enterprises. For strategic reasons, the navy wanted to encourage domestic shipbuilding to be less dependent on foreign yards. Several Spanish factories entered the competition, which was eventually won by José Martínez Rivas and his partner, the renowned British shipbuilder Charles Palmer.[129] Martínez Rivas's victory in the competition created high expectations in Bilbao, where he was touted as a visionary who would singlehandedly resuscitate the moribund Vizcayan shipbuilding industry.[130] In many ways, Martínez Rivas was trying to emulate the career of his British partner, who started his business as a mining company and eventually built up an industrial empire at Jarrow, near Newcastle, that included coal and iron ore mining, a steel mill, a shipyard, and a shipping company.[131]

Astilleros del Nervión, as the Martínez Rivas–Palmer company was called, was erected next to Martínez Rivas's San Francisco factory. It was planned on a grandiose scale, requiring an investment of at least thirty million pesetas.[132] Martínez Rivas did not wish merely to assemble the vessels from parts bought from subcontractors; he wanted to produce as much of the material as he could. Thus the shipyard included several metal workshops whose production ranged from simple screws to heavy artillery guns.

Unfortunately, Martínez Rivas's plan failed. A quarrel between him and Palmer over the nature of the partnership revealed the financial difficulties of the shipyard after the launching of the first vessel. Palmer believed that his association required only his technical expertise, while Martínez Rivas wanted him to contribute financially to the enterprise because the Basque businessmen could not bear the costs alone. In 1892, when Martínez Rivas halted work at the yard because of lack of funds, the government took over the shipyard to assure the construction of the three vessels. While Martínez

Rivas was pushed aside, the government retained Palmer as a technical consultant.[133] By the mid-1890s, the three vessels were completed at the staggering cost to the Spanish taxpayers of sixty million pesetas, around 33 percent higher than the original bidding price of forty-five million. Although the navy was proud of the quality of the vessels, a ruinous fate awaited them. All three were sunk in Cuban waters during the Spanish-American War.

Although a personal failure for Martínez Rivas, the construction of the three vessels established, at least in part, the objective of the Spanish government of promoting the national industry. Altos Hornos de Bilbao acquired new Martin-Siemens furnaces to provide steel armor plates for the vessels, and it must have become an important supplier because its president, José Vilallonga, was decorated by the Spanish government for services rendered.[134]

Similarly, the needs of the shipyard created a market for metallurgical firms. Several were founded between 1888 and 1894. Companies like Tubos Forjados, Aurrera, Talleres de Deusto, Talleres de Zorroza, and Alambres del Cadagua produced machinery, boilers, railway cars, and all sorts of metal parts needed in the construction business.[135] Their appearance is usually connected with the protective tariffs enacted by the government in December 1891, although several of them preceded the implementation of the new tariff by one or two years.[136]

The new tariff did encourage the foundation of some of these new factories, but perhaps a more compelling reason was a glut in the international iron and steel market during the early 1890s. The exports of the Big Three (Altos Hornos, La Vizcaya, and San Francisco) decreased noticeably during those years. Thus, it was not by chance that many of the main stockholders in these three companies participated in the creation of the new metallurgical firms. For instance, the Ibarra family owned an important share of Tubos Forjados, Alambres del Cadagua, Basconia, and Euskaria.[137] Other industrialist-miners, such as the Chavarris and the Gandarias, invested in several of these factories, too.[138]

At the same time that these new metallurgical factories were established to create markets for the Big Three, Bilbao's businessmen also tried to assure its supply of coal by developing fields in León, Palencia, and Asturias. For instance, in the early 1890s, Martínez Rivas participated in the formation of the Coto del Musel company to exploit coal mines in Asturias. Similarly, Chavarri and other investors such as Pedro Gandarias from La Vizcaya formed the Hulleras del Turón company, also to exploit coal mines in Asturias. The Ibarras and some of their partners in Altos Hornos de Bilbao invested in Unión Hullera Asturiana.[139] Hullera Vasco-Leonesa, Hulleras del Sabero,

Carbones Asturianos, Matallana, and Euskaro-Castellana were all examples of coal mining companies developed by Vizcayan businessmen during the early 1890s.

Another railroad construction cycle accompanied the development of these coal mines. Lines connecting Bilbao with the León and Palencia fields were built mostly under the sponsorship of Vizcayan businessmen. During these years, too, the Bilbao-Santander line was inaugurated, and the rail network within Vizcaya continued to grow. In 1894, the Basque province had a ratio of seven miles of railroad per ten thousand inhabitants. This proportion doubled the ratio for Spain as a whole and was even greater than those of Germany, Belgium, France, and the United Kingdom.[140]

As in the previous investment period, shipping continued to attract vast amounts of capital. While the local fleet measured approximately 90,000 tons in 1885, ten years later it grew to 156,000 tons. The formation of Sota and Aznar, which eventually became Bilbao's biggest shipping company, dates to this period.

Ramón de la Sota, one of the partners of the shipping company and an important ore exporter, was among the main promoters of the Banco del Comercio, founded in 1891. This bank improved lending practices that had been restricted by the only two financial institutions that existed until then, the Banco de Bilbao and the local branch of the Banco de España. Bearing in mind Sota's interest in the shipping business, it was no coincidence that the Banco del Comercio pioneered in Bilbao the use of steamers as collateral for loans.[141] The advent of a new competitor forced the Banco de Bilbao to expand its activities; to do so, it doubled its capital in 1891, to ten million pesetas.[142]

Another financial turning point was the establishment, in 1890, of the Bilbao stock exchange. During its first five years, the nominal capital of the stocks quoted on the exchange increased almost 50 percent, from 64.3 million to 95.6 million pesetas. Very few new stock issues appeared during the mid-1890s, but toward the end of the decade, the stock market boomed as Vizcaya's economy underwent a new, intensive investment cycle.[143]

This third investment cycle lasted from 1898 until 1901. Although short-lived, this cycle surpassed the previous ones in the number of enterprises created and amount of capital invested. Traditional industries such as shipping, mining, and metallurgy continued to attract capital, while new banking, insurance, and hydroelectric companies were developed to support and diversify the older investments.

Shipping attracted the largest amount of capital during those years. Helped by a rise in freight rates due to the Boer War, Bilbao's fleet increased its tonnage by approximately 50 percent between 1895 and 1900.[144] While Barcelona's fleet, traditionally the largest in the country, remained stagnant at around 212,000 tons toward the end of the 1890s, Bilbao reached that level in 1898, and approached 300,000 tons only two years later. Bilbao's fleet thus became the largest in the country, accounting for approximately one-half of the total tonnage registered under the Spanish flag.[145] According to the British consul, in 1900, of a total of 5.3 million pounds sterling invested in new companies in Vizcaya, 1.2 million pounds were channeled toward shipping companies.[146] As the local fleet increased, its total share in the ore traffic also grew. Between 1894 and 1900, British bottoms lost 30 percent of the transport market because of the "vast increase of Spanish-owned ships employed in the trade."[147]

The large increase in the local fleet revived the insurance business after years of decay and the use of foreign firms as underwriters. Two insurance companies were formed, La Polar and Aurora, with a combined nominal capital of 130 million pesetas.[148] La Polar's capital seems to have been linked, at least in part, with that of the Sota and Aznar shipping company through a complicated financial arrangement.[149]

Also tied to the development of the fleet was the founding of a shipyard, Euskalduna, in 1901. Sponsored by shipping and mining magnate Ramón de la Sota and other local businessmen, the shipyard's main goal was originally to repair Sota's and other shippers' fleets. Soon, however, it started building ships, and eventually became one of the most important shipyards in the country.

Sota and his partners followed, albeit more cautiously, the same strategy pursued by Martínez Rivas in the late 1880s. Mines, a steel factory, a shipyard, and a shipping company all formed part of Sota's business empire. In the early twentieth century, as Bilbao's mines decreased production, Sota, together with Eduardo Aznar, Cosme Echevarrieta, and other partners, formed the Sierra Menera Mining Co. to extract ore in the provinces of Teruel and Guadalajara. In 1917, these mines started supplying his recently created steel company, Siderurgia del Mediterráneo, located in Sagunto.[150]

The Sierra Menera Mining Co. was not the only corporation formed by Vizcayan businessmen to exploit iron mines outside the province. Similar companies, such as the Minas Calas, Sociedad Minera Villaodrid, Minas de Teverga, Minas de Irún y Lesaca, were involved in the extraction of ore in Huelva, Lugo, Oviedo, and Guipúzcoa, respectively. In addition to these iron deposits, Vizcayan businessmen promoted the extraction of other minerals such as lead, sulfur, coal, and copper throughout Spain.[151] Investment in mining peaked in 1901, when of a total of 482 million pesetas in nominal capital used to finance new enterprises, approximately one-fourth (122 million) was allocated to this sector.[152]

Another sector that attracted almost one hundred million pesetas was banking. This industry was completely restructured at the turn of the century. In 1901, the Banco de Bilbao and the Banco del Comercio merged; but at the same time, new banks appeared to compete with Vizcaya's oldest financial institution. Among the new establishments were the Banco de Vizcaya, Crédito de la Unión Minera, Banca y Bolsa Bilbaína, Banco Naviero Minero, and Unión Financiera.[153] The seven local banks had a combined nominal capital of ninety-eight million pesetas.[154] Only three of the seven banks survived the financial crash that occurred toward the end of 1901. Two of the three surviving institutions, the Banco de Bilbao and Banco de Vizcaya, were consistently among the five largest Spanish banks until their merger in 1988, when the amalgamated bank became the biggest in the country. The third survivor, the Crédito de la Unión Minera, was dedicated to the needs of the mining industry. It was a profitable institution until a financial scandal involving its directors forced its liquidation in the 1920s.

As the banking industry increased its resources during the first years of the twentieth century, it became directly involved in the creation of new businesses. The Banco de Vizcaya, for instance, encouraged the development of several hydroelectric companies. Backed by Vizcayan financial resources, these power companies went beyond the boundaries of the province, acquiring rights to supply electricity to different regions using the energy of several rivers in Spain.[155] The Sociedad Hidroeléctrica Ibérica, with capital of twenty million pesetas, was among the largest of these companies. In the 1900s, it secured water rights to the Ebro River and supplied energy to Bilbao, Santander, Barcelona, Valencia, and Alicante.[156]

Although it attracted much less capital than banking, shipping, and mining, the metallurgical industry also continued to grow, remaining the backbone of the Vizcayan manufacturing sector.[157] Perhaps its most important development was the merger in 1901 of Altos Hornos de Bilbao, La Vizcaya, and La Iberia. Under the name of Altos Hornos de Vizcaya, the new company had a combined capital of 32.7 million pesetas, constituting one of the largest manufacturing companies in Spain.

Growth through combination was not restricted to metallurgy and banking. In 1901, two large paper mills founded in the early 1890s, the Papelera del Cadagua and Papelera Vizcaína, merged to form the Papelera Española, with a total capital of twenty million pesetas. Similarly, in 1898, Unión Resinera Española was born of the fusion of several producers throughout Spain under the direction of Víctor Chavarri and other investors.[158] The tendency toward industrial concentration continued during the first decades of the twentieth century, becoming a trademark of the Vizcayan economy. While the concentration might have increased the economic efficiency of the firms involved, it also had, in some cases, a deleterious effect on the economy as a whole. Many of the amalgamated firms became de facto monopolies, earning high profits at the expense of Spanish consumers. Their position became increasingly stronger as the Spanish government protected them from foreign competitors with high tariffs.

The third investment cycle ended with the crash of the Bilbao stock exchange in 1901. The contraction in the local economy was short-lived, however. Those companies that were only the result of sheer speculation disappeared, but many solidly established firms continued to prosper. At the dawn of the twentieth century, Vizcaya was the most vibrant economic region in Spain. Local entrepreneurs seized the opportunities created by the mining boom to modernize traditional industries and develop new businesses throughout Spain. In the twentieth century, despite the gradual depletion of the mines, Vizcaya's place of honor within the Spanish economy is reflected in the following statistics. In 1929, Basque capital constituted 25 percent of Spanish banking resources, 38 percent of the investment in shipyards, 40 percent of the stock in engineering and electrical construction firms, 68 percent of funds dedicated to shipping companies, and 62 percent of the monies invested in steel factories.[159] All this was generated by a region that barely included 3 percent of the population. At a time when the mines had ceased to be the engine of growth of the Vizcayan economy, such figures seem to support Unamuno's remark that the key to Bilbao's wealth was not in its subsoil, but among its people.

Chapter Four

The Formation of Bilbao's Modern Business Elite

The power attained by modern Bilbao has its solid foundation on the capital amassed in its warehouses and humble commercial establishments.
F. Echegaray, "Cuatro generaciones bilbaínas," in *Libro de actos conmemorativos del centenarío del Banco de Bilbao*

Chapters 2 and 3 analyzed the economic development of the Bilbao region; this one focuses on the entrepreneurs behind this process. Who were these men? What was their relationship to past generations of business leaders? What changes took place among their ranks during the fifty years analyzed here? To answer these questions, I have compiled and analyzed lists of top businessmen for the 1850–1900 period. A comparison of these lists will illustrate the changes and continuities in the composition of Bilbao's modern business elite. I shall try to show that commerce was the breeding ground from which most of the city's entrepreneurial families came. The industrialists who presided over the mining boom and the modernization of the regional economy were not so much a new class of businessmen as the continuators of the city's proud mercantile tradition. As a result of this common background and the way old and new members of the elite merged during the second half of the nineteenth century, it seems appropriate to paraphrase British historian E. P. Thompson in saying that Bilbao's modern business elite did not rise like the sun at a scheduled time; it was present at its own making.

Business Leaders during the 1850–1873 Period

As mentioned in chapter 3, trade was the Vizcayan economy's engine of growth during the period between the two Carlist Wars. Not surprisingly, then, the dominant figure in the local economic scene was the merchant. Unfortunately, most of the information available on traders during these years is fragmentary. While Román Basurto's study on Bilbao's commerce ranks the main merchants of the late eighteenth century according to their participation in harbor traffic, such an endeavor is, ironically, impossible for a period much closer to the present for lack of statistics.[1] By default, then, I have had to use more impressionistic sources to compile lists of Bilbao's top merchants.

MOST IMPORTANT MERCHANTS DURING THE 1850S

To identify the most important traders during the 1850s, I have used three types of documents primarily: the 1852 electoral list, the French consular reports (F.C.R.), and the list of members of Bilbao's commercial tribunal.

The electoral list provides the names of the 150 individuals within the Bilbao district who enjoyed full political rights. Unlike in the rest of Spain, where the political franchise was restricted to an open-ended number of electors who could muster certain property or tax qualifications, in Vizcaya the situation was different. During the 1850s and most of the 1860s, no matter how many might have met the qualifications required in the rest of the Spanish provinces, the number of voters in Vizcaya remained fixed at 150 per electoral district. To compile the electoral lists, each municipality in Vizcaya had to determine who the wealthiest individuals were within its jurisdiction. A second selection at the district level among the pool of names drawn by each municipality established the final list of 150 electors. Since Vizcayans did not pay direct taxes, the authorities had no reliable way of assessing a person's wealth; thus it was left to the individual to demonstrate the worth of his assets. Unfortunately, it is not known how the authorities evaluated each potential elector's claim to fortune; but whatever method was followed, the 150 electors probably constituted a fair sample of Bilbao's economic and political elite.

Although the electoral list shows who the richest and most powerful men in Bilbao were, it, unfortunately, does not indicate their occupations. By using notarial records, the F.C.R., and a list of officials of the Bilbao commercial tribunal, I have been able to classify roughly two-thirds of the electors according to two basic categories: merchants and proprietors.[2] The two categories are not mutually exclusive. By the mid-nineteenth century, the distinction between a wealthy proprietor and a large trader was often very subtle. In many cases, the two categories might reflect only different stages in the economic life of an individual, since he might have started his business career as a merchant and ended as a proprietor. Such was the case, for instance, of Bartolomé Arana, an elector on the 1852 list, who recorded his occupation in his will as a "proprietor and formerly a merchant."[3] Similarly, many active merchants were real estate holders. For heuristic purposes, however, I have considered as merchants all those who were actively involved in commerce at the time the electoral list was made, whether they owned real estate or not. The classification resulted in an almost equal number of merchants and proprietors among the electors.[4] Appendix II lists those belonging to the mercantile elite during the 1850s. In addition to forty-nine traders extracted from the 1852 electoral list, I have included seven who appear only in the F.C.R.

Interestingly, only three surnames (Bergareche, Echevarría, and Recacoechea) in my list of top traders during the early 1850s can be traced back to the lists of the largest exporters of wool and iron compiled by Román Basurto for the 1750–1800 period.[5] In addition to these three cases, the Uhagóns and the Orbegozos might also be mentioned, since, despite their absence from the late-eighteenth-century lists of top exporters, they do appear on a list of twenty wealthy local merchants compiled by Napoleonic troops to tax the local population.[6] Similarly, although the Epalzas are conspicuously absent from all of Basurto's lists, they seemed to belong to a long dynasty of traders who occupied prestigious positions in Bilbao's *consulado,* the old merchants' guild.[7]

While these six merchant families may demonstrate continuity among the top ranks of the mercantile community since at least the early nineteenth century, the vast majority of the traders on my list seem to point toward a renewal of the top merchants' ranks during the 1800–1850 period. The Olaguivels, the Ibarras, the Aguirres, the Ayarragarays, the Uriguens, and the Abaituas replaced names such as Gómez de la Torre, Villabaso, Quintana, Busturia, Gardoqui, Arechaga, Ugarte, Azuela, and Uria Nafarrondo, names that had dominated the city's trade until the early 1800s but that had disappeared completely from the mercantile world by the 1850s.

The changing composition of the ranks of the top local businessmen paralleled a switch from an almost exclusive concentration on the export of iron products and wool to a trade centered around colonial staples (sugar and cocoa from the Caribbean), codfish, flour, and grain. Traders had to adapt to changes stemming from a radically different economic and political reality. Houses like Epalza and Son or Errazquin and Orbegozo, for instance, had been involved in the wool and iron trade during the first decade of the 1800s.[8] Yet, by the middle of that century, they had abandoned wool altogether and the iron trade became less important as they focused on colonial staples, wines, oil, grain, and manufactured products from northern Europe.[9] In fact, by the 1850s, among the fifty-six major traders cited in table II-1, Ibarra Mier and Co. seems to have been the only one still primarily involved in all aspects of the iron trade.[10]

It is tempting to explain the modifications at the top of Bilbao's mercantile world by arguing that new merchants adapted better than established traders to the new economic and political situation. There is no proof, however, that the old mercantile dynasties of the eighteenth century failed to adapt to the new trading environment. Nor is there evidence that the older mercantile dynasties fell prey to a so-called Buddenbrook effect, a rule that postulates the decline in the economic power of families after one or two generations at the top. The reasons for the gradual renewal of Bilbao's top businessmen will remain a mystery until the subject is properly studied.[11]

Still, a working hypothesis may be ventured to explain the change. The general expansion of trade in eighteenth-century Europe ended during the 1790s as a result of the French Revolution and the wars it spawned. In addition to the adverse international climate, technological reasons caused a precipitous drop in the demand for Bilbao's main export staples, wool and iron products.[12] This double calamity might have resulted in the ruin of many merchants. While other areas recovered after the Napoleonic Wars, the situation persisted in Vizcaya as the political turmoil of the 1820s and the war between 1833 and 1839 discouraged the growth of Bilbao's trade. Many of the merchants who weathered these storms might have divested their capital away from trade, seeking more secure investments in land and urban real estate. Such a process would have mirrored a similar trend among late-eighteenth-century French merchants in the town of La Rochelle.[13] In Vizcaya, a few examples of this process have been documented. González Portilla has shown how the Pérez de Nenín family gradually abandoned trade and acquired several urban properties.[14] A further illustration of the transfer of merchant capital into real estate is provided by a list of the fifty largest Vizcayan landowners, published in 1860.[15] Such names as Gómez de la Torre, Villabaso, and Mazarredo and other eighteenth-century trading families appear prominently on the list. Presumably, while old, powerful merchants wanted to protect their fortune from the perils of trade, up-and-coming traders could not afford such a luxury because their capital resources were not large enough. Nevertheless, given the state of our knowledge of this issue, it is hard to say how much of this transfer of capital from commerce to real estate was due to a crisis and how much was due to an apparent practice of abandoning active trade after a certain level of wealth had been accumulated.[16]

Although the disappearance of many important late-eighteenth-century traders cannot be explained fully, aspects of Vizcayan society assured the renewal of the merchant ranks. Unlike the cultures of other Spanish regions, which treated commerce as an activity unfit for the social elite, Basque custom and tradition actually encouraged trade and industry.[17] Commerce was perceived as a necessity to provide sustenance to the many sons who could not inherit land because of the practice of not partitioning family lands. Furthermore, the lack of a formal regional aristocracy guaranteed that those merchants who grew rich from trade did not have to overcome any legal barriers to reach the higher echelons of society. Indeed, the accumulation of wealth through commerce had its social and political rewards. Merchants such as the Uhagóns, for instance, became mayors, and marriages between traders and landowning families were not uncommon.

Bilbao's commercial vitality also helped renew the merchant ranks by attracting ambitious individuals from other areas within Vizcaya, from neighboring provinces, and even from abroad. Although the geographical origin of many of the merchants remains unknown, their names indicate that the vast majority had a Basque ethnic background. For example, Sebastián Ayarragaray and Pedro Errazquin came from the Basque province of Guipúzcoa, Romualdo Arellano was born in northern Navarre, an area heavily populated by Basques. Similarly, the Barroetas and the Ibarras seem to have settled in the city during the early nineteenth century, and both families had their roots in the Vizcayan interior. Although the Uhagóns came from France during the late 1750s, they emigrated from one of the Basque provinces beyond the Pyrenees.[18] Yet, not all foreigners could claim Basque origin; the MacMahons and the Mowinckels immigrated from Ireland and Scandinavia, respectively.[19]

In addition to their commercial activities, how did these traders encourage the economic development of the region? As chapter 3 points out, during the 1850s, they promoted a variety of businesses, such as the Bank of Bilbao and the Tudela Railroad. While shares in these businesses were widely distributed among the local population, the main Bilbao merchants monopolized the boards of directors. For instance, the Banco de Bilbao's first directors included J. P. Agüirre, V. Arana, Pablo Epalza, P. A. Errazquin, B. Escuza, G. M. Ibarra, F. Uhagón, E. Uriguen, L. Violete, and M. Zabalburu. With the exception of the last two, all the other directors appear on my list of the merchant elite.[20] Similarly, P. Epalza served as president of the Bilbao-Tudela line, and several other members of the mercantile elite joined him on the board of directors of the company, occupying two-thirds of the seats.[21]

Some of the top merchants also developed manufacturing companies. During the 1840s, for instance, Arellano, Epalza, Ogara, and Olavarri pioneered the first modern iron factory in Vizcaya, Santa Ana de Bolueta. The Ibarras soon followed, founding a larger plant on the banks of the Nervión. A third factory, founded in the mid-1850s, Santa Agueda, also stemmed from a project of an important local merchant, P. Errazquin.

The strong presence of large merchants among the developers of manufacturing plants seems to be in sharp contrast to the experience of England during the first decades of the Industrial Revolution. According to François Crouzet, the "'merchant-princes' of the main seaports" could not be found among factory owners.[22] Since the early English factories required little founding capital, the first generation of industrialists did not have to come from a wealthy background. In Bilbao, however, the iron industry needed large capital investment to prosper. The local merchants not only possessed the required funds but also were familiar with the industry through their long involvement in the iron trade. These considerations must have eased the transfer of capital from trade into manufacturing once Vizcaya fell under the umbrella of Spanish protective tariffs after the First Carlist War.

The participation of Bilbao's merchants in the industrial development of the region had parallels in other centers of heavy industry in Europe. For instance, in the Ruhr region, coal traders built some of the first iron factories during the early nineteenth century.[23] Yet, what is striking in Bilbao's case is the presence of merchants in all types of industry to the exclusion of other socioeconomic groups. Unlike their counterparts in other long-established European manufacturing centers, Vizcayan industrialists tended not to rise from the artisan class. While craftsmen may have provided a pool of skilled workers for the new blast furnaces, in general, the technical experts that ran the plants came from other regions. In the case of Nuestra Señora del Carmen, it was a Catalan, José Vilallonga, who ran the factory.[24] The region's other major factory, Santa Ana de Bolueta, hired French technicians to run the operation at the start.[25]

Thus, during the 1850s, a unique feature of Bilbao's economic world was the preeminent position of its mercantile elite. Trade still remained the local businessmen's main occupation, but they also promoted factories and other enterprises that supported the city's commercial activities. Although heirs to a long commercial tradition, the members of the mercantile elite in the 1850s could not boast an extensive history of leadership in the local business world. Political turmoil and economic changes were the background against which the Bilbao business elite was changed significantly during the first half of the nineteenth century. How well established did the new elite become during the 1850s? Did new traders continue to join its ranks during the next decade?

THE BUSINESS ELITE DURING THE 1860S AND THE EARLY 1870S

The main sources available for determining the composition of Bilbao's business elite during the 1860s are five lists of prominent Vizcayans that were published in 1860, 1861, 1863, 1865, and 1867 in the B.O.V.[26] Each list was divided into three economic categories: rural proprietors, manufacturers, and merchants. In theory, only the fifty wealthiest individuals in each of the three groups were listed for the purpose of selecting representatives to a provincial economic council (the Junta de Agricultura, Industria y Comercio—Agricultural, Industrial and Commercial Council).[27] As in the case of the 1852 electoral list, the process of selecting the wealthiest individuals in each category remains a mystery. Yet, in spite of the omission of a few important merchants, the lists seem to be a good sample of the regional business elite.[28]

Due to the preponderance of Bilbao in the commercial life of the province, it is not surprising to find that around 80 percent of the merchants listed resided in this port. An overwhelming number of the largest landowners also seem to have lived in Bilbao; in fact, one could almost say that all did, since the remaining ones were settled in Abando, an area separated from the city only by the Nervión River and incorporated into Bilbao's municipal boundaries in 1869. The city also included the largest number of manufacturers, whose share of the provincial total grew from 18 percent in 1861 to 46 percent in 1865.[29]

Of the thirty-seven Bilbao merchants who appeared on the 1860 and the 1861 lists, twenty-eight can be directly connected to my list of main traders in the 1850s.[30] The overlap between these lists seems to indicate a certain stability in the ranks of the top traders. Yet, soon after 1861, many new names started replacing older ones. Such changes were first reflected in the 1863 B.O.V. list. By 1867, of the forty-one names listed for Bilbao, only nineteen had appeared on the 1860–1861 rosters, while eighteen merchants, who had been listed in the early years, were eliminated.

Natural attrition and bankruptcy were the main reasons for the disappearance of most of those eighteen merchants. Ten traders died without being immediately replaced on the rosters by other family members.[31] At least four of the other eight merchants who disappeared from the list went bankrupt during the recession of the mid-1860s.[32] One of the remaining four, Eugenio Aguirre, appeared on the 1865 list under the industrial category, perhaps as a result of his operation of a flour mill. I could not document the other three cases.[33]

Not all of the twenty-two merchants added between 1861 and 1867 represented a completely new trading dynasty. On the contrary, several important families cited as part of the mercantile elite during the 1850s resurfaced after what may have been a period of inactivity or an omission in the sources. For instance, the MacMahon family did not appear on the 1860 list, but had two representatives on the 1867 list, brothers Pedro and Francisco.[34] Including the previous example, I have documented seven cases that represent the substitution of a son for a father, or a son-in-law for a father-in-law.[35] If these cases are added to the nineteen merchants on the first B.O.V. list from 1860, then slightly more than 60 percent of the top traders survived the recession of the mid-1860s.

It is hard to generalize about the other fifteen new merchants added to the lists, since I have not been able to obtain information about all of them. Yet, the few documented cases would seem to hint at the open character of the local business elite, since they show examples of merchants rising from the middle or low ranks of the local commercial world. The Gurtubay family provides a good illustration.[36] Simón Gurtubay settled in Bilbao during the early decades of the nineteenth century and earned a modest living as a small-scale codfish trader. According to local mythology, he amassed a fortune in a serendipitous way.[37] In 1833, he received by mistake an extremely large shipment of fish. Under normal circumstances, such a large quantity of codfish would have been impossible to sell profitably, but the outbreak of the First Carlist War made Gurtubay a millionaire. His oversupply became a precious commodity that fed Bilbao's inhabitants during the prolonged siege of the city.

Although the story may have a kernel of truth, it is doubtful that the Gurtubays' ascent was so sudden. None of my sources mentions any Gurtubay until Simón's son José appears on the 1863 B.O.V. list. It is likely, however, that Simón's origins were humble; a document shows that he was illiterate, something very unusual among the merchant community, and probably indicative of a modest social background.[38] In 1869, Simón's commercial partnership with his two sons and his son-in-law showed net assets totaling almost 1.25 million pesetas, a large amount of capital, which certainly put him among the top ranks of the mercantile community.[39] Fortune opened the doors to social and political prominence, and by the late 1860s, José Gurtubay appeared among the members of Bilbao's municipal government. Not as swift an ascent as the local mythology indicates, but it was still a remarkable rise for the son of an illiterate merchant.

Like the Gurtubays, most of the merchants added to the B.O.V. lists between 1863 and 1867 continued to trade in codfish, staples from the colonies, oil, wine, flour, grain, wood, and manufactured products from northern Europe. With the exception of the Ibarras, no other iron ore trader appears on the lists.

The B.O.V. lists also permit us to analyze the group of landowners and manufacturers. Throughout the five lists, the landowner group is the one that shows the greatest continuity. The names cited include prominent old Vizcayan families, such as Salazar or Novia de Salcedo, some of which date to the time of feudal chieftains during the Middle Ages; prominent eighteenth-century traders such as Gómez de la Torre or Mazarredo; and recently retired merchants such as José P. Aguirre or Juan Echevarría y La Llana.

Some of the landowners were engaged in small manufacturing activities, which were usually extensions of their real estate holdings. For instance, some owned *ferrerías,* which were fueled by wood from their private forests; or they took advantage of streams on their land to power flour mills. In general, the landowners did not operate these businesses themselves, but often rented them to merchants. Similarly, although they invested in some of the large provincial enterprises such as the Bank of Bilbao and the local railroad line, they rarely participated on the boards of these companies.

Although manufacturers were listed as a separate category on the B.O.V. lists, this classification does not seem to have been widely used in Vizcaya until around the 1890s. Since prominent merchants owned the most important factories, manufacturing activities were perceived as being subsumed within the larger concept of commerce. Thus, it was rare to find an important entrepreneur who called himself an industrialist or a manufacturer in the business contracts kept in the local notarial archive.[40] For instance, Gabriel María Ibarra, despite his vast investments in mines, railroads, and factories, consistently used the term "merchant" to indicate his occupation.[41]

Since the regulations required the presence of fifty of the "wealthiest manufacturers" in the province, the B.O.V. lists included what must have been small workshops together with the large factories owned by some of the main merchants. The lists did not specify the type of industry owned by the persons cited. Yet, through notarial documents and published sources, I have been able to match most of the names with their manufacturing activities. Flour mills and metalworks predominated among the documented cases. Of a total of eighteen manufacturers listed in 1863, five owned flour mills and six possessed metal factories.[42] Among the other five cases I verified, two involved tanneries; the third, food processing (preserves); the fourth, a glass factory; and the fifth, a paper mill.[43] Except for the flour mills and the iron and glass factories, the other manufacturing establishments were small-scale enterprises that did not reach 250,000 pesetas in capital. The paper mill's net assets were about half that amount, and the food-processing company's capital did not even total 62,500 pesetas.[44] Accordingly, the label of "wealthiest manufacturers" should be taken with a grain of salt for many of those listed in this category.

From 1863 to 1867, the number of Bilbao's manufacturers listed increased from eighteen to twenty-eight. Most of those cited in 1863 still appeared at the later date. The added manufacturers did not seem to alter the local industrial panorama much. In general, they owned small forges or other types of iron workshops. None of these served as springboards for future large firms. On the contrary, their lackluster performance during the mid-1860s and their demise in the early 1870s, when the Second Carlist War was the coup de grâce for most of them, dampened the interest of many local investors in risking their capital in manufacturing enterprises.

The late 1860s and the early 1870s represented a period of political turmoil and the tail end of a recession that had plagued the country since the mid-1860s. The effect of these events on Bilbao's business elite is hard to determine. In 1873 the municipality of Bilbao levied a tax on its inhabitants to defray the costs of the Second Carlist War. Basas has published a list of traders who were asked to pay the highest contributions to the war effort. Of thirty-three wholesale merchants who paid more than 6,250 pesetas in taxes, only eighteen appear on the 1867 B.O.V. roster under the "manufacturer" or "trader" categories.[45]

These numbers would seem to indicate that the ranks of the business elite were profoundly shaken during the years prior to the war. Yet, two considerations demand more caution before drawing such a conclusion. First, it is possible that many wealthy traders tried to hide their worth in order to be assessed a lower tax rate. Second, the timing of the assessment could also have skewed its results. Many people fled the city as the outbreak of war became certain, and, consequently, the tax rolls may have been incomplete. The absence of some of the richest merchants (i.e., Romualdo Arellano, Gabriel Ibarra, and Juan Gurtubay, to name just a few) leaves no doubt that the tax rolls were incomplete.

Despite its incompleteness, the tax list is worth mentioning because it does indicate the incorporation of a new group of merchants into the local business elite. The sizable tax of more than 6,250 pesetas imposed on these traders in itself bespeaks their economic prowess, since the taxes levied on the more modest merchants rarely exceeded 250 pesetas. In addition, for the first time on any available listing of top businessmen, two ore traders not related to the Ibarra family appear among the new group of merchants as paying the highest tax rate of 10,000 pesetas. The presence of Tiburcio Chavarri and Cirilo Ustara signal the first fruits of the nascent mining boom.

Thus, during the 1860s and the early 1870s, the business elite continued to show signs of openness. As in the 1850s, traders maintained their dominance of the local economy. In general, the city's commerce was still centered around colonial goods, wine, codfish, flour, and manufactured products from northern Europe. In the late 1860s, ore exports increased noticeably, and, consequently, some ore traders joined the ranks of the local economic elite. Yet, the full impact of the mining boom was not felt until after the Second Carlist War. During the last quarter of the nineteenth century, mining, manufacturing, and shipping replaced commerce as the engine of regional economic growth. The effect of these changes on the local business elite is the subject of the next section.

The Business Elite, 1875–1900

The rapid expansion of the economy during the 1875–1900 period complicates the task of defining the local business elite. As local businessmen diversified their investments, their classification according to occupation became increasingly difficult. To keep the analysis focused, therefore, I have concentrated the discussion on four of the region's main industries: mining, metallurgy, shipping, and banking. When I classified the businessmen discussed in this section, I tried to establish the activity that caused their original economic rise, regardless of subsequent investments in different areas.[46] In addition to determining the business leaders in these four sectors, I analyzed the relationship between these entrepreneurs and the business leaders of the interwar period and attempted to establish how much continuity and change characterized the economic elite. Finally, I tested the hypothesis that has caused a historiographical debate by postulating that mining entrepreneurs played a crucial role in the development of other regional industries.

MINING

Who were the main mining entrepreneurs? As discussed in chapter 2, the ownership and operation of the mines was concentrated in the hands of a few families and companies. Within the Triano-Somorrostro district, for instance, the mines owned by nine families accounted for almost 80 percent of total ore production in Vizcaya. These families included the Ibarras-Zubirías (and their partners, the Miers-Gorostizas), the Chavarris, the Ustaras, the Aranas, the Amezagas-Yandiolas, the Bellidos, the Allendes, and the Durañonas-Gandarias.[47]

Like the Ibarras, most of these mining families had been involved in the ore trade since the early 1800s. During the 1850s and the 1860s, they solidified their control of the mines as export restrictions on ore were lifted, and new legislation changed the provisions regulating property rights, permitting the appropriation of large deposits. Thus, even before the mining boom started, these nine families already owned the lion's share of Vizcaya's mineral wealth.

Except for the Ibarras and the Miers, the other seven families had to wait until the early 1870s to become part of Vizcaya's economic elite. Since, according to F.C.R., the ore trade constituted less than 2 percent of the value of Bilbao's exports during the 1850s, it is not surprising that the majority of the businessmen involved in this type of commerce did not form part of Vizcaya's mercantile elite.

Prior to the 1870s, few ore traders actually resided in Bilbao, where the most powerful merchants were concentrated. For instance, Chavarri, Arana,

Bellido, and Durañona were originally from the port town of Portugalete, at the mouth of the Nervión River. Before the mining boom, these families were not completely specialized in ore trading; like other Vizcayan merchants, they also dealt in such items as staples from the colonies and grain. Yet, their participation in this trade never reached the scale of that of the largest Bilbao merchants during the 1850s and the 1860s.[48]

Although those seven mine-owning families were not part of the provincial economic elite during the 1860s, they were certainly not poor, either. José Chavarri, for instance, left a sizable estate when he died in the late 1850s. There were no rags-to-riches stories during Vizcaya's mining boom; no peasant or mine worker is known to have become rich overnight by discovering an iron deposit.

With only a few exceptions, the leading merchants of the period prior to the Second Carlist War did not manage to obtain productive mining concessions in the Somorrostro-Triano district.[49] In the other districts, however, some of the old merchant houses fared better. In Galdames, for instance, the firm of Uriguen Vildosola registered the most productive mines in the early 1870s.[50] Similarly, in Ollargan, the old mercantile houses of Epalza, Arellano, and Olavarri indirectly owned rich deposits through their participation in the Santa Ana de Bolueta iron factory. These mines were registered during the 1850s to supply the needs of the factory. As the mining boom progressed, and especially during the late 1890s, when the Triano-Somorrostro mines showed signs of exhaustion, they were intensively worked with the purpose of exporting much of the production. Yet, even in their best years, in the late 1890s, the Galdames and the Ollargan-Abando-Begoña districts combined produced only about 20 percent of Vizcaya's ore.

The Ollargan-Abando-Begoña area also created business opportunities for another small group of entrepreneurs with varied backgrounds. The most important members of this group were Luis Levinson, Juan Aburto y Azaola, and Manuel Lezama Leguizamón. While the first two were relative latecomers to the mining industry, Lezama Leguizamón had been involved in the business for a long time prior to the export boom. Although the Lezama Leguizamón family seems to have belonged to a long line of Vizcayan landowners, the name does not appear on the B.O.V. lists published in the 1860s.

Levinson, a Danish entrepreneur, registered his mines during the late 1860s, but they were not intensively worked until the late 1890s, when they seem to have been leased to a local businessman, Luis Núñez.[51] Aburto, on the other hand, entered the mining industry only in 1885, when he bought the Malaespera mine, which became very productive soon after. Aburto came from the middle ranks of Bilbao's commercial community, having been involved in the wine trade prior to his acquisition of the Malaespera deposit.[52]

Family connections may have helped him initially in the mining industry. He bought the lease to the Malaespera mine from his brother-in-law, José María Martínez Rivas, one of the most important ore extractors in the region.[53]

The mining boom also established the dominance of a group of fourteen companies that extracted between 75 and 80 percent of all Vizcayan ore between 1880 and 1900. Among these fourteen, the Orconera, the Franco-Belge, and Martínez Rivas accounted for about two-thirds of the total. The first two were multinational companies that, in partnership with the Ibarras, exploited the richest Somorrostro-Triano deposits. The third seems to have been the creation of a local entrepreneur, José María Martínez Rivas.

Unlike the Ibarras, Martínez Rivas did not have a background in the ore trade or the mining industry prior to the early 1870s. The bulk of his ore production came from the Unión and Amistosa mines, which he acquired in 1871 for a British company, the Somorrostro Iron Ore Co. Martínez Rivas's relationship with this company remains unclear. There is no doubt that he worked these two mines, but I could not determine whether he did so as an agent, a lessee who had to pay royalties, an owner, or a partner of the British company.[54]

What was Martínez Rivas's background? Although his early career is something of a mystery, it seems probable that his uncle, Francisco de las Rivas, the first Marqués de Mudela, helped him with his early business deals.[55] The marqués's patronage was recognized symbolically by altering the family name, which was Martínez de Lejarza. José María's father, Santiago Martínez de Lejarza, who married the marqués's sister, was not a wealthy man himself. The marqués was the family's benefactor, provided generous dowries for his nieces, and employed his brother-in-law as the manager of his estates in Vizcaya. Given his disposition to help the family, the marqués must have noticed early his nephew's knack for business and may have financed his first deals.[56]

Among the other eleven mining companies, which controlled about a quarter of the ore extracted in Vizcaya, four belonged to some of the main mine-owning families: the Ibarras, the Chavarris, the Durañonas-Gandarias, and the Aranas. Foreign businessmen who had come to Bilbao either prior to the mining boom or attracted by the ore deposits led another four of those eleven companies: José MacLennan, Rochet and Co., Otto Kreizner, and Parcocha Iron Ore. The remaining three top producers were the Vizcaya steel company, Compañía Explotadora de la Demasia Ser, and Echevarrieta and Larrínaga.[57]

Echevarrieta and Larrínaga represents one of the most successful examples of local entrepreneurs who took advantage of the mining boom without actually owning any deposits. Cosme Echevarrieta and Bernabé Larrínaga, the only two partners, came from the middle ranks of Bilbao's mercantile community.[58] Family connections must have eased the transition from commerce to the extraction of ore. Larrínaga was married to a sister of Juan Aburto, the owner of the Ollargan mines that the two partners first worked. Similarly, it was Aburto who first leased other mines to them in the Somorrostro area.

A special case in Bilbao's mining elite was that of Ramón de la Sota, whose total mineral production was not included in the Vizcayan statistics because he extracted the bulk of his ore from a group of mines in Santander. Like those of other large Vizcayan mine owners, Sota's family was originally from Portugalete. His father, Alejandro, had been involved in the mining business since at least the early 1860s and formed part of the middle ranks of the Vizcayan business world during those years.[59] Through leases and the direct exploitation of some of his own deposits in Triano, Alejandro Sota managed to extract around 50,000 tons of ore annually during the 1880s.[60] Although large, this quantity was not enough to place Sota among the ranks of producers like the Chavarris, who consistently surpassed 150,000 tons per year during that decade. Yet after 1886, Sota's production reached the level of that of the large Vizcayan producers if one takes into account the output of his mines in Santander. In that year, Ramón Sota seems to have taken the business over from his father, establishing the Setares Mining Co. with his cousin Eduardo Aznar, Benigno Salazar, José del Cerro, and Norberto Seebold to work the extremely rich Ceferina mine in Santander (on the border with Vizcaya).[61] During the early 1890s, this company was extracting around 150,000 tons of ore per year.[62]

Thus, the expansion of mining permitted the incorporation of a new group of businessmen into Bilbao's economic elite. Thanks to their long involvement in the iron trade, nine mine-owning families from the Triano-Somorrostro district controlled a large share of the industry by leasing or working their deposits directly. Another small group of foreign and local mining companies also managed to accumulate great profits through their leasing of rich deposits. The majority of the larger Vizcayan mining entrepreneurs seem to have come from the middle ranks of the merchant community. Montero is right in stressing the fact that the rise of this group of mining entrepreneurs constitutes an important development in the Vizcayan business world.[63] Yet, considering the socioeconomic background of many of these businessmen, and the direct participation of some members of the old mercantile elite in the mining industry, Montero's division of Bilbao's business world during the 1880s into mining entrepreneurs and traditional merchants seems artificial. If the old business elite did not have a greater participation in the industry, it was not so much due to their lack of interest or to a traditional mentality averse to risk taking as to the fact that most concessions had already been granted by the time the mining boom started. Nevertheless, some members of the old mercantile elite still managed to participate indirectly in the industry. For instance, several powerful merchants provided capital to improve the infrastructure of the deposits and financed many aspects of the export trade.[64] Thus it would seem more appropriate to speak of a progressive fusion of old and new entrepreneurs. During the 1890s, when the local deposits were showing signs of depletion and Vizcayans promoted the development of mining companies in other regions, the ties between the two groups were further strengthened, since they both founded and managed these enterprises.

METALLURGY

The iron and steel factories were the backbone of Vizcaya's manufacturing sector. The mining entrepreneurs' involvement in the development of these factories has become a contested issue in Vizcayan historiography. This debate appears to be an extension of the theory that postulates that it was the capital accumulated through mining that permitted the modernization of the manufacturing sector. If this were the case, it would not merit further analysis, since in chapter 3, I show that the capital accumulated by the mining industry was not sufficient to explain total investment in the Vizcayan economy during the 1880–1900 period. The question may be analyzed in a different way, however. Did the mining entrepreneurs play a role that was out of proportion to their actual investment in the metallurgical sector? Historians have been trying to answer this question through indirect methods, since the sources available do not always provide the exact amount of capital contributed by each investor.

For instance, Jesús Valdaliso has used the Vizcayan Mercantile Register to try to measure the participation of mining entrepreneurs in the metallurgical industries. According to his computations, 25 percent of all the investors in metallurgical companies also had interests in mining.[65] He concludes that, although important, the participation of mining entrepreneurs was not decisive in the creation of a modern steel industry in Vizcaya. Since he does not give the percentage for other groups of investors, it is difficult to assess his conclusion.

Moreover, Valdaliso's article seems to be marred by methodological problems that stem from the source he used. First, since the Mercantile Register does not record the capital contributed by each investor to a new company, Valdaliso measures only the number of times a businessman appears in the different companies registered. Thus, those investing one or one million pesetas have equal weight on his scale. Second, although he recognizes that the classification of businessmen is problematic because of their multiple investments, he does not mention what criteria he uses to classify someone as a metallurgist or as a mining entrepreneur when that person had interests in both sectors. Third, given the high number of metallurgical companies in his sample, he does not seem to take into consideration the vast difference between a large mill and a small workshop.[66]

To avoid these problems, I use here a different strategy. I analyze only the largest metallurgical companies by looking at the boards of directors. I trace the socioeconomic backgrounds of the directors prior to their involvement in the new companies. The assumption behind my strategy is that membership on a board indicates ownership of a large number of shares and also active participation in the management of the company. Fourteen newly established companies form part of my sample: Altos Hornos de Bilbao, Aurrera, Alambres del Cadagua, Astilleros del Nervión, Basconia, Chavarri-Petrement, Euskaria, Euskalduna, Iberia, San Francisco de Mudela, Santa Agueda, Talleres de Deusto, Talleres de Zorroza, and La Vizcaya. The search of board members resulted in a group of sixty-two businessmen.[67]

Businessmen with mining interests were the most numerous among the board members of the metallurgical companies. Fourteen of the twenty I classify as mining entrepreneurs belonged to four families: the Ibarras, the Chavarris, the Martínez Rivases, and the Gandarias.[68] The Ibarras were especially active in promoting this type of industry; members of this family appear on the boards of seven of the fourteen companies I analyzed. Similarly, the Chavarris are represented on four of the fourteen boards. Slightly less active, the Gandarias participated in the founding and on the board of La Vizcaya, Basconia, and Talleres de Deusto. Martínez Rivas was involved in only one manufacturing company, Astilleros del Nervión, but this shipyard exemplifies one of the most ambitious industrial undertakings in Bilbao. Collectively, members of these four families acted as directors in twelve of the fourteen companies I studied. It is this high profile in the Vizcayan industrial landscape that had led some historians to believe that mining businessmen in general spearheaded the modernization of Vizcayan industries.

Yet, despite their high profile, these families did not act alone. The group of entrepreneurs I label as "outsiders" on table II-2 also played a crucial role in the development of Vizcaya's metallurgical industry. I include three types of businessmen in this group: non-Vizcayan entrepreneurs who acted as investors but for the most part did not reside in the province; Vizcayans whose fortunes were made outside the Basque region; and foreigners who settled in Bilbao.

Among the non-Vizcayan entrepreneurs, a group of financiers contributed almost 60 percent of the capital of Altos Hornos de Bilbao. This group of investors acted on behalf of large financial institutions from Madrid, such as the Banco de Castilla, Crédito General de Ferrocarriles, and Crédito Mobiliario (a branch of France's Crédit Mobilier, headed by the Pereire brothers).

The group of investors of Basque origin who made their fortunes outside their native region includes entrepreneurs such as the Marqués de Urquijo and the Marqués de Mudela (Francisco Rivas), and the Olano and Larrínaga families. The Marqués de Urquijo made a fortune as a financier in Madrid during the 1840s and the 1850s. In the 1880s, he not only was among the founders of Altos Hornos de Bilbao, but also participated in the creation of at least three other metallurgical companies in the Bilbao area. The Marqués de Mudela followed a similar career. His financial deals in Madrid made him a rich man, and in the late 1870s, he bought an iron factory in Bilbao, which became the third largest in the region. The Olanos and the Larrínagas offer a slight variation from the previous two cases, having made their fortunes in England. Their business base was the port of Liverpool, where they headed a large shipping company named after the two families. In 1882, José Antonio Olano, representing the Olano and Larrínaga Co., became the largest stockholder and first president of La Vizcaya steel factory.[69]

Finally, there was a small group of foreigners among the outsiders. Investors such as J. B. Davies, who had come to Bilbao to work in the mining industry, participated in the founding of some metallurgical plants. Also, engineers such as the Frenchman Enrique (Henri) Disdier, who had initially come to the region to direct the new steel factories during the early 1880s, became partners in several companies founded in the early 1890s.[70]

Although a smaller group than the mining entrepreneurs or the outsiders, the number of businessmen with mercantile backgrounds who served on the directorates of the metallurgical companies was not negligible. Six of these ten directors were descendants of the mercantile elite of the 1850s and the 1860s.[71] The rest seem to have been newly enriched merchants, since their names do not appear in any of the sources I studied for the interwar period. Unlike some of the most active entrepreneurs in the two other groups, each of these mercantile families restricted their investments to just one metallurgical company. For instance, the Gurtubays and the Uriguens appeared to have invested only in Altos Hornos de Bilbao. There was, however, one exception: the Echevarría y Rotaeche brothers. Their father, José, started as a trader of nails and iron products; by the early 1870s, the partnership between father and sons had prospered and gone into manufacturing. Without abandoning their family manufacturing business, they participated in the founding and were on the boards of directors of two large local companies, La Vizcaya and Iberia. By the 1890s, the family's manufacturing business included two medium-sized plants in the Bilbao area and had gone from producing nails to special steels, pioneering the use of an electrical laminating system in the region.[72]

Among the rest of the directors whose backgrounds I could determine, there were only two who belonged to landowning families, the Marqués de Casa Torre and Enrique Gana. While Casa Torre was involved in just one company, Gana, an engineer, participated in the founding of five companies. With the exception of José Orueta, a lawyer and son of an architect, the other directors in this group were involved in the shipping business, and, with the exception of Arrotegui, they were directors of the Euskalduna shipyard, which was founded in 1901.

Thus mining entrepreneurs, especially four families within this sector, did play a key role in the development of Vizcaya's metallurgical industries. The contribution of the mining industry, however, went beyond the participation of those families in the creation of local factories. Without the economic opportunities created by the mining boom, it is doubtful that the outside investors would have poured their capital into local industry. Consequently, it is not farfetched to claim that almost 60 percent of the directors of Vizcaya's metallurgical companies were somehow connected to local mining. At the same time, the presence of local merchants who had made fortunes trading goods from the colonies, codfish, or wines among the directors of these companies was a sign of the continued willingness of the local mercantile community to support new economic projects. Whatever their socioeconomic background, the majority of the directors of these factories did not owe their economic rise to the development of the metallurgical industry. For the most part, they were already established businessmen who expanded their economic horizons by promoting these enterprises.

SHIPPING

Bilbao's merchant fleet grew at a rapid pace during the 1875–1885 decade. In the early 1870s, the local fleet of steamers totaled around six thousand tons; in 1885, it reached ninety thousand tons and represented a total investment of approximately thirty million pesetas. During this period, the first large shipping companies were organized as corporations. Four companies led the field in the renewal of the local fleet: Marítima Vizcaya, Bilbaína de Navegación, Vapores Serra, and La Flecha.

I obtained the names of the main investors in these four companies from the partnership contracts in the local notarial archive. Fifty-seven businessmen appear in those contracts. To simplify the analysis, however, I considered only the thirty-four largest investors. Since they represent more than 90 percent of the capital invested in those four companies, the omissions will not adversely affect the analysis.[73]

Among these thirty-five investors, three stand out for their large contribution to their respective companies. José Serra y Font, president and founder of Vapores Serra and La Flecha, was a Catalan businessman who appears to have been involved in shipping prior to moving to Bilbao in the late 1870s. He owned approximately 50 percent of Vapores Serra and 25 percent of La Flecha. Juan Bautista Longa, president of Bilbaína de Navegación, shared a similar background with Serra. A native of Mundaca, a small port on the Vizcayan coast, Longa's early business career developed in England, where he was a partner until 1880 in the Olano Larrínaga shipping company based in Liverpool.[74] On his return to Vizcaya, his ownership of 11.5 percent of the shares in Bilbaína de Navegación made him the largest stockholder in the company. Finally, Fernando Carranza, president of Marítima Vizcaya, may also have had a background in shipping prior to the 1880s, although the issue is clouded in the partnership contract, where his occupation is listed as proprietor.[75] With slightly more than 40 percent of the shares, Carranza was by far the largest stockholder of Marítima Vizcaya.

Among the other thirty-two investors associated with Carranza, Longa, or Serra, twenty-one had a mercantile background; four belonged to mining families, one had been an army officer; three fell under the generic term of proprietors; one was a lawyer; and two could not be classified. Practically all of these investors were native Vizcayans.[76]

Forty percent of the investors with mercantile backgrounds represented either a direct line or only a single generational change from the economic elite of the 1850s and the 1860s. Such was the case of Eduardo Coste y Vildosola, Ezequiel Uriguen, Felipe Abaitua, José María Olavarri, the Escuza brothers, and the Rochelt brothers.[77] The remaining 60 percent of investors with mercantile backgrounds represented commercial partnerships that had not been among the largest mercantile houses during the period between the two Carlist Wars.[78]

Despite their small contribution to the creation of these four large shipping companies, some mining entrepreneurs formed their own fleets. For instance, in 1883, the Ibarras owned three ships (*Ibarra* Nos. 2, 3, 4), collectively measuring forty-three hundred tons. Jose María Martínez Rivas also possessed a three-vessel fleet (*Rivas, Marqués de Mudela,* and *Somorrostro*), totaling five thousand tons; and Manuel Taramona and Juan Durañona each owned one steamer of more than one thousand tons.[79] None of these private fleets, however, grew beyond their initial size, and during the late 1890s, most were absorbed by newly formed shipping corporations.

During the 1880s, other local businessmen, not involved in the mining industry, also ran large private fleets, which rivaled the size of companies such as La Flecha or Bilbaína de Navegación. For instance, Manuel María Arrótegui owned a nine-vessel line, representing almost twenty-two thousand tons. I could not determine Arrótegui's early business career or social background, although I did find that he was a native Vizcayan. His rapid ascent seems to have started in the 1870s. Toward the end of that decade, he bought several steamers in Liverpool in highly leveraged deals usually financed by the same British sellers, Dashper Edward and Walter Glyn.[80] Around the same time, similar deals were struck in Liverpool by another Vizcayan, Dionisio Eizaga, who also became prominent in the Bilbao shipping business.

The expansion of Vizcaya's merchant fleet continued during the 1890s, and toward the turn of the century, feverish investment in this sector increased the capacity of the local fleet to almost three hundred thousand tons, making it the largest in the country. Unfortunately, I could not find lists of investors in all the new companies. Nevertheless, the names of those who headed the most important firms appear in some economics publications. The companies founded by the businessmen discussed later together with companies created during the 1880s and still operating in the late 1890s constituted at least 75 percent of Bilbao's merchant fleet.[81]

At least three of the businessmen who led large shipping companies in the late 1890s had participated in the founding of Bilbaína de Navegación or Marítima Vizcaya in the previous decade. Eduardo Aznar, one of the managing directors of Bilbaína de Navegación, went on to create, with his cousin Ramón Sota, the largest of Bilbao's mercantile fleets, under the name of Sota and Aznar. In addition, Aznar managed other companies such as Naviera Bat and the Vasco Cantábrica de Navegación without the direct participation of Sota. Prior to the Second Carlist War, Aznar was a shipbroker, forming part of the middle ranks of Bilbao's commercial world. As the mining boom progressed, Aznar invested with Sota and other partners and acquired rich deposits in Santander during the late 1880s. The entrepreneurial spirit of these cousins rivaled that of the Chavarris and the Ibarras. During the 1900s, Sota's and Aznar's business empire expanded all across the peninsula, including interests in shipping, mining, shipbuilding, metallurgy, banking, and insurance.

Juan Bautista Astigarraga, Aznar's comanager in Bilbaína de Navegación, created Naviera Bachi around the turn of the century. During the 1860s and the 1870s, Astigarraga was Aznar's partner in the shipbrokerage business. Continuing his association with Aznar, he also participated in the mining business in Santander.[82]

Martínez Rodas, also a founder of Bilbaína de Navegación, formed two other shipping companies in the late 1890s, Marítima Unión and Marítima Rodas. Martínez Rodas, not a native Vizcayan, was a military man who came to the region during the Second Carlist War. He married into the Arana family of wealthy mine owners, but he did not appear to get directly involved in the extraction of iron ore in Vizcaya. By the turn of the century, however, he had invested in mining ventures in different regions throughout

Spain, participated in the founding of the large Aurora insurance company in Bilbao, and promoted banking ventures.

The expansion of the shipping industry also created opportunities for other small merchants, shipbrokers, and sea captains to join the ranks of the local economic elite. The Abásolo family provides an example. In the 1860s, Modesto Abásolo was an employee of a large Bilbao merchant, Benito Escuza.[83] Later he established himself as a shipbroker, and by the turn of the century his son Félix was managing one of the largest Vizcayan shipping companies, Naviera Vascongada.[84]

In general, the most active shipping businessmen in the late 1890s had few ties to the mercantile elite of the 1850–1875 period. With the notable exception of Sota, the main mining entrepreneurs kept a modest presence in this sector. This does not mean that the old mercantile elite and the principal mining entrepreneurs were completely excluded from this economic sector. Their participation, however, seems to have been mostly indirect, as they became passive investors in the companies managed by the likes of Aznar, Martínez Rodas, and Abásolo. During the early 1900s, the distinction between those involved in shipping and mining became more blurred, as the shipping magnates invested in the mining industry in different regions around the peninsula, and some of the large mining entrepreneurs created and managed large shipping companies.

BANKING

For almost two decades after its founding in 1857, the Banco de Bilbao was the sole financial institution in the city. Immediately after the Second Carlist War, the Banco de España opened a branch office. The conservative lending practices of these two institutions permitted the survival of several private banking houses, headed usually by wealthy descendants of the leading merchants of the 1850–1870 period. In 1891, the Banco de Comercio was formed, immediately becoming a strong competitor of the two existing financial institutions. During the late 1890s, a combination of high ore prices, hefty freight rates, and monetary transfers from other regions (especially Cuba) fueled an investment boom that led to the creation of several financial institutions. Many of these institutions did not survive the speculative frenzy of the period. Yet, two banks founded in 1901—the Banco de Vizcaya and Crédito de la Unión Minera—had a lasting impact in the region.

Although it would be important to study both successful and failed institutions, the sources I analyzed yielded information only about the financial companies that thrived. Thus the discussion of the leading banking entrepreneurs is limited to the five largest and most prosperous institutions in the city during the early 1900s: the Banco de Bilbao, the Banco de Comercio, the Banco de España, the Banco de Vizcaya, and Crédito de la Unión Minera. Since I could not obtain the lists of stockholders for all these banks, my analysis mostly focuses on the directors of these institutions. Table II-4 gives the names of the directors and the sources I used to compile the list.

Businessmen with a mercantile background and entrepreneurs from mining families had an almost equal number of representatives on the banks' boards. Together they accounted for slightly more than half of the total number of directors. The rest of the directors represented various business sectors. There were at least two landowners (Villabaso and Galíndez San Pedro), an engineer (Alzola) with multiple connections to the local business and political worlds, two doctors (Soltura Urrutia and Vicuña y Lazcano), one lawyer (Vallejo Arana), three important shipping entrepreneurs (Martínez Rodas, Carranza, and Longa), and a Basque who had made a fortune in Mexico (López de Letona). Unfortunately, I could not determine the background of about a third of the directors. Yet, a quick glance at the last name of most of those unidentified directors shows that they were of Basque origin. It is possible that some of these unidentified bankers were returned émigrés who made fortunes abroad and thus did not appear prominently in the local sources I analyzed. Others had family ties with some of the major entrepreneurial clans. For instance, R. Chapa, a director of the Crédito de la Unión Minera, was related to Sotera Mier, who as an early partner of the Ibarras owned large mines in the Triano area.[85] Regrettably, I could not establish Chapa's occupation.

The picture that emerges obscures some differences in the configuration of the board of each bank. For instance, the Banco de Bilbao's board of directors was composed almost exclusively of descendants of the mercantile elite of the 1850s and the 1860s and a few traditional landowning families. The only exceptions to this pattern occurred in 1891, when Cirilio María Ustara, a mining entrepreneur, and Enrique Aresti, a merchant who made part of his fortune in Mexico during the 1860s and the 1870s, were appointed to the board. This strong control over the board was kept until 1901, when the Banco de Bilbao merged with the Banco de Comercio. The merger did not actually displace the descendants of the old elite, but it did dilute their power on the board as many directors of the other bank became board members of the merged company.[86]

During its independent existence, the Banco de Comercio seems to have had a more heterogeneous board of directors. Its first president, Domingo Toledo, appears to have come from Castile and settled in Bilbao during the late 1860s or the early 1870s. In 1871, he established a mercantile partnership with his father and another local merchant. This company could not have been among Bilbao's largest, since the renewal of the partnership in 1880 established its capital at around 250,000 pesetas.[87] Yet, during the next decade his capital must have grown rapidly, since he became a large stockholder in the Banco de Comercio. Joining Toledo on the bank's board were important mining entrepreneurs such as Ramón Sota and Benigno Chavarri; shipping businessmen such as Fernando Carranza and León Longa; an industrialist, Lópēz de Letona, who had started a textile mill in the Bilbao region during the 1890s, but whose fortune had been made originally in Mexico; and a prominent Vizcayan landowner, Luis Villabaso. Although I could not identify about half of the board members' backgrounds, an analysis of forty-two of the bank's largest shareholders also showed varied backgrounds among those investors. This group of the forty-two largest shareholders in 1892 consisted of members of the old mercantile elite such as M. Ayarragaray, P. MacMahon, and A. Echevarría; important mining entrepreneurs, such as P. Gandarias and G. Yandiola; merchants with private banking houses and mining interests, such as C. Jacquet; and investors from outside the region, such as the Bank of Castile and Urquijo and Co.[88]

The Banco de Bilbao remained a bastion of the old commercial elite, but the Crédito de la Unión Minera was founded in 1901 by the mining industry for the mining industry. Among the fifteen board members presiding over the institution shortly after its founding, at least ten had interests in mining. These directors included members of important mine-owning families such as the Chavarris, the Gandarias, the Allendes, and the Ibarras, who shared the direction of the institution with ore extractors such as Otto Kreizner and Luis Núñez.

Also founded in 1901, the Banco de Vizcaya was conceived as an investment bank. As such, it attracted investors with various backgrounds. The mining industry was represented by three branches of the Ibarra clan, by P. Allende, and by A. Ustara. Descendants of the old mercantile elite, such as J. Basterra, P. MacMahon, and B. Uriguen, also sat on the board. Joining these two groups were entrepreneurs such as E. Borda, T. Urquijo, and P. Maíz, who must have prospered with the general economic expansion of the region.

Finally, the local branch of the Banco de España was administered by a director named in Madrid, but local investors also participated in the direction of the institution through a board based in Bilbao. The composition of the board changed every few years to permit the participation of the largest stockholders. According to Montero, the shares of the Banco de España together with government bonds were considered safe investments, attracting mostly the capital of members of the old mercantile elite and of rentiers who for the most part did not participate in the promotion of other business ventures. Indeed, Montero is right in stressing the connection of those stockholders to the old Vizcayan economic elite of the period prior to the Second Carlist War. Yet, not all of them displayed the cautious investment practices Montero claims. During the 1880s and the early 1890s, many of the stockholders Montero cites to illustrate his point had embarked on economic enterprises beyond commerce and the acquisition of government bonds. For instance, J. B. Longa was the promoter of Bilbaína de Navegación, one of the largest local shipping companies; Aureliano Schmidt also invested heavily in the creation of the Banco de Comercio. Similarly, the MacMahons were among the large investors in the Banco de Comercio and sat on the board of the Bank of Bilbao. The Uriguens had investments in metallurgy, mining, and banking; the Gurtubays participated in the founding of Altos Hornos de Bilbao and pioneered the oil industry in the country by building one of the first refineries in the harbor.

The direct participation of the old business elite in banking, then, seems to have been proportionally larger than in the other industries studied. They maintained control of the old Banco de Bilbao, which they had promoted in the late 1850s, and participated in the founding of at least three of the other four institutions. It was a logical step for those who had accumulated wealth through the trade of staples to recirculate their capital through financial institutions during a period when that type of trade had lost its paramount importance for the economic well-being of the region. During the 1890s and the 1900s, they were joined on the bank boards by mining entrepreneurs who had experienced a rapid accumulation of capital during the last quarter of the nineteenth century. In particular, among the group of important mining entrepreneurs who also played a key role in the development of the local metallurgical sector, the Ibarras, the Chavarris, and the Gandarias accounted for seven of the eighteen board memberships from the mining industry.[89] As the economy kept expanding, newly enriched businessmen became bank directors, too. In the end, a seat on a bank's board became the ultimate recognition of the success of the members of the local economic elite.

Conclusion

After the Second Carlist War, the merchants' tight control of the local economy came to an end. The mining boom and the concurrent general economic expansion of the region permitted the incorporation of a new group of entrepreneurs into Bilbao's business elite. The four industries studied show that the members of the old elite did not disappear from the business world, but their relative importance diminished as other entrepreneurs began to rise. In many cases, those rising entrepreneurs shared with the old elite a common trading background, yet their rapid economic ascent was not so much connected to commercial activities as to the accelerated growth of the mining and shipping industries. Revisionist objections to the idea that the mining entrepreneurs played a key role in the industrialization of Vizcaya are only partially justified. If one focuses only on the number of directors with a mining background, then the conclusion that this group played a key role should be tempered, since they never reached an overwhelming percentage of the total. Nevertheless, it cannot be denied that, in metallurgy, four families of mining entrepreneurs stood out as the most dynamic group of investors in the industry by promoting twelve of the fourteen firms analyzed. Three of these four families also participated actively in the development of Vizcaya's banking industry.

In shipping, the direct involvement of mining entrepreneurs, with the notable exception of Ramón Sota, seems to have been less dynamic until at least the early years of the twentieth century. Yet, this analysis tends to obscure a crucial fact in the evolution of the local business elite: there were no ideal types of mining entrepreneurs, merchants, and shipping businessmen. Although these categories may explain the initial economic rise of the businessmen, they do not fully capture the dynamic diversification of investment that took place among the business elite. By the late 1890s, when the capital invested in Vizcaya reached unprecedented levels, it was almost impossible to classify the main entrepreneurs according to just one economic activity. The closely interlinked boards of banks, shipping, metallurgical, and mining companies during the early twentieth century attest to this fact. The result of this investment strategy was the emergence of a cohesive elite that had the ability to bridge the potentially different business interests of the various Vizcayan industries.

Chapter Five

Business, Culture, and Society

[Bilbao] is a convent of merchants.
Miguel de Unamuno, "Madrid y Bilbao"

It is not rare to find in our mountains those who live worried about the great business of our salvation *in a truly puritan spiritual state.*
Miguel de Unamuno, "Mi raza"

From what kind of society did Vizcaya's entrepreneurs emerge? Can cultural factors help us understand their success in developing the local economic resources? Did success itself alter the economic and social behavior of the succeeding generations of Bilbao's businessmen? This chapter attempts to answer these questions. It focuses on certain themes, such as family, religion, education, wealth, and lifestyle, to try to establish some of the parameters that shaped the attitudes of the entrepreneurs. These topics are discussed within the context of two well-known scholarly works: Max Weber's thesis on the connection between Protestantism and capitalist development, and Arno Mayer's study of the persistence throughout nineteenth-century European history of an aristocratic hegemony.

Unfortunately, the documentation for this kind of study is not as complete as that available for other countries. For instance, there is nothing comparable to the material compiled in some German yearbooks, which served as the basis for a prosopography of the late-nineteenth-century business elite.[1] Unlike entrepreneurs in England, the United States, and Germany, Vizcayan entrepreneurs left behind practically no published memoirs or diaries.[2] Nor have any of Vizcaya's leading businessmen been the subject of biographical study, as have the Rockefellers, the Fords, and the Morgans. Contemporary nineteenth-century literature, put to good use in a study of Barcelona's elite, is not a rich source either.[3] Except for the novel *El intruso* by V. Blasco Ibáñez, Vizcaya's economic growth produced no Dickens or Zola to decry its problems, nor a Samuel Smiles or a Horatio Alger to praise the entrepreneurial spirit that underlay the industrialization process. Such literary silence should not be interpreted as a lack of interest in socioeconomic matters. The Vizcayans were proud of their economic achievements, but were not inclined to rhapsodize about them. In their silence, they gave credence to a trait captured by seventeenth-century Spanish playwright Tirso de Molina, who claimed that the Vizcayans were "short on words but long on deeds."[4] A similar reserve has unfortunately also limited access to private collections of documents.

Although the absence of sources is significant and renders conclusions more tentative, the notarial archive has provided a small window into the lives of Vizcaya's businessmen. Wills and marriage contracts yield information on attitudes concerning the family, religion, and business. Because personal documents cannot be consulted for one hundred years after filing, the window for the 1890s was closed. I was able to partly circumvent this last restriction, however, since several wills of prominent business families written during the 1890s were located at the archive of the Bank of Spain. While the notarial records form the backbone of the chapter, I complemented them with a few documents from the private archive of the Ibarra family, the largest mine owners in the region; local genealogies; travel guides; the available writings of contemporary observers; and other miscellaneous items.

Religion and the Businessman: A Basque Work Ethic

Throughout much of Basque history, business and piety complemented one another. One of the most visible links between the two activities was the confraternities that gathered Basques within their home territory and in different centers of the Spanish empire from the sixteenth century onward. Curiously, only the Basque confraternities outside the home region have received any mention in the historical literature. In Seville, around 1540, the Basque colony formed the Cofradía de los Vizcaínos, which was partly financed by a self-imposed contribution on its members' commercial transactions.[5] During the early decades of the eighteenth century, the growing number of Basques in Madrid formed the Royal Congregation of Saint Ignatius, named after Ignatius of Loyola, the Basque founder of the Jesuit order.[6] This association soon developed ties with similar groups created in the New World, especially with one in Mexico, which was known as Nuestra Señora de Aranzazu.[7] In addition to providing religious services, the confraternities preserved the ethnic identity of their members and served as trade, banking, and labor networks for Basque businessmen. Rare was the emigrant from these provinces who did not rely on these networks to ease the difficulties of starting a new life away from the home territory.

During the second half of the nineteenth century, the Bilbao business elite also displayed the same mixture of piety and economic activity that characterized earlier Basque generations. For instance, the use of religious names for some of the mines (e.g., Nuestra Señora de Begoña, San Miguel, San Antonio) and local iron factories (Nuestra Señora del Carmen, Santa Ana de Bolueta, or Santa Agueda) provides one of the most overt signs of piety.

Wills provide a more intimate look into the religious feelings of businessmen. Although after 1850 the religious function that wills had fulfilled in earlier periods was declining and they were becoming simply instruments to designate heirs, some wills left by members of the business elite still expressed elements of piety.[8] We should not assume, however, that those businessmen who did not add religious clauses to their wills beyond common formulas shared by practically all wills were indifferent or uninterested in this matter.[9] For instance, in 1899, Ramón Ibarra wrote that he "left up to the executors of the will all the arrangements related to the burial, funeral, and masses . . . for the eternal rest of his soul."[10] This was the most usual wording used to settle these matters in a will. Unfortunately, it does not reveal much, if anything, about Ibarra's religiosity. Yet, he may have been a devout Catholic, since the inventory of his estate shows that he had a private chapel in his home.[11]

Ramón Ibarra's father, Juan María, left clear indications of his religious feelings in his will, written in the late 1880s. He expressed a wish to have the chapel of the steel factory at Baracaldo expanded. The same document shows that among the recipients of his charitable bequests were his "spiritual adviser" (a priest who served as his personal confessor), religious groups such as the Conference of Saint Vincent de Paul, several communities of nuns, the Jesuits, and parish priests, who were instructed to distribute funds among the local indigents.[12]

Another member of the Ibarra family, Rafaela, demonstrated a piety that bordered on sainthood during the 1880s and the 1890s. Although this may seem an exaggeration, in Rafaela's case it is supported by the Vatican's imprimatur, since she was officially beatified in 1984.[13] Among her pious works, Rafaela created an order of nuns and donated very large amounts of money to charitable causes. In 1900, for instance, she gave 1.2 million pesetas for the founding of a religious school, the Colegio de los Santos Angeles Custodios. An idea of the size of the donation is given by the fact that Rafaela seems to have inherited from her father around 1.3 million pesetas in 1892.[14]

According to her biographer, Rafaela tried to "bring up [her children] in the holy fear of God, teaching them how to pray."[15] The lessons were not lost. One son, Gabriel Vilallonga, became a Jesuit priest, and a daughter, Rosario, entered an order of nuns. Rafaela's other children did not become ecclesiastics, but they nevertheless showed deep piety. Her oldest son, Mariano, requested in his will to be buried in the habit of the Company of Jesus and advised his children to "always seek the Fathers of the Company for the direction of their soul."[16]

The Ibarras' religious attitude was not unique. Other prominent members of the business elite expressed similarly pious sentiments. For instance, according to his will, Juan Durañona, a founder of La Vizcaya, donated money for the construction of a church in Santurce, a town on the estuary of the Nervión River.[17] Sotera Mier, a partner in many of the Ibarras' enterprises, bequeathed funds for the establishment of religious schools for poor girls in Portugalete and Sestao.[18] The joint will of José Echevarría and his wife, Victoriana Rotaeche, founders of a family factory, which their son Federico expanded into one of Vizcaya's largest mills, specifically requested the celebration of one hundred masses for each of their souls after they died.[19] A similar provision for daily masses after her death appears in the testament of Simón Gurtubay's widow, the mother of two founders of Altos Hornos de Bilbao.[20] Large merchants such as Pedro F. Olavarria requested in their wills to be buried wearing the habit of religious orders or the shroud of the Virgin, as did the widow of trader Francisco Uhagón.[21] Other prominent business families saw some of their children become priests or nuns.[22] Finally, the inventories of the businessmen's estates reveal that some of them had private chapels in their homes.[23]

Although religion surfaced in many areas of the entrepreneurs' life, it was perhaps in the educational field where it was most influential. After all, religion passed from one generation to the next through educational efforts. The Jesuit order handled most of those efforts in the region because of the deep admiration that its founder, the Basque Saint Ignatius of Loyola, inspired as a favorite son of the country.[24] The Jesuits' links with the local elites were especially close. In 1883, for instance, several prominent members of the Bilbao business community (Juan María Ibarra, Tomás Epalza's widow, Pedro F. Olavarria, and José Vilallonga, among others) established a partnership under the name "La enseñanza católica" with the purpose of promoting "teaching according to the principles of Catholicism, establishing colleges or universities." The charter expressly mentioned that the partnership would grant the direction of the institutions to the Jesuits.[25] Within a few months of the creation of "La enseñanza católica," the University of Deusto was founded, which is still managed by the Jesuits. In 1916, a similar association served as the steppingstone for the organization of the Deusto Business School (Universidad Comercial), the first institution of higher learning in Spain exclusively dedicated to the study of economics. Throughout the twentieth century, a large number of Basque business and political leaders studied in these two institutions.[26]

It is clear, then, that in their private lives and in their support for religious education, many Vizcayan entrepreneurs who modernized the region's economy displayed a profound piety. As a devout group of Catholics, these businessmen seem to defy Max Weber's idea of a link between Calvinism and the spirit of capitalism. Diehard Weberians may argue that by the second half of the nineteenth century such ethical principles had been assimilated by non-Protestant European countries. Yet, Basques did not display entrepreneurial spirit only during the nineteenth century; they had a long history of capitalist undertakings. Was there, then, a connection between their religiosity and their economic development akin to what Weber perceived to be a strictly Calvinist phenomenon? Or did development occur independently of their religiosity?

For Vicente Blasco Ibáñez, the influence of religion in general and of the Jesuits in particular was nefarious. In his novel *El Intruso,* set in Bilbao around the early years of the twentieth century, Blasco Ibáñez portrays the priests as a deleterious force that intrudes into the households of the wealthy businessmen through their wives' honest spiritual cravings. In short order, the priests take command of the life of the whole family. Finally, completely under the sway of the Jesuits, the prototypical local entrepreneur, Sánchez Morueta, makes decisions related to his business that reflect religious priorities that clash with the economic well-being of his enterprises. His new ultrareligious friends begin to interfere with his affairs, alienating his trusted advisers who, unwilling to mix business with religion, leave the company.

Blasco Ibáñez, an anticlerical writer, undoubtedly exaggerated the negative influence of religion. Contrary to what Sánchez Morueta's story suggests, the spiritual behavior of the businessmen does not appear to have been devoid of pragmatism. This is perhaps most clearly indicated in the language and clauses of certain wills. Although it was common among the members of the business elite to distribute their fortunes equally among their children, those heirs who joined religious orders were generally given lesser shares than were those who did not do so. Such measures were taken because, as the widow of Luis Zubiría explained in her will, "daughters and sons who preserve their lay status will have greater needs in their lives than those who join the church."[27] Similarly, donations to religious charities never reached levels that could significantly reduce a family's fortune. Admittedly, inheritance laws that limited the testator's freedom to give away his or her fortune prevented such extreme cases. Yet, although the law allowed the free disposition of up to 20 percent of a person's estate, charitable contributions rarely approached even 1 percent of the testator's net assets. This was true even in such strongly religious families as the Ibarras. Ramón Ibarra's executors decided to donate 13,500 pesetas to charity, which was less than 1 percent of the total estate. His brother Juan Luis left a similar portion of his fortune to strictly religious charities. Their father, Juan María Ibarra, left 22,000 pesetas in bequests, which represented about 1.5 percent of the income generated by his estate in the first year of his death. Only Rafaela Ibarra, an exceptional case, probably came close to the 20 percent limit.[28]

Nor does it seem that the Jesuits played as detrimental a role as Blasco Ibáñez describes in his novel, where some of the characters under their influence appear as opponents of economic development. Julio Caro Baroja mentions that during the eighteenth century the Jesuits supported the enterprises of the regional bourgeoisie, "praising the virtues of work against the aristocratic conception of life symbolized by Madrid."[29] That exaltation of work combined with an iron discipline and a tough competitive environment characterized the Jesuit-run Deusto Business School in the early twentieth century.[30] As much as the Calvinists did, the Jesuits believed in the sanctification of lay life, which Weber saw as key for the development of the capitalist ethic.[31] The writings of a Vizcayan Jesuit, Uria-Nafarrondo, who served as the chaplain of the Saint Ignatius Confraternity in Madrid during the late eighteenth century, clearly express such a view: "The conscience of merchants active in their trade is as pure as those of the best Christians: to treat them as usurers or thieves is a classic slander."[32] Uria-Nafarrondo not only denounced that misconception, but proudly proclaimed among his ancestors many merchants who had been leaders of Bilbao's *consulado.*

The idea that Jesuit teachings gave moral support to capitalist development is not new. It was advanced by H. M. Robertson during the 1930s as part of a critique of Weber's thesis.[33] After examining the writings of prominent Jesuits, Robertson concluded that they blessed business activities with much greater zeal than did any Protestant sect.[34] Yet, unlike Weber, Robertson did not think that the spirit of capitalism was linked to religious factors. Turning the argument on its head, he believed that capitalist ideology had a secular root, and that the actions and writings of the Jesuits merely reflected their frequent exposure to the aims and difficulties of the lay world.

Robertson and other critics of Weber have been faulted for misinterpreting the latter's work, especially their reading that religion was the only initiator of the capitalist ethic. On the contrary, Weber's defenders claim, the work of the German scholar uses multiple causes—among which religion was singled out—as contributing cultural elements that permitted sustained economic development. As to the objection that the Protestants were not the only ones to advocate an ethical system favorable to business activities, Weber's defenders answer that other groups that might have espoused a similar ethic throughout history did not enjoy sufficient influence to be able to impose their values.[35]

The defenders' first point is a sensible one, since Weber did not maintain that religion was the only cause of modern capitalism. In addition, whether one agrees or not with the details of Weber's thesis, his broader point—that cultural factors such as religion do influence economic matters and can encourage or discourage business activities through their moral sanction—cannot be disputed. Yet, the idea that the capitalist work ethic is an exclusive development of certain Protestant doctrines cannot be accepted, nor can the argument that other religious groups that advanced similar moral teachings lacked following or influence be conceded. In the Basque case, at least, the Jesuit message was heard among large sectors of society. Thus it may not be farfetched to infer that their ethical teachings did have an effect on the economic behavior of local businessmen.

Whether the Jesuits' ethics had a religious or a secular origin is a chicken-and-egg problem. However, the important point is that, whatever their source, Jesuit moral teachings helped encourage Basque business activities. The Jesuits preached hard work, discipline, and diligence, and their system of ethics found a receptive audience among the Basques. Echoes of their ethical teachings may be found in several written sources. For instance, José María Ibarra y Gutiérrez de Cabiedes recommended in his will that his sons "avoid idleness [huyan de la ociosidad], and never stop working no matter how many goods you might own, but don't be greedy."[36] Similarly, Mariano Vilallonga urged one of his sons to live up to his inheritance by being a "model of religiosity, honesty, and industriousness [*laboriosidad*]."[37]

The same motive of hard work recurs in funeral tributes to Vizcayan personalities. Pablo Alzola, an engineer who participated in many Vizcayan industrial projects, eulogized the Marqués de Casa Torre as a man who, in spite of his fortune, which would have allowed him to live as a rentier, chose to pursue a career, achieved a doctorate of jurisprudence, became involved in politics, and even participated in the establishment of several businesses in Bilbao.[38] A further indication of devotion to work appears in an anecdote involving industrialist Víctor Chavarri. Unwilling to put off business engagements because of the happy occasion of his first son's birth, he went back to work, visiting the mines with an associate after learning that the infant and the mother were in good health.[39] I could not confirm the veracity of the anecdote, but what is important is the respect and admiration that industriousness could evoke.

If industriousness deserved praise, its opposites, laziness and dereliction of duty, merited opprobrium. A report written during the late 1860s for a French sociological society explained that the number of beggars in Vizcaya was extremely low because of "the profound repugnance to ask for alms, because it seems that such an action implies the confession of sloth," a mark of dishonor.[40] This work ethic did not tolerate perfunctory activities either.

For instance, the testament of Juan Antonio Villabaso Echevarría, a very rich proprietor who could afford to let his children live the life of rentiers without many worries or great exertions, specifically condemned such habits. He purposely mentioned that if his sons, who at the time were studying in the university, were "not applying themselves to their studies and due to that reason were to flunk the classes," they should be charged for all the expenses incurred in their education.[41]

Such a strong work ethic supported by the religious beliefs of the entrepreneurs contributed to their economic success. Yet, other social and cultural factors also helped make their achievements possible. The next section analyzes some of them.

Business and Society

Social mores, ethnic solidarity, and family structure also supported business activities. Inheritance practices in the countryside sanctioned by the local *fueros* forced many Vizcayans to earn a living outside the agricultural sector.[42] Farmsteads were passed on intact from one generation to the next to a single heir. Parents had complete freedom to choose their successor from among all their children. The rest of the siblings were given small cash settlements to help them get started in their adult lives. In general, Basques perceived these inheritance rules as promoting social and economic stability in the countryside. Given the small size of the farms, division among all heirs would have rendered their exploitation unfeasible.

Interestingly, the inherent bias of the inheritance system did not undermine the strength of family ties, nor did it diminish rural emigrants' strong sense of ethnic solidarity and attachment to the region. On the contrary, all these feelings seem to have been heightened as a way of reducing the hardships of those forced to leave their farmsteads and even their homeland. A report written in the 1860s explains how Basques prospered by relying on kinship or ethnic networks: "all the young people who go overseas are called by a brother, an uncle, a relative, or the childhood mate of their parents, who is well established as a merchant, . . . who finds him a place in his house or in a friend's."[43] On many occasions, those who grew rich outside their home territory returned to the Basque country, becoming in turn protectors and sponsors of other relatives.

The story of Pedro Aguirre Basagoitia exemplifies these social practices.[44] Pedro was sent by his father, a small landowner, to a nautical school to become a ship's pilot. Sometime during the 1860s, following an interlude as a navigator on the Bilbao-Cuba route, Pedro was sent for by an uncle who had established a commercial house in Mexico. A few years later, his brother Domingo joined him there. By the early 1880s, having made a considerable fortune, both brothers were back in Bilbao participating in the social and economic life of the city.[45] As childless bachelors, the two brothers lived with a widowed sister and her son, Pedro Icaza. Probably thanks to the Aguirres' financial assistance, the nephew became an engineer and took part in the construction of important hydroelectric projects funded with Vizcayan capital. In his will, Pedro asked Icaza to create a charitable foundation with a portion of his fortune. The nephew fulfilled the request, and, among its many contributions, the Aguirres' estate promoted the Deusto Business School.

The Aguirres' story shows the importance of the Basque network not only for the emigrants, but also for the economic development of Vizcaya itself. Many of Bilbao's richest inhabitants started their careers abroad, usually in the Americas, but sometimes in the Philippines or in Great Britain. Among a list of eighty-five Bilbao millionaires around 1873 published by Manuel Basas, at least fourteen had made part of their fortune overseas.[46] When they returned to Vizcaya, several of these millionaires invested part of their fortunes in local business.[47] Several of these returned émigrés did not just bankroll enterprises, but sometimes actively worked in them. For instance, Enrique Aresti, who worked during the late 1860s first as an agent and then as a partner in a trading company in Mexico, became the president of a large paper mill in the Basque region during the 1890s.[48] Some Vizcayans, such as bankers Cristóbal Murrieta and J. L. Uribarren, even lent their financial support from abroad. From their respective bases in London and Paris, they became partners with the Ibarras in the Baracaldo factory.[49] In Murrieta's case, family links reinforced the business relationship with the Ibarras, since an Ibarra was married to Cristóbal's brother. Further examples of the importance of the Basque diaspora to the development of the home country are provided in chapter 4.

Paradoxically, a completely different set of legal rules governing inheritance matters in cities like Bilbao also encouraged the formation of social networks that influenced business practices. City dwellers could opt to divide their estates according to Castilian or *foral* dispositions. The main difference between the two systems was that the Castilian laws did not permit the selection of a single heir and the exclusion of the rest of the siblings, as the *fueros* did.[50] The Castilian legislation was not completely egalitarian, however. The testator was granted the right to improve by up to a third of his or her total estate the share of one of the heirs, but the remaining two-thirds had to be divided equally among all the successors. Thus, in a family of three children, for example, one heir could receive five-ninths of the estate; the other two could receive two-ninths apiece.

TABLE 19
Number of Children per Family, 1850–1875

						Total
No. of children	2–3	4–5	6–7	8–9	10–12	
No. of families	8	11	8	5	0	32
%	25	34.4	25	15.6	0	100

Source: A.H.P.V.

While *foral* inheritance provisions were designed to regulate the transfer of land from one generation to the next, they included hardly any rules on how to distribute other forms of wealth. As a result, families whose fortunes consisted mainly of personal property or urban real estate distributed their estates according to Castilian laws.[51] In general, such distribution was strictly egalitarian, since the right to improve the share of one heir was rarely exercised.

Although Malthusianism to reduce the possible number of heirs might, to our eyes, seem a logical response to such inheritance practices, the Bilbao business elite did not follow that path. My analysis of wills permitted me to reconstruct thirty-two of the fifty-six largest mercantile families during the 1850s.[52] As table 19 shows, 75 percent of those families had four or more children; slightly over 40 percent of that group actually had six or more offspring.

Did the entrepreneurs who presided over the economy after 1875 behave in a similar fashion? Using local genealogies and some wills, I was able to analyze twenty-nine of the most prominent business families of the period. As table 20 shows, the number of families with four or more children increased to almost 83 percent from the 75 percent recorded for the earlier period. Three-quarters of the 83 percent actually had more than six children.

Although personal fortunes grew rapidly as a result of the economic boom, the increase could not keep pace with the growth of large families and the custom of equal division of estates among all heirs. Nowhere can this quandary be better appreciated than in the case of the Ibarras, who, as the largest mine owners of the region, benefited most from industrialization. During the late 1870s, revenue from the mines and other family assets was divided among three heirs; as a new generation took over in the 1880s, income and assets had to be shared by eighteen family members. By the turn of the century, the eighteen heirs had given birth to at least seventy-two children with rights to the family property. By the early 1900s, an ever-ballooning number

TABLE 20
Number of Children per Family, after 1875

						Total
No. of children	1–3	4–5	6–7	8–9	10–12	
No. of families	5	8	7	5	4	29
%	17.25	27.6	24.1	17.25	13.8	100

Source: Pérez Azagra y Aguirre, *Títulos.*

of Ibarra relatives had to deal with the fact that their main asset—the mines—could not be renewed and its output was inexorably declining. Putting so much demographic weight on a limited resource meant a sure formula for downward social mobility. To counterbalance this force, the entrepreneurial elite followed different strategies.[53]

In many instances, the combination of large families and egalitarian inheritance practices provided an incentive to maintain an active presence in the business world. In this sense, many Vizcayan entrepreneurial dynasties defy the so-called Buddenbrook effect, or Law of Three Generations, which postulates the economic decline of families after two generations at the top.[54] The Ibarras' active participation in regional business, for instance, spans almost two centuries and more than four generations. The founder of the dynasty, José Antonio Ibarra y de los Santos, involved the family in the iron ore trade during the early decades of the nineteenth century. Two of his sons, Juan María and Gabriel María, and a son-in-law, Cosme Zubiría, expanded the business, assuring their position as the largest mine owners of the region. Most of Zubiría's and the Ibarra brothers' sons went into business, organizing and presiding over the vast resources of the family. By the early 1900s, members of the fourth generation of the family were leaving their mark on the local business world, as the presence of Gabriel María Ibarra y Revilla among the founders of the Bank of Vizcaya attests.[55] The uninterrupted participation of the family in business is felt even today: Emilio Ibarra, a sixth-generation representative of the dynasty, leads the largest bank in Spain, the recently merged Bilbao-Vizcaya, and four other Ibarra relatives sit on its board of directors.[56]

Other families survived well into the twentieth century. The Sotas, for example, contributed at least three generations of businessmen before the family's property was seized by Franco after the Civil War.[57] Alejandro Sota y Alvarez was the first member of the family to achieve some prominence, becoming a significant exporter of iron ore during the early 1880s. With Alejandro's support, his son Ramón Sota y Llano created a much larger mining company after leasing a rich deposit in Santander. He then went on to found, together with his cousin Eduardo Aznar, the largest shipping company in Vizcaya. By the early 1900s, his son Ramón Sota y Aburto had become a partner in the family's businesses. As evidence of the remarkable resiliency and entrepreneurial drive of this family, a fourth-generation member of the clan founded a prosperous shipping company in Argentina after the Sotas were forced into exile by the Civil War.[58]

I could cite further examples of families with two, three, or more generations in the business world. The permanence of a family among the business elite was not so much a question of generational sequence as of the interplay of personal aspirations, entrepreneurial ability, socioeconomic climate, and political circumstances. For the business elite, the family was the arena in which all these factors played to determine career choices. Ultimately, however, the decisions were filtered by individual desire. Industrialist Federico Echevarría, for instance, sent his son Juan to study engineering in Great Britain, with the intention of preparing him to lead the family's metallurgical factory near Bilbao. Although Juan finished his studies and worked in his father's firm for a few years, he soon abandoned the business world to become a distinguished artist.[59] Juan's choice contrasted with that of his brother Luis, who followed the example of their father without straying, leading the family firm and passing it on to his son Arturo, a fourth-generation entrepreneur.[60]

Unfortunately, because of the difficulty of probing into private family matters, I have not been able to study in depth the profession of each son of the top businessmen between 1850 and 1900. Obviously, not everybody wished to pursue business careers. Yet, with the end of the Second Carlist War in 1875 and with the rapid expansion of the mining and metallurgical sectors, a business fever encouraged many young Bilbao dwellers to become entrepreneurs. José de Orueta, founder of a local metallurgical factory, expressed in his memoirs how the economic success of many of his former schoolmates and friends during the 1880s and the 1890s made him abandon his work as a lawyer shortly after finishing his studies to try to emulate his comrades.[61] In the same vein, Basque journalist Ramiro de Maeztu wrote in 1910: "Ten years ago Bilbao's conversations referred only to millions [of pesetas]. Those who were not rich admired the wealthy and felt confident that they too would soon be affluent."[62] According to José F. Lequerica, a Basque economist who became one of Franco's foreign ministers during the late 1940s, this attitude toward wealth was more than a result of temporarily favorable economic conditions. For him, the Vizcayans had always ascribed a positive moral value to efforts dedicated to the acquisition of wealth, and such an attitude paved the way for the rapid growth of the local economy toward the end of the nineteenth century.[63] Thus a combination of cultural, social, and economic factors seems to have encouraged many young men to pursue business careers.

Active participation in the business world was not the only way used to maintain a family's economic clout. Marriage alliances among powerful families provided another method of counteracting the tendency of inheritance practices to dilute accumulated capital. The existence of these alliances during the 1850–1875 period seems to contradict the perceptions of foreign visitors and local observers, who noted the lack of a rigid social hierarchy in Bilbao prior to the mining boom. For instance, a British woman who toured the country at the outbreak of the Second Carlist War wrote: "The young man, who here [at a shop] serves you in the morning, will claim your acquaintance in the evening at the club, and challenge you to a game of billiards or cards; and you will find him on a footing of perfect equality with the merchants, officers, and others that compose the upper classes."[64] Remembering his childhood, the son of an important merchant of the interwar period wrote: "We all knew each other. . . . As people passed by, we could have drawn the family tree of each one of them, showing the dates on which grandparents died and parents got married. At the Arenal and Albia, the child of the most prominent and distinguished family used to play . . . with the sons of shopkeepers and carpenters."[65]

Physically small and with a population of around twenty thousand during the 1860s, Bilbao certainly provided an intimate setting for the social intercourse of its dwellers. Despite the preference of some rich merchants to cluster around certain streets, residential patterns did not physically separate the rich from the poor. Although European capitals were by then building exclusive neighborhoods for their elites, in Bilbao social differentiation had a vertical component rather than a horizontal one. Wealthy businessmen and artisans resided in the same buildings, but while the former inhabited the lower floors, the latter lived in the upper stories.

Yet, when it came to selecting spouses, all the intermingling of the different social classes appears to have been of no consequence. Of all the marriage contracts recorded by the local notaries, none connected an artisan or a small shopkeeper with a member of the business elite. I have not been able to determine precisely whether marriages were arranged by parents, but given the residential patterns and the mores described by contemporary observers, the predominance of marriages within similar social groups indicates that partners were probably carefully selected.

Marriage sawed the bottom rungs off the ladder to help the social ascent of the local population. Not all its rungs were missing, however. Marriages between the business elite and what may be termed middle-class groups did occur. Such marriages involved prominent families like the Ibarras and the Zubirías. In 1866, for instance, Rosario Ibarra married Adolfo Urquijo y Goicoechea, a lawyer and son of a local notary.[66] Similarly, in 1872 Mariano Basabe y Solaún, a co-owner with his sister of a Bilbao pharmacy, married María Pilar Zubiría, one of Cosme's daughters, who would inherit mining and industrial properties.[67]

Although in some cases marriage served as a step to gain entrance into elite groups, it usually consolidated the positions of those at the top. Links among business families were extremely common. Among the fifty-six largest merchants in the city cited in table II-1, thirty-four had at least one family tie with one of the other traders listed. Yet, as a group, the businessmen did not form an endogamous caste. Family bonds also connected them to some important Vizcayan landowners.[68]

The fact that marriages usually linked those at the top does not mean that the business elite became a closed caste. Once a man reached the top through any means, he was accepted into the elite and his children became potential spouses for other members of the group. Those who made their fortunes abroad had no trouble marrying into prominent Bilbao families.

Similar marriage patterns continued during the last quarter of the nineteenth century. If anything, the ties among the main entrepreneurial families became more pronounced. As household size increased, families pursued strategies that characterized wealthy groups in other countries.[69] Marriages between cousins, uncles and nieces, or involving several brothers and sisters from two families helped reconstitute the multiple divisions of an inheritance within one or two generations. For instance, three of Mariano Vilallonga's sons married three of Ramón Sotas's daughters; two of Vilallonga's daughters became the wives of two men who were cousins, Agustín Pombo Ibarra and Ignacio Careaga y Urquijo.[70] It would be cumbersome to provide more examples of this kind of family alliance; suffice it to say that by the early decades of the twentieth century, major entrepreneurial clans such as the Martínez Rivases, the Ibarras, the Vilallongas, the Zubirías, the Aranas, the Martínez Rodases, the MacMahons, the Olavarrias, and the Chavarris were all connected through marriage.

These social links among the most prominent entrepreneurs left a distinctive imprint on the structure of some of the largest Vizcayan companies. Not only were firms interconnected by members of boards of directors, as mentioned in chapter 4, but in many cases these directors were related to one another. Appendix III establishes the social links among the directorates of the Banks of Bilbao and Vizcaya around 1914. The kinship connections facilitated and strengthened many of the company mergers that took place in Vizcaya during the early twentieth century. The resulting socioeconomic cohesiveness also helped present a unified front in the campaigns launched by the entrepreneurs to obtain government protection for their industries during the 1890s and the early 1900s.[71]

Business and Education

Education was also stressed as a way to improve life. It was not a coincidence that the literacy rate in the Basque region was among the highest in Spain during the nineteenth century.[72] Even earlier, many Basques had received at least a rudimentary instruction, since education opened the door to careers in commerce or in the royal bureaucracy, where Basques found plenty of opportunities for their excess population.

The merchants who led Vizcaya's economy during the period between the two Carlist Wars received mostly a practical training in countinghouses. Others, like Aguirre Basagoitia, graduated into the higher ranks of the mercantile world after stints as ships' pilots or sea captains.

The richest Bilbao merchants tended to send their sons to England, France, or Germany to learn the techniques of international trade and to acquire proficiency in foreign languages.[73] The concern to provide their children with a good education appears in several traders' wills. For instance, J. B. Mendezona left clear instructions for the education of his son: "I beg the [legal] guardians to have my son follow a useful career [carrera provechosa], and if he shows a preference for commerce, send him to Germany, to a school there until he learns the language well, and then England to finish [his studies]."[74] Special provisions were included in wills to ensure that younger sons would have the same educational opportunities as elder brothers whose schooling had been financed during the lives of their parents. Thus the widow of the merchant Francisco Uhagón allotted an extra 5,000 pesetas per year to her youngest son, Federico, until he came of age to cover his own educational expenses.[75]

For some business families, the evolution from commerce to industry involved a change in their sons' educational choices. Many of the businessmen born between the two Carlist Wars had university degrees. Engineering and law seem to have been the two most popular fields of study. Business families like the Ibarras and the Chavarris, who had entered the ore trade early in the nineteenth century, must have felt the need for technical expertise to be able to expand their family enterprises. By the 1870s, at least two Ibarras, Ramón and José Antonio, and the two Chavarri brothers, Víctor and Benigno, held engineering degrees from the University of Liège in Belgium.[76] The sons of José Vilallonga, the first president of Altos Hornos de Bilbao, studied industrial engineering in Barcelona. Others, such as Pedro D. Arana, a large ore extractor, obtained degrees in mining from schools that I could not determine.[77]

Whereas the older generation provided the capital and the ownership of the mines, the young businessmen with their university degrees presided over the modernization of the mining and steel industries. According to the will of Juan María Ibarra, his eldest son, José Antonio, together with his cousins Fernando Luis Ibarra and Tomás Zubiría, negotiated the contracts from which sprang the Orconera and the Franco-Belge mining companies. And although the older generation of Ibarras subscribed most of the family shares in the revamped Altos Hornos de Bilbao in 1882, their sons actually worked as directors of the firm. Because of his technical expertise, José Antonio Ibarra in particular played a key role in the purchase of the machinery needed to produce Bessemer steel.[78]

The technical expertise of the generation of Vizcayan entrepreneurs who modernized the local economy has been mostly overlooked, and the oversight has led to factual errors. For instance, Gérard Chastagnaret claims that, in contrast with the Asturian case, in which several major businessmen held engineering degrees, in Vizcaya technical skill and capital ownership were clearly separated.[79] As the foregoing examples make clear, this was not the case. The importance of technical knowledge did not escape the local entrepreneurs. Fathers understood the need for such training for their sons and prompted them to study engineering. For example, Mariano Vilallonga's will shows how he politely nudged his sons in that direction.[80] By the early twentieth century, a second generation of businessmen with engineering degrees included members of such prominent entrepreneurial families as the Sotas, the Martínez Rivases, the Echevarrías, the Allendes, and the Chavarris, among others. Several of these engineers received their degrees from universities in England, Belgium, or France.

The disposition toward a utilitarian education that prepared the businessman to manage his enterprises seems to have been in sharp contrast with what happened during the late nineteenth and early twentieth centuries in Catalonia, the other major Spanish industrial center. For the Catalan upper bourgeoisie, university education seems to have been more a matter of prestige than a skill required to make business run efficiently. As a result, Catalans tended to study law at the University of Barcelona, where they could obtain their degrees with only minimal effort. Furthermore, at a time when Bilbao's entrepreneurs were founding a business school, such institutions were shunned by the Catalan elite.[81] Nor does it appear that Barcelona's successful businessmen sent their children to universities abroad. The Vizcayans' study in foreign countries not only improved managerial skills, but also exposed them to new technologies and encouraged business contacts with northern Europe.

The Vizcayans' faith in education as a business tool also seems to have assured easier economic transitions from one generation to the next than in the Catalan case. According to McDonogh, prolonged and excessive parental authority in Catalonia stunted the development of young managers in family firms.[82] In contrast, youthfulness distinguished the early decades of rapid economic growth in Vizcaya. For instance, the Chavarri brothers must have been right out of engineering school when, only in their mid-twenties, they participated in the founding of the Vizcaya steel factory. Similarly, when he was heading an ever-expanding shipping and mining enterprise, Ramón Sota, a lawyer by training, was only in his mid-thirties. Also young and technically trained were the members of the Ibarra-Zubiría family, who, in the early 1870s, were sent to London by their parents to negotiate the creation of the Orconera Iron Ore Co.

The "Aristocratization" of the Bourgeoisie?

In *The Persistence of the Old Regime,* Arno Mayer argues that throughout the nineteenth century the European aristocracy managed to retain social, political, and economic power. Although he recognizes some institutional changes, he believes that the aristocracy remained the most important social group due to the diffidence of the bourgeoisie, its major potential challenger. According to Mayer, no challenge ever took place because the bourgeoisie's main aspiration was to become part of the nobility.

A few years before the appearance of Mayer's book, a prominent Spanish historian, Manuel Tuñón de Lara, anticipated the argument as it applied to the Iberian context. According to Tuñón, the main feature of Spanish history of the second half of the nineteenth century was the amalgamation of the nobility and the upper middle class. In this process the aristocracy supposedly imposed its cultural and social values on a bourgeoisie mesmerized by its splendor.[83] The mechanisms that permitted the merger of the two groups included the ennoblement of the upper echelons of the bourgeoisie, marriage alliances, and common economic interests. As it became part of what Tuñón calls the "power bloc," the upper bourgeoisie drew closer to the old nobility, cutting itself off from other middle-class groups that were denied leadership positions in Spanish society. It is interesting to note also that many of Tuñón's examples of the upper bourgeoisie are extracted from the Basque business elite.

Undoubtedly, there was a fusion of elites in nineteenth-century Spain. Whether this process took place as Tuñón and Mayer describe it is another matter. Perhaps too much emphasis has been put on certain outward signs that lend themselves to various interpretations. It is undisputed that many prominent Vizcayan families such as the Aznars, the Chavarris, the Costes, the Ibarras, the MacMahons, the Sotas, the Vilallongas, and the Zubirías were granted titles of nobility during the 1890s and the early 1900s. Admittedly, the acceptance and, in some cases, perhaps the aggressive pursuit of such titles reflected a disposition to seek prestige, honor, and status. Yet, such an inclination did not necessarily indicate a betrayal of bourgeois values as such. The newly ennobled understood that, to a large degree, their titles were a direct result of their business success. The fact that many of them continued to be actively involved in business instead of enjoying an aristocratic retirement on a country estate should caution us about inferring changes in attitudes.

Indeed, if the aristocratic mentality is based on the idea that inherited privileges override any sense of personal merit or accomplishment, there are further indications that the ennobled Vizcayan businessmen did not share such views. Witness, for instance, the wishes expressed in the will of Count Mariano Vilallonga regarding his first-born son, who was to inherit the title: "if his health and capacity allow him, [he should try] to study to obtain a university degree either in Spain or abroad, which the testator would prefer to be a degree in engineering. Such a degree would bestow more honor than the noble title, since it must be obtained through one's own work, with which he will fulfil his moral obligation to be a role model for his younger brothers."[84] Clearly, these are not the thoughts of a person blinded by the splendor of aristocratic customs. Personal merit, hard work, and education are emphasized over the title. Contrary to Tuñón's assertion, then, it would seem that as the businessmen became incorporated into the nobility, they did not jettison the values that aided their social ascent; equally important, they tried to pass on those values to their children. In this regard, they continued to uphold the traditional Basque idea that nobility and business enterprises were not incompatible, a concept that appears often in Spanish literature of the sixteenth and seventeenth centuries. In one of his comedies, Lope de Vega captures this notion and the suspicion it raised in other Spanish regions in the protest of the daughter of a Vizcayan merchant whose nobility is questioned by Castilians after he has made a fortune trading in the Indies:

> Es de mi padre el solar
> el más noble de Vizcaya
> que a las Indias venga o vaya
> Qué honor le puede quitar?[85]

Nor is it clear that marriage alliances isolated the recently ennobled from other bourgeois groups. True, many members of Vizcaya's business elite married into noble families from other Spanish regions. Mariano Vilallonga is once again a good example, having married, one after the other, two Andalucian noble women. Pedro Zubiría, who was given the title of Marqués de Yanduri, became the husband of the Count of Aguiar's daughter. Several other noted alliances could easily be cited. Yet, these marriages seem to have been the exception rather than the rule. Of a sample of 116 marriages between 1890 and 1939, taken from a genealogy of ennobled Basque families, only 19 cases involved weddings with noble families from other Spanish regions. In 70 of the 116 cases, the families linked by marriage lived primarily in Bilbao, with a few cases of bonding with important families from the other Basque provinces. These marriages did not seem to divide the bourgeoisie at all. In fact, of the examples Tuñón uses, it is easy to point to many links between those inside and outside the "power bloc." For instance, Tuñón considers Ramón Sota to be an outsider mostly because of his support for the Basque Nationalist Party. Yet, as has been shown, several of Sota's daughters married into the Vilallonga-Ibarra family, a pillar of what may be termed the "aristocratized" establishment. Cosme Echevarrieta and his son Horacio are also perceived as outsiders because of their Republican sympathies. In fact, Horacio is said to have rejected a title that the king was willing to grant him because of his intervention in the rescue of Spanish soldiers in northern Africa during the early 1920s. Yet, the Echevarrietas belonged to the same social club, La Bilbaína, as many other wealthy Vizcayans and shared economic interests, and Horacio's sister eventually married into a Vizcayan family with a Castilian title.[86]

The fact that Tuñón did not include Sota and Echevarrieta among those who betrayed bourgeois values had more to do with their political ideas than with their social or economic behavior. In Tuñón's and Mayer's view, a conservative bourgeois was an aspiring aristocrat no matter what he or she did. Only thus can Mayer write that "[the other industrial titans who remained in the third estate] prided themselves on being as much the masters in their own industrial domains as the Junker proprietors' were on their estates . . . none of the great industrialists ever joined any of the . . . bourgeois-progressive parties."[87] To consider pride in ownership a mere reflection of aristocratic behavior is stretching the idea too far. Similarly, political conservatism was not necessarily an aristocratic or bourgeois quality. In Vizcaya, for instance, the true defenders of conservative political values were the Carlists, whose main supporters were the peasantry and not wealthy Bilbao businessmen.

Although nobody can dispute that old and new elites were fused during the nineteenth century, this fusion did not involve the cultural, social, and political conquest of an established group over a rising one. On the contrary, it meant a complex weaving of old and new values. As a result, the ennobled businessmen of the late nineteenth century seem to have had more in common with twentieth-century English actors, rock musicians, and businessmen, who, after being knighted, live their lives much as they did prior to receiving the honor and are not necessarily considered paragons of Ancien Régime society.

Family Fortunes and Lifestyles

Another way of approaching the problem of aristocratization or gentrification of the bourgeoisie is by studying the evolution of its fortunes and standard of living. Investment patterns can not only reveal the private economic behavior of businessmen but can also shed some light on their social aspirations. In order to discuss the business elite's standard of living, one must first determine what assets they were able to command and how they managed them, since consumption patterns have a direct relationship to personal fortunes. In addition, a point of reference is needed to establish whether the economic boom of the last quarter of the nineteenth century did in fact alter the businessmen's lifestyle. The fortunes and consumption patterns recorded for the 1850–1875 period will thus be compared to those displayed as the economy grew at a more rapid pace during the last quarter of the nineteenth century.

Inventories of estates and dowries recorded in marriage contracts permit us to establish individual fortunes with some precision. Registered by notaries, these inventories and marriage contracts were private documents designed to clarify family financial affairs. Since there was no inheritance tax in Vizcaya during the second half of the nineteenth century, assets did not need to be hidden to evade payments to the state. If some property was not recorded or appraised in the inventory because of arrangements among the heirs, this was usually noted. And although certain items could have escaped the scrutiny of the executors of a will, in general the estate inventories are a reliable indication of a person's fortune.

Many commentators, such as Miguel de Unamuno and Ramiro Maeztu, believed that there was widespread economic equality among Bilbao's inhabitants prior to the mining boom.[88] Yet, an analysis of marriage contracts for the 1850–1875 period belies such a notion. As table 21 indicates, there was a wide gulf between the dowries contracted by members of the business elite at the time of their marriages and those of artisans and small shopkeepers—not to mention those whose total lack of assets made marriage contracts useless.

The gulf separating the artisans and shopkeepers from the business elite was perhaps even wider if one considers that the latter were able to disburse a great portion of their dowries in cash or other types of highly liquid assets. In contrast, artisans and shopkeepers included mostly objects such as tools, household furniture, and clothing among their contributions to their marriages. In addition, the gap between the groups had started to widen even before the prosperous decades of the mining boom.

Among the elite, males' contributions appear to have been greater than females' dowries.[89] Usually, such disparities reflected the fact that grooms were older than brides. As a result, men had already established themselves in the business world by the time of their marriage and possessed assets that were earned through their own efforts. Sometimes parental donations or inheritances increased their capital. Since elite women did not work outside the home, their contribution depended completely on the dowries that their families could grant them. For example, Mariano Vilallonga and Rafaela Ibarra's 1866 marriage contract illustrates the differences in age and in capital between grooms and brides. Twenty years older than his wife, Vilallonga had already inherited property from his father and had been in business for some years by the time of his marriage. His contribution of 475,000 pesetas was about five times greater than the 88,400 pesetas that his wife's family handed over as her dowry.[90] Still, for brides such as Rafaela Ibarra, dowries represented only an advance of their total capital contribution to the marriage, since the possibility of a larger inheritance loomed in the future.[91] Thus the difference between the groom's and the bride's dowries did not indicate a family practice of granting women less money, or the lower economic standing of brides' families.[92]

While table 21 establishes significant economic differences among Vizcayans, a comparison between the dowries of Barcelona's and Bilbao's elites may help set the problem in a broader perspective. In a pioneering analysis in Spanish historiography, Gary McDonogh has attempted to quantify the average dowry of the several Catalan social groups. Unfortunately, his data appear to be somewhat problematic.[93] The most perplexing and unexplained trend in his analysis is that, while all the other occupational groups analyzed increased their dowries between 1858 and 1878, the merchants/industrialists suffered a significant drop, from 49,400 to 40,200 pesetas.[94]

Bearing that problem in mind and the fact that McDonogh records only women's dowries, the Catalan businessmen's 1858 figure of 49,400 pesetas was of the same order of magnitude as sums given to the brides of Bilbao's business elite during the 1850s. Yet, casting doubt on McDonogh's 1878 figure, the Catalan merchants' 40,200 pesetas recorded for that year were only about a third of the amount given to Bilbao's elite women during the 1870s. It is very

TABLE 21

Male and Female Contributions to Marriages (pesetas)

	1850–1859		1860–1869		1870–1879	
	Male	Female	Male	Female	Male	Female
Business elite	100,000 (7)	52,750 (7)	205,750 (25)	82,450 (27)	263,175 (13)	121,175 (17)
Merchants and professionals	10,000 (25)	17,900 (25)	23,350 (17)	33,100 (18)	17,500 (23)	32,100 (23)
Artisans and shopkeepers	3,400 (20)	4,500 (20)	3,600 (33)	4,350 (33)	7,800 (33)	6,000 (33)

Source: A.H.P.V.

Note: The numbers in parentheses indicate the number of cases for which information was available.

unlikely that during that decade Bilbao's businessmen were three times as rich as their Catalan counterparts, especially when, two decades earlier, their daughters' dowries were about the same. Yet, Bilbao's dowries during the 1870s surpassed even the Catalan peak of 85,000 pesetas, which McDonogh established as the average sum given to landowners' daughters in 1878.[95] Thus the dowries seem to indicate that Bilbao's business elite did not lag behind Catalan elite groups in capital resources. It is possible, however, that, in contrast to the Catalan cases, the Basque dowries represented a higher proportion of a family fortune. Unfortunately, at the moment, that possibility cannot be explored because there are no studies of Catalan fortunes for these years.

If the marriage contracts provide a first approximation to the question of the capital resources of Bilbao's entrepreneurs, their estate inventories permit us to analyze the amount and composition of their property in more detail. I examined all the inventories in the notarial archive for the 1850–1875 period, and it appears that, with one exception (R. Uhagón), thirty major entrepreneurs left estates of at least 250,000 pesetas. The upper boundary of wealth among this group seems to have been Romualdo Arellano's 4.43 million pesetas. Arellano was a merchant who participated in the formation of the Banco de Bilbao and the iron factory Santa Ana de Bolueta. At the time of his death in 1875, Arellano's personal fortune surpassed by one-third the capital of the Banco de Bilbao.[96]

How big were these fortunes compared to those of other elite groups? Within Vizcaya, their wealth seems to match that of the largest landowners, which also ranged from about 250,000 to approximately 2.75 million pesetas during the 1850–1875 period. For instance, José Landecho, whose family had possessed entailed estates since at least the sixteenth century, left a fortune of almost 2.75 million pesetas in 1874.[97] That same year, the inventory of José María Jusué, who had also inherited entailed estates and occupied high positions in the Vizcayan government, recorded his wealth at 1.2 million pesetas.[98] In 1865, Braulio Zubia's fortune of 300,000 pesetas represented the lower end of the wealth range for landowners.[99] In other regions of Spain, there are only sporadic indications of levels of fortune in the historical literature. For example, Juan Guell, a powerful Catalan businessman, left an estate of 7.05 million pesetas when he died in 1872.[100] That was considerably larger than the top Bilbao fortune recorded around the same period (4.43 million pesetas). The Basque businessmen's fortunes were also well below the 16.75 million pesetas left in 1858 by a Madrid banker, Juan Gaviria.[101] Francisco de las Rivas, a Basque financier who grew rich in Madrid during the mid-nineteenth century and participated in the process of Vizcaya's industrialization, is said to have amassed 50 million pesetas by the time of his death in 1882.[102]

It should not be surprising that Madrid, home of financiers and great nobles, concentrated the greatest fortunes in the country.[103] Although there are no estimates of the assets accumulated by the high Spanish nobility, the grandees, their income levels at the beginning of the nineteenth century seem to indicate vast resources. In several instances, a grandee received more than 1.5 million pesetas per year in rents. The richest among these select nobles, the Duke of Medinaceli, had an annual income of 3.15 million pesetas in 1808.[104] These annual revenues alone surpassed most of the fortunes amassed in Bilbao during the 1850–1880 period.

Table 22 shows the composition of some of the Basque businessmen's fortunes between 1850 and 1875. In general, the inventories tend to distort the distribution of assets. For instance, Damaso Escauriaza's estate shows a large amount of cash, which came from the sale of bonds and stocks after his death. Similarly, the "business investments" column, which includes merchandise or capital placed in noncorporate partnerships, tends to reflect the mature businessman at the end of his career. In some instances, the "credits" column indicates other aspects of the active life of a trader by recording positive balances in accounts held with other merchants inside and outside the Iberian Peninsula.

Whether active or retired, all businessmen diversified their fortunes into investments that could serve as a cushion in times of trouble and could easily be transferred to their heirs after death. In Bilbao, this type of investment seems to have been channeled primarily into stocks and bonds. Table 23 shows in detail the kinds of securities that the entrepreneurs listed on table 22 purchased during the years 1850–1875. Given the legal restrictions in effect during most of those years, joint-stock companies were limited to banks and railways. Highways and other public works encouraged by the Vizcayan provincial government also sometimes sold shares to the public. The reforms introduced in the Commercial Code in 1869 permitted the creation of joint-stock companies in other economic areas. Yet, the corporate structure began to be used for industrial companies in Vizcaya only during the 1880s, and the Bilbao stock exchange did not open until 1890. Thus the fortunes described on table 23 were affected by these business and legal practices.

Among the financial securities, Spanish state bonds drew the most funds. These bonds usually offered an interest rate of 3 percent on their nominal value, but since they were greatly discounted, the real interest could be several points higher. Such a high return for the time, and the expectation that the state would honor its debts, made the bonds attractive. Yet, this investment must have given nightmares to more than one investor, as the value of the bonds depreciated rapidly, reflecting the financial and political troubles of the Spanish state. Although the bonds rose during the relative stability and prosperity of the 1850s and the early 1860s, they plunged after 1866 to hit an all-time low during the height of the Second Carlist War in 1874.

The second most attractive securities were foreign stocks and bonds. These French, British, Latin American, Russian, and Ottoman state bonds were usually bought on the London or Paris exchanges through the agency of Basque banking houses established in those centers. These investments, in conjunction with money channeled into foreign (mainly French) railroad shares, constituted a sizable portion of the Vizcayans' portfolio. Finally, about an equal amount of money was allotted to banks and to Spanish railroad companies. The bulk of the capital in financial institutions was directed toward solid institutions, primarily the Banco de España and the Banco de Bilbao. The Tudela and the Norte lines constituted the favorite railroads.

If the Vizcayans' pattern of investment in stocks and bonds reflected the general behavior of the Spanish public at large, then the thesis that railroad construction starved other economic sectors of capital should be revised.[105] In the Vizcayans' portfolios, Spanish state bonds represented slightly more than twice the capital invested in national railroad companies. It would thus appear that the relentless borrowing needs of the state were much more of a hindrance to economic development than was the accelerated construction of the railroad lines during this period. Yet, the state's needs do not explain the situation completely. The fact that even investment in foreign securities was preferred, by a considerable margin, over Spanish railroad shares seems to indicate a certain lack of confidence in the Spanish economy as a whole during those years.

Real estate provided the second most popular alternative for the diversification of fortunes. Urban properties attracted the bulk of the investment in this sector. According to some inventories, houses in Bilbao rented annually for about 4 or 5 percent of the value of the property.[106] Thus rental income roughly equaled that obtained from state bonds, but, unlike bonds and stocks, real estate assets also required maintenance expenditures.

Rural estates in Vizcaya appear to have yielded lower annual rents, between 2 and slightly over 3 percent of the land value.[107] Such low returns probably explain their more modest place in the portfolios of Bilbao businessmen. With the exception of Máximo Aguirre, who possessed some rural properties in the Burgos province, none of the other businessmen I studied owned land outside the Basque region. Even a former merchant such as Juan Echevarría y La Llana, who retired from active trade relatively early in life and called himself "proprietor" in notarial documents, had around 80 percent of his fortune invested in stocks and bonds.[108] Although there were noneconomic incentives for owning real estate (i.e., the acquisition of voting rights

TABLE 22

Composition of Basque Businessmen's Fortunes, 1850–1875

Name	Gross Assets (pesetas)	% Stocks & Bonds	% Real Estate	% Credits	% Business Invest-ments
Arellano, R.	4,449,488	66.77	17.04	4.42	0.00
Casas, J.	2,699,603	52.22	22.74	21.14	0.00
Sanginés, M.	1,885,748	17.06	31.00	22.02	18.76
Aguirre, M.	1,827,675	15.28	13.85	23.06	40.90
Escauriaza, D.	1,533,183	32.70	10.14	0.02	23.66
Aqueche, J. R.	1,387,400	85.79	6.98	0.00	0.00
Gorbena, B.	1,357,925	31.60	7.81	11.92	14.29
Ayarragaray, S.	997,100	44.43	4.11	18.51	21.50
Uhagón, widow	989,335	46.10	39.68	0.00	0.00
Briñas, J.	969,430	23.36	30.16	22.00	20.32
Abaitua, S.	825,430	11.54	2.89	53.93	28.82
Blanchard, D.	801,960	3.28	61.34	3.65	15.97
Urien, C.	730,190	66.57	32.68	0.00	0.00
Escuza, M.	699,933	75.95	7.86	0.00	0.00
Mendezona, J.	602,745	49.19	1.63	19.40	25.30
Barua, S.	566,368	38.79	12.54	24.78	0.00
Artenana, A.	505,225	11.68	3.56	77.05	0.00
Urquiza, E.	499,905	36.74	8.38	28.76	0.00
MacMahon, D.	484,040	37.44	0.00	7.45	0.00
Gorocica, S.	383,843	24.02	26.53	16.89	9.77
Escuza, B.	326,593	53.89	28.82	8.90	5.74
Galíndez, P.	316,900	7.02	25.56	65.05	0.00
Solaún, J.	284,331	52.41	0.30	36.45	0.00
Uhagón, R.	257,722	16.94	30.42	4.33	43.65
Total					
pesetas	25,381,199	10,864,759	4,599,988	3,956,588	2,605,700
%	100.00	42.81	18.12	15.59	10.27

Source: A.H.P.V.

TABLE 22
Continued

% Cash	% Dowries	% Furniture, Jewelry & Clothing	% Miscel-laneous	Debt	Net Assets
0.11	11.24	0.35	0.07	14,600	4,434,888
2.28	0.00	1.16	0.47	0	2,699,603
0.88	7.40	0.57	2.32	281,150	1,604,598
5.28	0.00	0.65	0.98	667,715	1,159,960
32.12	0.00	1.34	0.00	84,300	1,448,883
6.01	0.00	1.03	0.19	0	1,387,400
32.54	0.00	1.84	0.00	159,775	1,198,150
1.31	10.03		0.00	88,000	909,100
0.03	11.12	3.09	0.00	0	989,335
2.43	0.00	1.71	0.03	128,250	841,180
0.02	0.00	2.80	0.00	365,200	460,230
0.89	12.47	2.35	0.05	108,240	693,720
0.75	0.00		0.00	0	730,190
15.2	0.00	0.99	0.00	4,250	695,683
0.13	0.00	1.04	3.31	49,650	553,095
7.35	16.54		0.00	0	566,368
5.05	0.00	2.66	0.00	13,975	491,250
23.88	0.00	2.24	0.00	21,350	478,555
32.95	20.65	1.49	0.00	0	484,040
9.21	11.28	1.84	0.46	0	383,843
0.46	0.00	2.19	0.00	26,585	300,008
0	0.00	2.37	0.00	12,750	304,150
7.55	0.00	3.29	0.00	1,050	283,281
0.08	0.00	3.80	0.77	26,325	231,397
1,758,728	1,186,500	304,670	104,267	2,053,165	23,328,904
6.93	4.67	1.20	0.41		

TABLE 23

Types of Stocks and Bonds Owned by Basque Businessmen, 1850–1875

	Total (pesetas)	% Spanish State Bonds	% Foreign Bonds	% Spanish Railroads	% Foreign Railroads	% Bank/ Insurance	% Public Works	% Vizcayan Bonds	% Miscel-laneous
Arellano, R.	2,970,738	10.74	31.16	15.44	13.31	16.60	0.00	0.00	12.75
Casas, J.	1,409,600	69.18	18.83	7.22	0.00	4.77	0.00	0.00	0.00
Aqueche, J. R.	1,190,250	22.40	71.44	1.88	4.27	0.00	0.00	0.00	0.00
Escuza, M.	531,590	33.31	16.00	6.96	14.43	16.43	9.63	0.00	3.24
Escauriaza, D.	501,290	31.20	0.00	53.04	0.00	0.00	0.00	0.00	15.75
Urien, C.	486,090	24.43	0.00	6.34	0.00	68.81	0.00	0.41	0.00
Uhagon, widow	456,125	42.99	8.81	23.94	24.26	0.00	0.00	0.00	0.00
Ayarragaray, S.	443,000	12.79	83.00	0.33	0.00	0.22	0.00	0.00	3.65
Gorbena, B.	429,125	42.67	18.77	28.18	9.60	0.00	0.00	0.77	0.00
Sangines, M.	321,625	59.11	0.00	0.00	0.00	35.08	5.81	0.00	0.00
Mendezona, J.	296,485	40.47	35.57	19.90	0.00	4.05	0.00	0.00	0.00
Aguirre, M.	279,255	0.00	2.75	43.13	25.35	17.67	0.45	0.00	10.65
Brinas, J.	226,370	54.06	0.00	0.00	0.00	0.00	45.94	0.00	0.00

Barua, S.	219,688	7.00	0.00	0.00	0.00	13.81	33.47	42.26	3.45
Urquiza, E.	183,690	34.35	0.00	0.33	0.00	65.32	0.00	0.00	0.00
MacMahon, D.	181,245	13.45	0.00	21.64	0.00	57.24	3.97	0.00	3.70
Escuza, B.	176,015	14.80	0.00	13.80	0.00	51.05	5.60	0.00	14.75
Solaun, J.	149,031	35.42	48.30	0.00	0.00	0.00	16.27	0.00	0.00
Abaitua, S.	95,275	45.69	0.00	0.21	40.28	5.77	0.00	0.00	8.05
Arteñana, A.	59,000	0.00	0.00	95.21	0.00	0.00	0.00	0.00	4.78
Uhagon, R.	43,663	0.00	18.24	27.08	0.00	54.68	0.00	0.00	0.00
Blanchard, D.	26,300	0.00	0.00	14.45	0.00	0.00	0.00	0.00	85.55
Galindez, P.	22,250	0.00	0.00	30.11	0.00	20.11	0.00	49.78	0.00
Total, pesetas	10,697,699	3,109,085	2,808,155	1,470,539	783,918	1,534,580	289,120	108,163	594,140
%	100.00	29.06	26.25	13.75	7.33	14.34	2.70	1.01	5.55

Source: A.H.P.V.

and the concomitant social prestige that such rights granted), these considerations did not prevent Bilbao's businessmen from funneling most of their capital into other, more profitable investments.[109]

The "furniture, jewelry, and clothing" column on table 22 gives a rough indication of consumption patterns and of the standard of living. Expenditures on these items was relatively small, only 1.2 percent of the total. In twelve of the twenty-one cases I analyzed, these expenditures were below 2 percent of the individuals' fortunes; only three businessmen appear to have held more than 3 percent of their estates in this type of assets. Daumard calculates that large Parisian businessmen ("négociants") around the middle of the nineteenth century held about 2.25 percent of their fortunes in personal items ("meubles meublants") such as furniture and jewelry, which is 1.8 times higher than the average for Bilbao.[110]

Not all the inventories give a complete list of the personal objects left by a person; sometimes just the total value of those objects is recorded. Yet, among the estates of the twenty-one businessmen who appear on table 22, two cases in particular—the estates of Mariano Sanginés and of the widow of Francisco Uhagón—offer more details and can illustrate some aspects of their way of life.[111] Both as a percentage of her fortune and in absolute value, the widow had the most expensive collection of furniture and other personal objects among those listed. Sanginés represented the opposite extreme, with one of the most modest household sets.

In Sanginés's case, the 10,850 pesetas recorded as furniture and personal objects actually included three carriages and several animals required to pull them, which were appraised at 4,500 pesetas. The household furniture added up to 2,500 pesetas. Most of the pieces were valued at fewer than 12.5 pesetas each, with the exception of some armoires and desks, a table with a jasper topping, and a set of upholstered chairs and a sofa in mahogany, each of which was appraised between 75 and 150 pesetas. The listing of furniture also included twenty-five decorative paintings with religious, nautical, and historical motifs. As an art collection, it was rather humble: no painting was assessed at more than 25 pesetas. His personal wardrobe, which consisted of two black frock coats, several pairs of trousers, vests, shirts, hats, and a few pairs of boots, was valued at only 250 pesetas. Assessed at 10 pesetas each, the frock coats were the most expensive articles among his clothes. To put this in perspective, in 1858, it would have taken an apprentice at a local iron factory four days of work to earn enough money to buy one of those frock coats, while some of Sanginés's more elaborate furniture would have required up to sixty days' wages to buy.[112]

Sanginés seems to have put more emphasis on a comfortable sleep than on his personal wardrobe, since bedding items such as mattresses and sheets were three times more valuable than his clothing. A few gold pins and some silver objects such as cutlery and other table utensils accounted for all his jewelry, which was valued at 625 pesetas. The rest of the articles in the list consisted of the religious objects in his private chapel (a silver chalice, a crucifix, and a few other items of religious paraphernalia assessed at 525 pesetas), kitchen and table utensils, a small armory with a few pistols and shotguns, a library, and a cellar with beer, sherry wine, chacolí (a Basque wine), and cider. Although the library did not have a high monetary value (only 75 pesetas), it is interesting because it reflects a completely utilitarian bent in Sanginés's readings. There were no novels, no history books, and, strangely enough for a man with a private chapel, not even a Bible. All the volumes were commercial treatises of different kinds, legislative compilations, French-Spanish and English-Spanish dictionaries, and a few books on home medicine.

The furniture and other personal objects belonging to Francisco Uhagón's widow were appraised at 30,600 pesetas, which represented about 3 percent of her total fortune. Undoubtedly, the widow lived in a more refined household than did Sanginés. The value of her furniture—including fifty-four paintings, decorative china objects, and a marble bathtub—reached 15,275 pesetas. Her paintings alone were estimated at 1,750 pesetas; four frames with religious motifs accounted for almost half of that amount. Bedding items, tablecloths, china, and kitchen utensils were valued at 6,425 pesetas. Unfortunately, her husband's personal wardrobe, which the widow donated to her two daughters, was not included in the inventory. Neither the will nor the inventory mentions any jewelry at all, but the widow owned a large number of silver items, which were appraised at 6,450 pesetas. The remaining 2,425 pesetas involved a cellar with some five hundred bottles of "fine wines" (1,775 pesetas), and a library with more than five hundred volumes (650 pesetas). Unfortunately, only a partial listing of the books is given, and it does not always include titles. Among the works cited are *Don Quixote,* several religious biographies, what appears to be historical studies ("Wars of Europe" listed in English but without mentioning the author), reference works (a French dictionary and an atlas), legal compilations (the Napoleonic Code, and *Ordenanzas del Consulado,* which were the old commercial rules, no longer in effect by the time the inventory was drawn), and forty-seven tomes of what probably was a periodical ("Annales de Artes y Manufacturas").

In short, the inventories tend to show that Bilbao's business elite lived comfortably, but well within their means during the 1850–1875 period. Their households do not appear to have been extravagantly luxurious. On the contrary, compared to the lifestyle of one of the richest of Madrid's financiers, the Marqués de Salamanca, the Bilbao business elite lived rather modestly. Next to Salamanca's collection of Goyas and El Grecos, Uhagón's paintings cannot impress, nor can Sanginés's carriages, which were rare in Bilbao, be considered opulent when one learns that Salamanca had a private railway coach equipped with gold plate.[113] Similarly, the residences of the Bilbao elite were usually large and comfortable, but not monumental. They generally lived in apartment buildings that were five or six stories high; the ground floor was commonly occupied by an office or shop, while each of the upper levels contained one or two apartments. Floor plans found in notarial documents display apartments with many small bedrooms, a studio or office area, a dining room, a living room, and a kitchen. Such a distribution of space seems to have been well suited for the large households of the business elite.

In the early 1870s, a British traveler left a written description of the interior of those dwellings that matches and complements the floor plans examined in the notarial archive:

> The principal room in the house is the salon, which is only used on state occasions. . . . The arrangement of this room is invariable in all houses . . . ; the furniture consist[s] of a long sofa, two arm chairs, and six other chairs, all of the same make, and covered with the same stuff, a console table with a glass over it, and two large candelabra; the floor is polished, and in the center of the room is a large square rug. . . . Opening out of the salon are usually two alcoves, . . . these are used as bedrooms. . . . The remainder of the bedrooms are small, and generally contain two beds apiece. . . . The only comfortable room, and that in which the family live, is the 'comidor' [*sic*], or dining-room.[114]

From what can be inferred from this description, the inventories, and the plans, the apartments were spacious but rather unpresumptuous.

If the elite's houses and personal possessions were relatively modest, their lifestyle reflected a similar simplicity. Although the fictional Vetusta in Clarín's famous novel *La Regenta* is supposed to be a realistic depiction of life in a small provincial town in the Asturias region, much of what the author describes could easily be applied to Bilbao. Despite being one of Spain's busiest harbors, the Vizcayan city offered few urban amenities. Like Vetusta, it had only one theater, where popular operas, *zarzuelas,* and plays were staged during a six-month annual season. Yet, the scarcity of this kind of entertainment did not seem to bother the local elite, possibly because the church still saw the theater at best as morally dubious. Although her parents selected carefully the shows she could see, Rafaela Ibarra recalled with much remorse going to the theater twice a week while she was a teenager during the late 1850s. When she became a mother, she permitted her children to attend only on rare occasions.[115] Such behavior seems to have been quite generalized; as one observer noted, "distinguished people went to the theater only on holidays and were characterized by a stern puritanism."[116]

Still, many members of Bilbao's elite enjoyed music passionately and formed small societies in which they played instruments or sang.[117] As in other small Spanish towns, the musical associations together with clubs or coffeehouses provided the setting for most social intercourse. For instance, at the Sociedad Bilbaína, a sort of gentlemen's club, members gathered to eat and read newspapers, periodicals, or books from an ever-increasing collection, to play billiards or cards, and to discuss current events. Social events like elaborate balls were virtually unknown. Beyond these simple pleasures, the routine of managing their business kept the members of the elite busy. Occasionally, they rewarded themselves and their families with trips to Paris or Madrid. Overall, however, their lifestyle was anything but glamorous during the interwar period.

How did the post-1875 economic boom affect the local entrepreneurs' habits and fortunes? Unfortunately, the one-hundred-year restriction rule on notarial documents prevents us from analyzing inventories taken after about 1887. Still, some records I found in the archive of the Banco de España and several inventories that were kindly made available to me from the private archive of the Ibarra family have allowed me at least to outline the boundaries of wealth for the period after 1875. The documents from the Banco de España include partial inventories from which I was able to extract the total value of fortunes, but not their composition. In addition, a series of reports produced by the bank about the personal financial conditions of the main depositors and directors of the Crédito de la Unión Minera in 1925 also helped me establish the boundaries of wealth in Bilbao.

During the 1880–1900 period, the wealthiest of Bilbao's businessmen seem to have had fortunes that ranged from one million to ten million pesetas approximately, or roughly two to four times the 0.25–4.25-million-peseta range in the 1850–1875 period.[118] Because they were the largest mine owners and manufacturers, the fortunes amassed by Juan María and Gabriel María Ibarra (10 million pesetas each) by the time they died in the early 1890s probably constituted the upper boundary of wealth in Bilbao. Thirty-five years later, theirs were still large fortunes, surpassed only by those of V. Chavarri's widow, B. Chavarri, and H. Echevarrieta.[119] Considering inflation, especially acute during the First World War, ten million pesetas in the 1890s represented a fortune at least as large as that of Chavarri's widow in 1925.

Within Spain, the 10-million-peseta mark could be considered literally a princely fortune. According to a study of the private wealth of Alfonso XIII, the king's fortune at the time of his coronation in 1902 reached 8.9 million pesetas.[120] Some grandees may have had larger fortunes, but it would seem that the Vizcayan industrialists were closing the gap. On the international scale, considering the exchange rate between the Spanish and the British currency during the 1890s, ten million pesetas translated into approximately one-half million pounds, the lower boundary of what W. D. Rubinstein considers the largest fortunes in Great Britain.[121] Similarly, the Vizcayan industrialists' fortunes did not reach the heights achieved by some of their French counterparts. For instance, one of the members of the Wendel family, the largest steelmaker in the Lorraine region, left a fortune in 1858 of 28 million francs (approximately 28 million pesetas).[122] Other important French steel manufacturers amassed fortunes between 1.5 million and 12 million francs during the last quarter of the nineteenth century, however, which paralleled the wealth of Vizcaya's industrialists.[123] All European fortunes paled when compared with those of U.S. industrialists such as Andrew Carnegie, who in 1900 received 480 million dollars (close to 100 million pounds) for the sale of his steel company.

Unfortunately, I have not been able to analyze the composition of most of the Vizcayan fortunes I found data on. Still, a comparison between Juan María Ibarra's estate, drawn in 1888, and that of his son Ramón, recorded in 1903, illustrates how investment patterns changed after 1875. Tables 24 and 25 demonstrate how the Ibarra family coped with the challenge of distributing its capital to successive generations and permit us to understand some aspects of their lifestyle.

In Juan María's case, the mines constituted his most valuable asset, accounting for slightly more than half of his total fortune.[124] Stocks and bonds were his second most important form of investment. His shares of Altos Hornos de Bilbao and Orconera represented almost 70 percent of his securities portfolio. Half of the remaining 30 percent was invested in Spanish railroad issues; 6 percent in bank shares; a tiny 1.3 percent in foreign railroad stocks; 3.4 percent in Spanish state bonds; and 4.3 percent in shares and bonds of the Franco-Belge mining corporation and of Ibarra and Co. (the shipping company managed by a Seville branch of the family). As was the case before 1875, his most valuable real estate holdings consisted of four houses on the outskirts of Bilbao, where the family lived.[125] Perhaps reflecting recent mine royalty payments, Juan María had a considerable amount of cash, which had been deposited in his business partnership with his sons and then transferred to the Banco de Bilbao.

The executors of Juan María's estate also provided an account of the income that the fortune generated during the twelve months prior to the distribution of the inheritance. The 1.46 million pesetas included the cash deposited in the family partnership (576,000 pesetas) and Juan María's capital

TABLE 24

Composition of Two Ibarra Family Estates

Family Member	Assets (pesetas)	% Cash	% Miscel-laneous	% Furniture, Jewelry & & Clothing	% Mines	% Dowries	% Real Estate	% Stocks & Bonds	% Credits	Debt	Total
Juan María	10,135,150	8.55	0.16	0.63	53.71	4.44	11.65	20.86	0.00	0.00	100.00
Ramón	8,267,950	0.14	2.10	1.39	6.17	0.00	13.55	75.10	1.55	0.00	100.00

Source: I.F.A.

TABLE 25

Detail of Stocks and Bonds of Two Ibarra Family Estates

Family Member	Total (pesetas)	% Spanish Railroads	% Foreign Railroads	% Banks	% Metal-lurgy	% Mining	% Ship-ping	% Other Industry	% Spanish Bonds	%Foreign Stocks & Bonds	Total
Juan María	2,113,900	15.35	0.56	6.05	53.27	18.72	1.90	0.00	3.37	0.78	100.00
Ramón	6,209,690	17.27	0.00	15.65	16.34	18.95	6.32	15.21	9.80	0.46	100.00

Source: I.F.A.

share in the family business (166,700 pesetas), none of which accrued during the year after his death. Without these two sums, his annual income has to be reduced to 717,300 pesetas. The bulk of this revenue came from the leases and direct exploitation of his mines (263,000 pesetas), dividends and interest on his stocks and bonds (121,900 pesetas), and loans and matured bonds (263,600 pesetas). These figures, however, do not include the royalties paid by the Orconera and the Franco-Belge that he had donated to his children in 1882. Had such payments still belonged to Juan María, they would have added approximately 700,000 pesetas to his annual income in 1889. Yet, even without the royalties from the Orconera and the Franco-Belge, Juan María's revenue far exceeded the 500,000 pesetas per year assigned to the future king as his personal income during the 1890s.[126]

Unfortunately, Juan María's inventory does not give an account of his personal expenditures. The only broad indication of consumption recorded is the value of his furniture and jewelry, which were appraised at 63,850 pesetas. In absolute terms, this figure was slightly more than twice the sum spent by Uhagón's widow before 1875; yet in terms of Ibarra's fortune, it represented only 0.6 percent of the total, which was about the same percentage recorded for Sanginés's estate. It is also worth noting as a possible indication of Juan María's priorities that he spent more money on the education of his three sons than on household furniture and jewelry.[127]

Ramón Ibarra's fortune appears more diversified than his father's, reflecting the growth of the Vizcayan economy. During the 1890s, Ramón played a key role in the local business expansion: he was a vice-president of Altos Hornos de Bilbao, president of Tubos Forjados, founding director of Basconia, and member of the board of the Banco de Bilbao. While the mines represented slightly more than 50 percent of his father's fortune, Ramón's main investment consisted of stocks in the metallurgical, mining, and banking companies that flourished in Bilbao during the 1890s. Unlike businessmen before 1875, Juan María and Ramón Ibarra dedicated an almost negligible amount of their fortune to Spanish state bonds and foreign stocks. The mining boom had bred an unprecedented confidence in the Vizcayan economy, and the Ibarras' portfolios reflected that optimism. According to Ramón's inventory, his share of the old family mines in the Somorrostro-Triano district accounted for approximately 6 percent of his total fortune. Had he not diversified his assets, the family's economic preeminence would have been diluted by the next generation because of inheritance divisions and the progressive depletion of the mines.

Ramón's real estate holdings, only slightly less valuable in absolute terms than his father's, show some interesting differences from Juan María's properties. Whereas the elder Ibarra lived in a comfortable three-storied house built on a lot of 340 square meters, his son inhabited what appears to have been a much larger home with more than twenty-four rooms erected on a tract of land occupying 54,000 square meters. Ramón's residence alone cost as much as the four houses owned by his father in Bilbao. The younger Ibarra's other properties included an apartment building, which he constructed probably as rental property in one of the new and expanding neighborhoods of the city, and some land in Baracaldo, which he co-owned with his siblings as part of their inheritance.

Although they were not listed, the furniture and paintings in Ramón's house were appraised at 60,000 pesetas, and his jewelry at 54,775 pesetas. Combined, the two figures represented 1.4 percent of his total fortune, about twice as much as his father spent on these items. Immediately after his death, the guardians of Ramón's children estimated that the fortune left by their father would produce an annual income of 380,000 pesetas.[128] It is reasonable to assume that such an income was larger when Ramón was alive and received compensation from his participation on the boards of several companies. Yet, even with these additions, it would seem that Ramón's income was smaller than his father's during the late 1880s. About half of the heirs' income was still expected to come from the mines, while the different stocks and bonds were to provide the other half. In 1902, the guardians estimated the expenses of the household at 80,000 pesetas per year, which accounted for 21 percent of the annual income. In order to increase the wealth of Ramón's five children, the guardians decided to reinvest three-fourths of the annual income generated by the estate in safe stocks and bonds, allocating the remaining one-fourth to covering household expenses and the children's personal needs. Investment, then, took precedence over consumption, a wise policy considering that the minors would not be able to count for too many years on their mining income as a way to increase their capital.

As Ramón Ibarra's inventory hints, the lifestyle of the Bilbao businessmen became more opulent toward the turn of the century. During the 1880s and the 1890s, members of the business elite abandoned their apartments in the old section of the city (the so-called *casco viejo*) and built mansions in new and expanding neighborhoods such as Abando or Deusto, which until then had been more or less rural areas. During the twentieth century, the relentless growth of the city replaced most of these mansions with apartment and office buildings. As this process unfolded and with the parallel improvements in the means of transportation, many members of the business elite moved toward what used to be vacation spots such as Las Arenas and Neguri, which are only a few miles from Bilbao.

Some leisure activities also became more expensive and glamorous. During the early 1900s, businessmen became interested in sports such as yachting. They organized clubs such as the Real Sporting Club and the Club Marítimo del Abra, which sponsored regattas.[129] King Alfonso XIII, an avid sportsman, ministers, and other members of the royal court, frequently presided over these summer events. The regattas illustrate the fact that, although the businessmen were honored by the presence of the king and his court, such attitudes did not make them lose sight of their enterprises. On the contrary, the sporting and social events were also an opportunity to lobby ministers and apprise them of the conditions of the Vizcayan economy.[130]

Despite their more lavish houses, the summer regattas, and their concomitant social events, the businessmen seem to have continued to lead a relatively simple life. Writing about the Bilbao elite during the first decades of the twentieth century, Maeztu remarked: "There are no dances, banquets or reunions in the private mansions. The custom is to have late dinners . . . , and nobody goes out after the meal. As a result, there is no opportunity for noisy bad habits or flashy displays of wealth."[131] If Ramón Ibarra's inventory is an indication, the Vizcayans' expenditures on furniture and jewelry were still well within their means. Unlike notable U.S. businessmen such as J. P. Morgan, whose purchases of art objects appear to have affected his banking firm, the Vizcayans' penchant for consumer goods does not seem to have interfered with their enterprises and constituted only a small proportion of their expenditures.[132] Similarly, the attitude of the legal guardians of Ramón Ibarra's children also seems to indicate a careful handling of money affairs with priority given to investment and the growth of capital over consumption.

Conclusion

It seems fair to conclude that several aspects of Basque culture and society facilitated the success of the Bilbao business elite. From the point of view of the connection between religion and economic development, the Basque case presents a glaring counterexample to Weber's thesis. The devoutly Catholic Vizcayan businessmen displayed the same traits that Weber ascribed to followers of Calvinist ethical teachings. In this regard, the influence of the Jesuits was considerable in encouraging hard work, diligence, and lay activities as worthy of good Christians.

The ethical encouragement provided by the Jesuits was complemented by several social practices. First, family networks at home and abroad advanced the business careers of many Basque entrepreneurs. Second, the movement of people and merchandise that sustained the networks instilled a certain cosmopolitanism, which kept the Bilbao business elite abreast of new developments in economic techniques. This cosmopolitanism also encouraged several families to send their sons to study engineering and other disciplines abroad. On their return, the foreign-trained graduates played a key role in the modernization of the local steel factories and the mining industry. Third, the large size of the businessmen's families and the equal partition of inheritances among all children encouraged many sons to remain active in the business world in order to maintain a standard of living as high as that of the previous generation. Family size and inheritance practices also led to marriage alliances among the elite that avoided the dilution of capital, facilitated the company mergers that characterized the Vizcayan economy in the early twentieth century, and helped present unified positions in lobbying campaigns to improve the conditions of local industry.

Finally, success in the business world brought many entrepreneurs titles and, in a few cases, led to marriages with old noble houses. In addition, economic success toward the end of the nineteenth century brought a somewhat more lavish lifestyle than the relatively austere one that characterized the period prior to 1875. Despite these developments, which seemingly support Mayer's thesis about the persistence of the Old Regime, the Vizcayan businessmen do not appear to have jettisoned the bourgeois values that made them succeed: hard work, personal merit, education, and what seems to have been a disposition to invest rather than to consume. In this regard, Vizcaya's entrepreneurs came closer to Weber's ideal capitalist than did better-known symbols of the business world such as J. P. Morgan.

Chapter Six

Business and Politics

Dominion is property, real or personal.
James Harrington, *Oceana.*

Bilbao's factories and the Vizcayan mines created a plutocracy—that of the new steel counts.
Miguel de Unamuno, quoted in García de Cortázar, "La oligarquía vasca"

Historians have always wrestled with the connection between economic and political power. No scholar denies the existence of this connection. The consensus, however, usually breaks down as soon as someone tries to characterize the relationship in some specific way. Only a diehard economic determinist would argue today that the economy is the exclusive mover of the political process. Yet, even those who recognize that the relationship is more complex often quarrel over the degree of influence that the economy exerts on politics, and vice versa. In Bilbao's case, the connection has never been properly studied, since, in general, historians have focused on other issues. This chapter attempts to probe that link by analyzing the direct participation of the Bilbao business elite in Vizcayan and national elective institutions and by studying the economic measures adopted by those political bodies. My goal is to establish how the entrepreneurs interacted with the central and regional governments and to determine whether this interaction helped or hindered their economic activities.

Vizcaya's Political Elite, 1850–1873

Some historians assume that the *foral* regime was a hindrance to economic development because it was designed to perpetuate the privileges of a landed elite that dominated Vizcaya's political institutions.[1] This argument echoes the Marxian idea that new forms of production give rise to new political bodies. Thus in the Vizcayan case, the landed elite's domination of provincial institutions supposedly thwarted the political and economic aspirations of local businessmen. This section, after a brief discussion of the status of the *fueros* after the First Carlist War, tests this assumption by analyzing the social composition and economic policies of the provincial government.

As was mentioned in chapter 1, the First Carlist War ended with the promise to respect the Basque *fueros*. The promise was formally recognized when in October 1839 the Cortes approved a resolution to uphold them. The resolution itself became a source of dispute because it was awkwardly phrased to try to satisfy both the centralizing tendencies of the Spanish government and the expectations of the majority of the Basque people. Its first article called for the restoration of the *fueros* "without prejudicing the constitutional unity of the monarchy." The second article required the government to consult with the Basque authorities and then to present a project to the Cortes to adapt the *fueros* to the Spanish constitutional regime. While those negotiations and the drafting of the project were still in progress, the Spanish government was given the power to resolve provisionally any difficulties that might arise.

Immediately after the resolution was approved, Basque officials and the central government in Madrid differed over the concept of constitutional unity. For the national authorities, it meant a homogeneous set of laws to be applied equally to all Spanish provinces. According to this interpretation, a good portion of the *fueros* and the singular system of government they had created would have to be scrapped. For the Basques, constitutional unity meant accepting certain common institutions such as the crown, the Cortes, and a central judicial system, but the acceptance of these institutions did not require a radical modification of the *fueros*, if any at all.

During the early 1840s, the *fueros* became a contested issue not only between the Basque provinces and the central government, but also between the main political forces in Spain: the Moderate and Progressive Parties, the two groups into which the Spanish liberal movement had divided. Ironically, the Progressives, whose advocacy of elected municipal officials would seem to have made them more sympathetic to decentralizing ideas, launched the most devastating attacks against the *fueros*.[2] The Moderates, although certainly not friends of regional rights, adopted a more cautious attitude and were willing to let the situation simmer because they feared a renewal of social uprisings such as those that led to the Carlist War. While they were in power, such a cautious policy exposed them to the Progressives' attacks. They were accused of not fulfilling the parliamentary resolution of 1839. Both parties also had to deal with internal rifts, which diverted their attention from this issue. The survival of the *foral* question during the interwar period owes much to the chronic political turmoil affecting the different governments in Madrid.[3]

While the *fueros* were merely a nettlesome side issue for the central government as long as no further uprisings occurred in the province, in Vizcaya the matter monopolized the attention of all political forces. Although this set of laws and customs provided a sense of identity for the Basques and therefore enjoyed wide support among all social groups, a recurrent tug-of-war existed between those Vizcayans willing to adapt the local laws to Spain's new constitutional situation and those for whom such reformist ideas were anathema. Accordingly, the local branches of the two liberal parties supported the *fueros,* but while the Moderates firmly resisted modifications, the Progressives advocated changes without complete abolition. In their defense of the *fueros,* Vizcaya's Moderates counted with the support of the Carlists, who, despite their defeat in the national arena, still constituted a powerful political force in the province.[4] In fact, the constant fear of further Carlist uprisings against government in Madrid was one of the Basques' best bargaining tools in trying to convince the central authorities to respect the *fueros.*

The attitude of Vizcaya's Moderates and Carlists toward the *fueros* prevailed, since the negotiating strategy of the provincial authorities with the central government demonstrated a clear resistance to changes in the local laws.[5] This strategy exasperated the Progressive government of General Espartero, which took a draconian step by abolishing the *fueros* altogether in 1841. Two years later, after Espartero was overthrown, the Moderates formed a government that restored the Basque laws, although with important modifications. Customs collection became permanently established at the borders and the seacoast, ending the traditional duty exemptions enjoyed by the Basque provinces; the *pase foral*—the right to refuse to accept central legislation deemed contrary to the local laws—was not restored; the administration of justice was reorganized so that it would follow the same principles applied in the rest of the country; and the provincial government lost jurisdiction over the mining sector. Still, the Moderates permitted the restoration of the provincial ruling institutions, the Junta General of Guernica and the Diputación. Fiscal and military issues remained unresolved, and were the subject of long, inconclusive negotiations.

Historians seem to disagree over the character of the Vizcayan regime from 1844 to 1876. Rafael Mieza and María Larrea, for instance, believe that the loss of the *pase foral* and the incorporation of the region into the Spanish customs network diluted the power of the local administration.[6] In contrast, José M. Ortiz and J. M. Portillo argue that, in spite of those losses, local institutions were actually very effective in containing the power of the central government and even increased their authority within the province.[7]

Paradoxically, there seems to be evidence to support both positions. On the one hand, in its fight against administrative reforms enacted in Madrid,

Vizcaya's provincial government successfully claimed for itself an overseer function over municipal budgets that it had never enjoyed under the traditional *foral* regime.[8] Similarly, in taxation matters, the Vizcayans effectively resisted attempts by the central government to levy an annual contribution in the form of a negotiated lump sum, which was supposed to compensate for the absence of national taxes in the Basque region.[9]

On the other hand, Vizcaya's government unsuccessfully protested the enactment of a national education law in 1857 that suppressed its right to supervise the instruction of children and to appoint teachers within the province. It also failed to block the disentailment and subsequent sale of municipal lands in the province as prescribed by national law. Furthermore, local authorities were unable to stop the central government's imposing an excise tax on certain imported articles such as sugar, cinnamon, and codfish. Although the national administration returned the money collected from this tax (minus 10 percent) to local officials, the tax had not been levied in Vizcaya previously. For that reason, the provincial government must have felt that it set a bad precedent, and local merchants balked at paying higher taxes.[10]

Despite the central government's inroads into the *foral* regime, Vizcaya's government still enjoyed a degree of administrative autonomy and fiscal exemptions that no other Spanish province outside the Basque region could match. The main Vizcayan officials were the general deputies, who were elected by a provincial assembly, the Juntas Generales, which met in Guernica every two years. Although the Juntas had authority over the deputies, the latter actually ran the province, since the Juntas met for only short periods. A random drawing determined a group of electors from among the Junta's delegates. The delegates, according to tradition, were divided into two groups, the Gamboinos and the Oñacinos.[11] Each of these groups in turn had to propose six names from which the two general deputies and two teams of alternates were randomly selected.

The elected deputies served for two years. A rule prohibited reelection for two consecutive terms.[12] Eligible candidates did not need to be delegates to the Juntas, but eligibility was restricted to those Vizcayans able to draw a high income from real estate, which limited the pool to a small proportion of the population.[13]

Unfortunately, the official records of the Juntas do not reflect how the electors chose the candidates, nor are there documents describing political campaigns or attempts to try to influence the decisions of the electors. Thus it is difficult to determine the political affiliation of the deputies. Nevertheless, historians such as Vázquez Prada have claimed that, during the interwar period, the Diputación was in the hands of the local Moderates.[14]

Aguirreazkuenaga claims that Moderates and Carlists actually alternated in power between 1839 and 1876.[15] Larrea and Mieza do not perceive a clear political affiliation among those who served as general deputies. Still, they mention that, from the 1850s on, Liberals seem to have dominated.[16]

Given the policies of the general deputies in their negotiations with the central government, it is clear that they all opposed concessions on the *foral* issue. It would thus appear that the local Progressives never held the office, except for a brief period from 1841 to 1843, when they were imposed by Madrid. Whereas it seems possible to infer the absence of the Progressives in the provincial government, the exact allocation of power between Moderates and Carlists remains hard to establish.

Although it is difficult to determine the precise political inclinations of the deputies, their names appeared neatly summarized every two years in the books that compiled the resolutions of the Juntas at Guernica. During the 1850–1870 period, forty-one individuals were nominated to serve as general deputies or as their alternates.[17] Fourteen of those forty-one individuals were chosen on more than one occasion. Yet, only one of those fourteen, J. J. Jauregui, actually held the office of deputy twice. The rest were selected as alternates more than once without ever becoming deputies or held office on only one occasion and became alternates in succeeding terms.

Twenty-two deputies were real estate owners; five others were involved in manufacturing; and only three in the group were actively engaged or recently retired from commercial activities. The remaining ten deputies could not be classified.[18]

Sixteen of the twenty-two deputies identified as proprietors were among the largest real estate holders of the province. The proportion of large proprietors may have been even greater. Eight of the eleven unclassified officials were included among the 450 wealthiest Vizcayans with the right to vote in national elections; and given the income requirements for becoming a deputy, it is probable that they too were among the largest real estate owners in the province.[19] Furthermore, three of the five identified as manufacturers were primarily large landowners whose industrial activities were only a small extension of their estates. For instance, J. J. Jauregui owned a small iron factory, which probably served as an outlet for the wood collected from his forests.[20]

The economic classification of the deputies tends to support the claim that the province was ruled by a group of real estate holders. Contrary to what Larrea and Mieza believe, the largest proprietors seem to have had a very strong representation in the provincial government.[21] Thus the mines and the factories did not create a plutocracy, as Unamuno's remark in the epigraph suggests; government by the rich actually preceded the rapid industrialization of the region after the Second Carlist War.

It should not be surprising that a regime that privileged real estate ownership over other forms of wealth was ruled by proprietors. For the same reason, the small number of merchants among the deputies seems understandable. Yet, the property requirement could not have been an insurmountable barrier to the many rich merchants who constituted the business elite of the province. In fact, many of Bilbao's wealthiest merchants enjoyed high rents from urban and rural properties.[22] Nor does it appear to have been a problem of geographic concentration in the Bilbao area, which suffered from underrepresentation in the Juntas of Guernica. After all, almost a quarter of those elected as deputies during this period resided in or around Bilbao, a proportion that considerably surpassed the city's meager representation in the Juntas.

The precise reason for the small proportion of businessmen among the deputies remains elusive. It would be tempting to say that proprietors enjoyed more leisure time than the merchants, whose businesses might have required constant attention in order to prosper. Yet, this reasoning does not explain the merchants' participation in Bilbao's municipal government. Seven of the twelve mayors between 1850 and 1868 were active merchants; the other five could be classified as proprietors.[23] They ruled the city in conjunction with a group of elected aldermen called *regidores*.

The franchise to select municipal authorities was highly restricted. In 1853, for instance, out of a total population of about fifteen thousand, only 2 percent of Bilbao's dwellers had the right to vote, and only half of those were actually eligible to serve as city officials.[24]

Municipal government also seems to have been a prerogative of the affluent. The twelve mayors during this period were among the wealthiest proprietors or merchants in the province. Unfortunately, since the municipal archive of Bilbao was closed during my stay because of a flood that damaged its collection of documents, it is extremely difficult not only to determine the economic background of the aldermen but even to know their names.[25] Some notarial documents provide some of this information. Although I examined a large number of protocols, only three provided the names of members of the city government for the years 1850, 1861, and 1868.[26] Of a total of twenty municipal officials, the 1850 notarial document lists only fifteen, among whom were six merchants, five proprietors, and four unclassified aldermen. From a total of eighteen names in the 1861 document, there were eleven merchants, four proprietors, and three unclassified officials. Only twelve of the municipal leaders appear in the 1868 notarial record. Four of them could not be categorized; five were merchants, three, proprietors. If these three years can be taken as representative, then, just as in the case of the mayors, most of the aldermen were among the wealthiest merchants or proprietors in Vizcaya.

On the other hand, the merchants' participation appears to be lower at the national level, in the selection of Vizcayan representatives for the Cortes. During the 1850–1868 period, only two active local traders, P. Epalza and P. P. Uhagón, of a total of seven elected deputies, occupied Bilbao's seat in the Cortes, and only for five years.[27] No member of the mercantile elite was elected from the other two Vizcayan districts during those years. Although the businessmen's electoral record in Bilbao was not negligible, their participation paled when compared to that of other local notables, who dominated that district's seat for thirteen of the eighteen years. In general, Vizcaya's national representatives shared a similar socioeconomic background with their provincial counterparts. In several instances, the same person served at both levels, although not at the same time.[28]

In Vizcaya, national elections took place under one of the most restrictive franchises in Spain. Since Spanish electoral laws granted voting rights to those who paid direct taxes to the state, their application in the Basque region posed a problem because those taxes were not levied in the province. Aware of this anomaly, the legislators included a special provision in the law for this province that established an even tougher guideline than for the rest of the country: only the 150 wealthiest individuals in each of the three districts into which Vizcaya was divided for electoral purposes had the right to vote.[29] This meant that only 0.2 percent of the population could vote in these elections, one-fifth the number who could vote in Spain as a whole.[30]

It is unclear how Vizcaya's authorities evaluated the wealth of potential electors. Yet, whatever method was used to establish the 150 wealthiest individuals in every district, Bilbao's merchants figured prominently. Using diverse sources, I was able to determine the economic activities of 106 of the city electors who appeared on an 1852 list.[31] I identified more active merchants (55) than proprietors (51).

Despite the merchants' relative importance among the electors, they did not usually challenge the proprietors as candidates for the Cortes. In fact, with the possible exception of Bilbao's municipality, where merchants may have outnumbered proprietors (see Appendix IV), it appears that the political elite did not recruit its members from the large businessmen. Some scholars believe that businessmen's lack of participation in the elective institutions showed an inherent bias of the *foral* regime, which was designed to uphold the interests of a rural society. According to this view, the business elite saw its activities hindered by the provincial power structure and therefore welcomed the abolition of the *foral* institutions in 1876.

How accurate is that interpretation? Did the predominance of proprietors in government necessarily mean that businessmen felt that their interests were not heeded by the provincial authorities? Did the process of industrialization require the abolition of the *fueros,* as some historians suggest?

The fact that proprietors ran the provincial institutions should not exclude the possibilty that those institutions could have benefited other social groups, including businessmen. England during the eighteenth and most of the nineteenth century offers an example of an accommodation that permitted a landed elite in charge of the state's institutions to promote the interests of commercial and industrial groups.[32]

Furthermore, many of Vizcaya's property-owning families had either a background in trade or social ties that connected them to the wealthiest merchants. For instance, José S. Orue, a *foral* deputy for two terms between 1854 and 1858, was related through the marriage of one of his sons to the family of a wealthy Bilbao merchant, P. F. Olavarria.[33] The Urquizu brothers, who were both *foral* deputies between 1852 and 1870, seem to have been two generations away from direct mercantile ties. Their mother was the daughter of a wealthy Bilbao merchant, José A. Arriaga, a leader in the city's merchant guild in 1811.[34] In addition, two of their uncles, Juan Echevarría y La Llana and Tiburcio M. Recacoechea, were among the largest merchants in Bilbao during the 1850s.[35] Similarly, the Ibarras, the region's largest ore traders, developed kinship ties with the Zabálburus, the Jáureguis, and the Quadra Salcedos, families who served at the highest levels of the provincial administration.[36]

The proprietor/merchant dichotomy sometimes fails to capture the historical evolution of families during even one generation. In some instances, the two categories might have represented only different stages in the life of an individual. During much of the nineteenth century, it was not uncommon to find individuals starting their careers as merchants and ending as proprietors. Among those who served as deputies, J. Echevarría y La Llana illustrates the point.[37] The case of Pedro Novia Salcedo offers perhaps another example with the possibility of an interesting variation. A deputy and wealthy landowner in the 1850s, Novia had been a leader of Bilbao's merchant guild in 1825. It is difficult to ascertain from his position at the guild whether he was a merchant during the early decades of the nineteenth century. In some instances, the highest positions of the guild were conferred on "distinguished gentlemen from Bilbao" not directly involved in trade.[38] Yet, even if he had not been a merchant in 1825, Novia's willingness (and that of other proprietors like him) to become part of the clearest institutional symbol of the merchants' life in Bilbao tends to support the conclusion that the two social groups maintained close connections with each other.

The social connections were extended into the realm of business. It is not uncommon to find in the proprietors' inventories of their estates, loans to local merchants, or transfer payments from abroad handled by the large banking and trading families based in Bilbao. A few examples will help illustrate the point. In 1853, the division of the estate of José M. Escauriza, a proprietor from the Baracaldo area near Bilbao, shows that part of his liquid assets were held in deposit with Epalza e Hijo, Ibarra Mier and Co., and Uribarren e Hijo.[39] The inventory of Fernando Landecho, another large proprietor, who had inherited part of several entailed estates that had been part of the family since 1585, also shows a loan to Epalza e Hijo for a considerable sum.[40] The case of the trader Francisco Gaminde demonstrates how family ties between merchants and proprietors might lead to business connections. In 1850, his wife's sister, the widow of an important landowner, gave him a sizable sum to invest in secure enterprises.[41]

Capital did not always flow from proprietors to merchants. Sometimes proprietors obtained loans from traders by using their real estate holdings as collateral. In 1870, for instance, the daughters of B. Zubia, a wealthy landowner, refinanced a mortgage with the merchant house of Gurtubay e Hijos on a property inherited from their father.[42] Similarly, the dowries given to merchants' daughters for their marriages to proprietors were another form of capital flow between the groups. In addition, enterprising sons of proprietors used their connections with merchants to start business careers. For instance, Fernando Zabálburu obtained loans from his uncles Juan M. and Gabriel M. Ibarra that enabled him to build a sizable business capital of his own by the time of his marriage in 1865.[43]

Given the many interests shared by Vizcaya's business and political elites, is it still possible to think that the *foral* regime actually hindered the region's industrialization because of the political predominance of one group over the other? The answer is not simple. Usually, those scholars who believe that the *fueros* delayed the development of Vizcaya's economy point to the mining industry as their main example. Indeed, Vizcaya's laws had prohibited the export of ore to foreign countries since 1526, when the so-called Fuero Nuevo was compiled. Vizcaya's laws also granted local ironworks special privileges for the purchase of mineral; these privileges were designed to give them an edge over competitors from other Spanish provinces. Because of its bulk and relatively low price per unit of weight, the ore could not have been exported in quantities large enough to affect this sector of the economy in a meaningful way before the nineteenth century. As long as the traditional ironworks remained competitive, the mining industry probably did not suffer any major hardships from the export ban. During its three-century existence, the ban does not seem to have elicited any strong protest from merchants or mining entrepreneurs who might have felt harmed by it.

In the early 1840s, the central government gained jurisdiction over Vizcaya's mines. The prohibition on ore exports, however, remained in effect until 1849. During the 1850s, France became an importer of Vizcayan ore, but the volume and value of this trade was quite small. In contrast to what happened with other national laws that violated *foral* rights, this one did not originate any protest from Vizcaya's government. In fact, had the mines continued under provincial jurisdiction, the ban probably would have been lifted by the local authorities when international demand exploded as a result of the invention of the Bessemer converter in the late 1850s. After all, the Vizcayan government had tried to promote more efficient exploitation of the mines by tightening property rights and lifting traditional restrictions on the number of months during which extraction of ore was permitted and by improving the infrastructure of the Triano mining district.[44] Furthermore, when in 1868 the Vizcayan government attempted to revive the *foral* regime as it existed prior to 1841, the issue of the export ban was not even mentioned.

Although the change of jurisdiction clouds the issue about the Vizcayan government's intentions with regard to the mining industry, its support of other economic projects belies the idea that local institutions delayed regional development. For instance, in the mid-1840s, the local government launched a program to expand and repair the provincial road network.[45] In 1858, the Diputación agreed to subsidize the private railroad line joining Bilbao with the Castilian interior. At the same time, it decided not just to promote, but to build and operate at its own expense, the mining railroad that linked the Triano district with the Nervión River. Inaugurated in 1865, just as the mining boom was about to begin, the Triano railroad became so profitable that it developed into one of the main sources of income for the Diputación.

The Vizcayan government also tried to protect local businesses in other ways. For example, in 1863, when the national government was planning to reduce the duties on imported iron products, it supported efforts by the Ibarra family, owners of Vizcaya's largest iron factory, to prevent the approval of the new tariff law.[46] Similarly, the Diputación defended the Banco de Bilbao against the national government's attempt to collect a 5 percent tax on its earnings by arguing that the *fueros* exempted Vizcaya's financial institutions from such levies.[47]

Thus the Vizcayan government was dominated by politicians who espoused conservative ideas and stressed tradition and the importance of the *fueros*. Yet, their actions showed that their conservatism was not opposed to local business interests. Economic development seems to have been a common goal of the political elite. Such an objective went beyond personal interests and involved deputies without any known connection with the business elite. A good example of this group of deputies was J. M. Arrieta Mascarua, a Carlist sympathizer and landowner, who actively encouraged, from public office, the construction of the Triano Railroad.[48]

The elections held after universal male suffrage was instituted by the regime established in Madrid after the September Revolution of 1868 showed the conservative nature of Vizcayan society. The four seats allotted to the province in the national constituent assembly as well as all those in the Diputación went to Carlist candidates. As the rest of the country was beginning to embark on a more liberal path, Vizcaya continued to be firmly anchored in tradition. Inevitably, these divergent paths led to new confrontations.

Taking advantage of the political turmoil caused by the September Revolution in Madrid, Vizcaya's Diputación decided to restore the full *foral* regime as it existed prior to 1841. It abolished all the taxes and dispositions that the central government had forced on the province since the end of the First Carlist War.[49] According to the deputies' own report, only the establishment of customs collection at the coast and the administration of justice were left in place. Despite a long history of protests against these two measures, the Diputación felt it was better to postpone action until the two commissions appointed to study the situation recommended an appropriate policy. This prudent course of action regarding economic matters reflected a concern that drastic changes could ruin the provincial business community. Perhaps for similar reasons, the Diputación refused to reassert its power in a third area: it never reclaimed jurisdiction over the mines or tried to enforce the *foral* ban on ore exports. In fact, a new mining law enacted by the national government in 1869 permitted the mine owners in Vizcaya to enlarge their deposits.

Although the government that toppled the Bourbon dynasty in 1868 seemed more amenable to a decentralized administration, the Vizcayans did not take advantage of this opportunity to solve the *foral* problem once and for all. On the contrary, the internal peace that had been achieved after the First Carlist War broke down. The understanding that had permitted Moderates and Carlists to work together in local institutions collapsed not because of disagreements over the *fueros,* but because of the bellicose position that the Carlists adopted with respect to the central government. Indeed, the Carlists' animosity was primarily directed against the new king and the constitution enacted in 1869, which they saw as an attack on the church.[50] In 1870, a rebellion started by the two elected *foral* deputies, Piñera and Urquizu, was quelled rapidly. In response, the central government appointed an interim Diputación formed of local notables opposed to the Carlists. This intervention and the proposal made by the interim Diputación to modify the composition of the Juntas of Guernica to allow the cities greater representation fueled Carlist-inspired rumors that the new regime wanted to abolish the *fueros.*

As the country's political troubles increased in 1873 with the abdication of the new monarch, the establishment of a Republican regime, canton revolts in the South, and an armed rebellion in Cuba that began in 1868, the Carlists rose once again. They started a full-scale civil war, which had the Basque region as its main battlefield. Under these difficult conditions, a disproportionate number of Republican sympathizers, mostly residents of Bilbao, were elected to the Cortes from the different Vizcayan districts. Among those elected were several members of the middle ranks of the Bilbao mercantile community, such as Bernabé Larrínaga, Cosme Echevarrieta, Federico Solaegui, and José F. Vitoria. With the exception of Vitoria, who went bankrupt in the 1880s, the others expanded their businesses and joined the region's entrepreneurial elite during the Restoration.

Although Republicans were overrepresented in the 1873–1874 elections, there is no doubt that most urban dwellers, among them the Bilbao business elite, did not support the Carlist cause.[51] Indeed, as in the previous war, the Carlists' strength was in the countryside, where their message of religion, king, and *fueros* appealed to the pious and conservative Basque peasants. That same conservative message presumably attracted some members of the political elite, who joined the rebellion more out of fear of what they perceived as the radical policies of the republic than out of a deep commitment to Carlism. Although other members of the political and business elites perhaps shared those same fears, they deplored the seditious tactics of the Carlists. They thought it was the least effective way of defending the *foral* regime, which they did not see as being under attack, as the Carlists claimed.[52] Unfortunately, the sociological underpinnings of the Second Carlist War have not been studied yet; therefore, it is difficult to assess how deep a split the rebellion caused among the provincial elites. Nevertheless, those notables who did not join the uprising were proven right, since the war was used as an excuse to abolish the *foral* regime.

The Restoration of the Bourbons and the Abolition of the *Foral* System

After the short-lived Republican experiment, the Bourbon dynasty regained the Spanish crown. The architect of the Restoration was Antonio Cánovas del Castillo, the leader of the Conservative Party. A great admirer of the British parliamentary system, Cánovas sought to re-create it in Spain. In order to do so, he decided to lessen the risk of political uprising by agreeing to alternate in power with the leader of the moderate Progressives (or Liberals, as they were known during the Restoration), Práxedes Sagasta. Thanks in part to this political maneuver, to the disarray of radical groups after the chaotic situation of the early 1870s, and to tight control of the voters through electoral corruption, the Restoration inaugurated a long period of institutional stability.

Following the Carlist defeat on the battlefield in 1875, Cánovas did not waste any time in tackling the *foral* question. Outside the Basque region, there was strong anti-*foral* sentiment and a belief that the *fueros* had been the cause of the two civil wars. Despite repeated pleas from Basque deputies in the Cortes that the region was being unjustly punished for what had been a Carlist and not a *foral* uprising, Cánovas passed a law in July 1876 imposing fiscal contributions and military conscription in the Basque region. It was a lukewarm measure that did not satisfy those requesting the total abolition of the *fueros*, yet, it sufficed to alarm the Vizcayans, who believed that such a measure was the first salvo of an offensive to destroy the provincial laws. Despite Cánovas's attempts to reach a compromise with the *foral* authorities, the Basques stubbornly refused to make any concessions.[53] This intransigence cost them dearly. Cánovas does not appear to have intended to abolish the *foral* institutions, but the Vizcayans' refusal to accept the 1876 law led to elimination of those institutions one year later.

Cánovas's drastic measure in 1877 provoked a split in Vizcaya's political elite between those in favor of negotiating with the central government to try to salvage at least part of the *foral* regime and those who stuck to the intransigent position that had led to the abolition of the *fueros* in the first place. The attitude of the intransigents was hopeless. It was impossible to drum up large popular support for their cause after the Carlists' military defeat; moreover, a powerful national army still garrisoned in the region was ready to strike at the slightest attempt of rebellion. In addition, once Cánovas realized that there was a group of Vizcayans willing to negotiate, he was able to offer concessions that made the abolition of the old institutions more palatable.

The *foral* system of government was replaced by a Diputación that formally resembled the administrative institutions of the other Spanish provinces. The deputies were directly elected by the voters. Whereas previously only two officials were in charge of the administration, the new Diputación included sixteen members; after 1884, the number was increased to twenty.[54] Although the premise behind the creation of the new Diputación was to homogenize Basque provincial administration and that of the rest of the country, the practical results were quite different. Unlike what happened in other provinces, Vizcaya's Diputación retained control of certain taxes within its jurisdiction.

Fiscal autonomy was the consequence of a special agreement between the Vizcayan authorities and the central government that came to be known as the *concierto económico*.[55] Despite Cánovas's belief that the Basque region should contribute its share of taxes to the national treasury, he did not impose the same fiscal system that existed in the rest of the country. Instead, every few years the Basque provinces and the Madrid government negotiated the annual quota that each province had to turn over to the central treasury. The quota covered the direct taxes on commerce, agriculture, and industry that in the rest of the country were collected by the central authorities.

The Diputación's fiscal jurisdiction did not include all the impositions levied in the province, however. For instance, import and export taxes were directly collected by the national authorities. In addition, the central government reserved the right to create new fiscal impositions. Nevertheless, the Diputación had broad powers over local fiscal matters, since it not only collected the impositions included in the agreement, but also changed them by replacing them with consumption taxes. No direct impositions on individuals or businesses were actually levied by the Diputación until the early twentieth century.[56]

The fiscal prerogatives of the Diputación led in turn to broad administrative functions, similar to the ones the *foral* institutions had enjoyed during the interwar period.[57] Vizcaya's government had control over municipal finances and budgets; it ran hospitals, schools, and charitable institutions; it contracted loans and issued debt; and it had its own local police force. In short, the juridical status of the province changed with the abolition of the *fueros,* but its relative autonomy vis à vis the central government was preserved. This autonomy was not supported by a solid legal base, but stemmed from the central authorities' recognition that the Basque provinces required a certain administrative leeway to be able to meet the requirements established in the tax agreement of 1878. The de facto self-rule that emerged from the tax treaty created some confusion among the authorities in Madrid and Vizcaya over the exact functions of the local government.[58]

Yet, seemingly satisfied with what it had achieved with regard to the *fueros,* the central government did not seek to reduce the powers of the Diputación. In fact, when in 1880 the national parliament approved a new law concerning the administration of provinces and municipalities, it had to amend the legislation two years later to accommodate the Basque region. Had that law been fully enforced in Vizcaya, the Diputación, for instance, would have lost its supervisory authority over municipal budgets, which would then have fallen under the jurisdiction of the central government.[59]

Armed with ample fiscal powers, the Diputación set the tax rates in such a way that it retained a good portion of the money collected to finance its own expenses. Thus, Vizcaya's government managed one of the richest provincial budgets in Spain.[60] The forced savings that fed the Vizcayan treasury were mostly invested in infrastructure that supported and complemented the private initiatives of local businessmen. During the 1880–1881 fiscal year, for instance, 62 percent of expenses consisted of railroad subsidies, maintenance and construction of highways, and costs related to the administration of the Triano railroad line.[61] Although the quota paid to the national treasury increased noticeably as the fiscal agreement was renewed in 1887, 1894, and 1906, Vizcaya's payments seem to have been lower than the amount the central government could have collected had it applied the same principles as it did in the rest of the country. In 1901, for instance, despite the tremendous economic boom in the region, Vizcaya was paying almost 18 percent less than Teruel, a small, poor province on the Castilian plateau.[62] True, Vizcaya's Diputación had to finance certain expenses that in Teruel's case were paid by the central treasury. Yet, even taking those expenses into consideration, the contribution of the Basque province, one of the most prosperous in the country, seems to have been among the lowest in Spain.[63]

Little fiscal pressure had an important effect on the development of the region. Many companies, whose main economic activities were carried out outside the province, established their legal addresses in Vizcaya to take advantage of its fiscal privileges. In 1900, this abuse of the tax agreement was curtailed as the central government established that new companies whose main activities took place outside the Basque region were to be taxed like businesses in the rest of Spain.[64] Still, businesses inside the region continued to enjoy relatively low taxes. A Vizcayan entrepreneur understood the advantages of the administrative autonomy of the Basque provinces when he wrote: "the freedom to allot tax resources within our *diputaciones* and to use them according to the needs of the provincial economy is the true reason for the prosperity of the region.[65] The quote exaggerates the point, since there was not just one "true reason" that could explain that prosperity, but certainly low taxes and an efficient local government contributed to the local economy's growth.

Thus the Diputación's ample autonomy from the central government made it a coveted prize for those aspiring to rule the province.

The Political Elite during the Restoration, 1875–1900

Did the new system bring changes in the political elite of the province? Several historians have suggested that the abolition of *fueros* and the establishment of a new Diputación marked the advent of a new oligarchy in Vizcaya. For instance, García de Cortázar claims that the new Diputación became the personal fiefdom of the local business elite.[66] For Manuel Montero, too, "the mining, industrial, and financial oligarchy" controlled Vizcaya's government.[67] This interpretation has been widely accepted, but no detailed study of the workings of the Diputación or of its members has ever been attempted to determine whether the business elite did actually control provincial government.[68]

In a sense, the new institution did bring forth a new set of politicians. There were 101 provincial deputies between 1880 and 1900. Out of that group, only eight family names match those of deputies during the *foral* period.[69] These new politicians, however, did not behave like an entrenched oligarchy. Although there was no rule against reelection, only fourteen individuals held office for two terms, and just two deputies served during three administrative periods.[70]

I have not been able to establish the socioeconomic background of all the deputies. Only 46 out of the 101 officials could be classified.[71] Elite businessmen constituted somewhat less than half (22) of those whose profession I identified. Another 9 deputies were small entrepreneurs who were not connected to the major companies of the region. Eight proprietors and seven members of the liberal professions accounted for the rest of the classified deputies.

The fact that the elite businessmen had such a seemingly strong representation among the deputies should not be misconstrued. Given the documents I studied, they were the easiest to identify, and therefore their percentage would probably not increase even if we could determine the profession of all the deputies. If they in fact represented about 20 percent of all the deputies, then their participation in the Diputación increased only slightly from its proportion during the *foral* period.[72] Although they were far from a majority, members of the business elite held the presidency of this institution during sixteen of the twenty years I studied. Yet, it is hard to say to what extent this represented the advent of a new class, since four of the five businessmen who became president had sizable real estate holdings prior to their industrial investments.[73] In fact, if all the deputies who were classified as members of the business elite and who also had significant properties before their involvement in other economic activities were relabeled as proprietors, then the latter's representation would equal the number of top businessmen in the Diputación.[74]

Perhaps more notable than the active participation of top businessmen in the Diputación is the fact that I could not connect the unclassified deputies to the large real estate holders, the social group that had dominated the institution before the war. In addition, the lack of visibility of these unclassified deputies in the major business enterprises of the region could also suggest that the Diputación may actually have been less oligarchic in its composition than is commonly believed; there may have been room for representatives of different social classes.

The appearance of a new group of politicians was related to changes in the electoral laws. For the national parliamentary elections, the Restoration imposed a restrictive suffrage, which limited voting rights to about 5 percent of the population.[75] The law remained in effect until 1890, when the Liberal Party managed to reintroduce universal suffrage for males. At the provincial level, even prior to 1890, voting rights were extended to a larger segment of the population. An 1877 law enfranchised everyone who paid direct taxes, no matter how small their contribution. In Vizcaya, since those taxes were not collected, the law was changed to include those who could prove a relatively low level of income or property ownership. In 1882, the law added to the provincial voting pool all those who were literate but paid no taxes. As a result, for the next eight years, there were twice as many voters in provincial elections as in national elections. Finally, after 1890, universal suffrage for males also ruled in provincial elections. In the district of Bilbao, for instance, the proportion of electors to inhabitants rose from 11 percent under the restrictive 1882 law to 18 percent under universal male suffrage.[76]

During the 1880–1900 period, the Diputación was dominated by a coalition of liberals. In Vizcaya, this political label indicated opposition to Carlism and a general acceptance of the changes introduced to the *foral* regime after 1876. The Carlists remained an important political force until the 1890s. In the 1880 election for the Diputación, they won eight out of twenty seats to become the largest single party in the institution. By the late 1880s, they had started to lose ground for several reasons: divisions within the party that split into an ultra-Catholic wing (the Integrists) and the traditional monarchists; the gerrymandering of some rural districts, which diluted their strength; and electoral corruption.[77] The third largest force was the so-called Fuerists, former liberals unreconciled to the loss of the *foral* institutions. This group had some success during the mid-1880s, but never seriously challenged the liberal coalition.[78] Finally, leftist parties such as the Republicans and the Democrats obtained a minuscule number of seats during the 1880–1900 period.

Although the business elite did not act as a political party, fifteen of the twenty-two major entrepreneurs who served as deputies won their seats under the broadly defined Liberal banner. The remaining seven elected businessmen included two Fuerists (R. Sota and G. Vilallonga), one Republican (B. Larrinaga), one Carlist (F. Ibáñez Aldecoa), one independent (P. D. Arana), and two undetermined (R. Goyoaga and A. Uriarte y Arana).

Whatever the social origin or political orientation of the elected members of the Diputación, the institution pursued economic policies that did not differ from its *foral* predecessor. Indirect taxation and promotion of the economic infrastructure of the region had already become the rule during the

1840–1873 period, yet nobody claims that the Diputación was controlled by the business elite at that time. Of course, the policies it adopted benefited the entrepreneurs both before and after the war, but this does not mean that the Diputación was the pawn of a clique of businessmen intent on running the provincial government for their own interests.[79] No one has ever presented clear examples of such occurrences, and the arguments used to support this thesis are sometimes contradictory. For instance, it has been said that the income of the Triano Railroad, owned by the Diputación, was used to reduce the tax burden of businessmen.[80] This line of reasoning does not consider that, in such a case, mining entrepreneurs were being taxed indirectly by paying for the high profit margins of the line, which was subsidizing the rest of the taxpayers.

It has also been alleged that the business elite used its control of the Diputación to contract its own services by charging the provincial treasury to perform public works. One specific example is the construction of the regional railroad lines.[81] The construction of the railroads, however, cannot be considered a public project. Except for the Triano line, the other railway companies were private enterprises, free to contract with whomever they chose. True, they received subsidies from the Diputación, but this was not a conspiracy to raid the public coffers. It was the common practice of nineteenth-century governments to subsidize railroad construction, since the lines were perceived to benefit the general public. Although some deputies were shareholders in railroad companies and thus had a vested interest in the prosperity of those firms, it appears to have been the practice to exclude officials whose private affairs might influence their votes from decisions such as the allocation of railroad subsidies.[82]

To determine whether the Diputación blatantly favored members of the business elite in its public works projects would require extensive analysis of all its contracts and the practices used to distribute them. Time constraints did not permit me to undertake such a study; however, the records of the Diputación's sessions tend to show that the institution was not a syndicate for the welfare of businessmen. Even with wealthy entrepreneurs holding the presidency of the provincial government, the deputies did not hesitate to turn down projects that could have benefited private business, especially if they felt that those plans did not further the public interest. In 1894, for instance, the Diputación rejected a request from a group of mine owners to extend the Triano Railroad so that it could service mines in the Sopuerta region. The provincial government decided that, because of the mine owners' refusal to guarantee a minimum amount of ore as freight, it could not risk the province's money on a project that did not hedge public funds against potential losses.[83]

Although the notion that the Diputación was the personal fiefdom of a small group of wealthy entrepreneurs seems exaggerated, there is no doubt that businessmen took a more active role in politics after 1876. Their presence became most noticeable during the 1890s, especially in the national elections. In part, this was a result of the ethos created by the economic boom that followed the Second Carlist War. The business success of the captains of industry was easily transformed into political capital. In 1891, for example, the Bilbao newspaper *El Nervión* justified its support for J. Martínez Rivas as a candidate in the national elections not because of his political ideas, but because of his founding of the Astilleros del Nervión shipyard. In the paper's opinion, "it was the obligation of all Vizcayans . . . to applaud such a work, praise the author, and carry him to a position that would be a just reward for his merit and where perhaps he could continue to serve the region.[84]

As the businessmen became more visible, their reputations acquired mythical proportions as power brokers or, as they were pejoratively called, *caciques*. The quintessential Vizcayan businessman-*cacique* was Víctor Chavarri, whose career as a politician and businessman by the time of his premature death in his early forties had been meteoric. Admired and reviled at the same time, Chavarri embodied many of the qualities and defects of the business elite. Capturing this ambivalence, a Bilbao Socialist newspaper wrote the following about him after his death: "With half a dozen men like Chavarri, Spain would have easily been spared its industrial underdevelopment. . . . Vizcaya was his fiefdom; he almost singlehandedly selected full Diputaciones and municipalities. . . . We could say about Chavarri: 'As a man, we abhorred you; as an industrialist, we admired you.' "[85]

Undoubtedly, Chavarri was one of the most powerful men in Vizcaya, but his political and economic omnipotence is exaggerated.[86] He often met opposition from other members of the Bilbao business elite. In 1885, for instance, several of his partners refused to name him executive manager of the La Vizcaya steel factory.[87] In 1891, J. Martínez Rivas defeated him in the Cortes election for the district of Valmaseda, which Chavarri had been representing since 1885. Two years later, despite his power and an alliance with his old rival Martínez Rivas, he was not able to deliver Bilbao's seat in the Cortes to his candidate, F. Solaegui. A. Urquijo y Goicoechea won that election, counting with the support of his relatives, the Ibarras, the largest mine owners of the region.[88] As for municipal elections, Chavarri's presence could not stop the Left, represented by the Republicans, from winning the largest number of seats on Bilbao's city council in 1893 and 1895. Although in 1897 Chavarri and other members of the business elite formed an electoral coalition of Liberals that regained the municipal government, their victory did not last long. In 1901 and 1903, Republicans, Socialists, and Basque Nationalists won most of the seats on Bilbao's municipal council.[89]

Nevertheless, Chavarri seems to have been behind political deals that paved the way for the election of candidates to the Cortes and provincial offices. In addition to the coalition he helped form in 1897 for Bilbao's municipal elections, he also engineered electoral pacts with the Carlists in 1896 and 1898. Since the latter had a strong following in the Marquina and Durango districts, Chavarri agreed with them that in exchange for their support in national elections, they would share the representation of those districts in the Diputación.[90]

Although Chavarri's political omnipotence may be arguable, it is indisputable that some members of the business elite gained absolute control of Vizcaya's seats in the national parliament. This control was consolidated during the 1890s. Prior to that decade, the elite businessmen either had not run for office or had little success in getting elected. For instance, in 1884, J. Martínez Rivas ran and lost against R. Mazarredo, a member of a Vizcayan landed family. In 1885, Chavarri had more luck, winning Valmaseda's seat. Yet, this was an exception rather than the rule. During the 1880s, most of the national deputies were large proprietors such as the Allendesalazars, the Landechos, or the Ibargoitias. In contrast, from 1893 to 1918, the major entrepreneurial families, who had propelled the industrialization of the region, monopolized the six seats in the Cortes.[91] As I will show in the next section, this concentration on national politics came at a time when the businessmen aggressively lobbied the Spanish government for protective tariffs and state contracts.

Unlike the provincial deputies, who usually remained in office for only one term, national representatives retained their positions for long periods. For instance, J. T. Gandarias held Guernica's seat from 1896 until 1914; Benigno Chavarri held Valmaseda's seat from 1894 until he was replaced by his son José M. Chavarri in 1910; in Durango, the Marqués de Casa Torre was elected uninterruptedly from 1891 to 1914; and the Baracaldo seat was firmly held by the Ibarra family from the time it was created in 1896 until 1914.[92]

Electoral corruption characterized many of the political contests in which businessmen participated. Such practices preceded their deep involvement in politics and were widely used by all candidates. Nor was this type of corruption an exclusive Vizcayan phenomenon; on the contrary, it was common throughout Spain during the Restoration period. But whereas in the rest of the country, the central government manipulated the elections as it pleased, in Vizcaya the local businessmen had enough resources to defy the ruling party in Madrid and to impose their own candidacies. With the establishment of universal male suffrage, elections in Vizcaya became expensive affairs, as candidates tried to outbid each other to purchase the necessary votes. For instance, in 1893, M. Allendesalazar, a landowner, withdrew from the race in the Marquina district because he could not compete against his rival, shipping entrepreneur Martínez Rodas, who spent two hundred thousand pesetas on the race.[93] Certainly, it must have been these practices and the persistence of the same businessmen as the national representatives of the province that prompted Unamuno's remark about the existence of a plutocracy in Vizcaya.

The business elite did not present a unified political front in the elections. The example cited in which J. Martínez Rivas competed against V. Chavarri was not an isolated case. In 1896, two shipping magnates, E. Aznar and F. Martínez Rodas, fought for Marquina's seat in the Cortes; and F. Martínez Rivas and B. Chavarri faced each other in Valmaseda, as their brothers had done in 1891. In 1898, there was even a case in which two members of the Ibarra clan ran against each other in Baracaldo. As it became increasingly difficult to accommodate the political ambitions of the Vizcayan businessmen with the number of seats allotted to the province in the Cortes and the Senate, many entrepreneurs decided to run outside the Basque region. In 1899, P. Alzola, for instance, was elected to represent the district of Huesca; and E. Aznar and F. Martínez Rodas won senate seats in Burgos and Santander, respectively.[94] According to J. Varela, the electoral migrations of these businessmen were feared in other provinces because, armed with large campaign chests, they could easily buy themselves a seat.[95]

With rare exceptions, the businessmen who became national deputies during the 1893–1910 period belonged to the Conservative or the Liberal Parties.[96] Yet, political affiliations and personal rivalries were put aside when it came to the promotion of Vizcayan business interests. Chavarri, a Liberal in the early 1890s, became a Conservative after 1894 since Cánovas's party promised more vigorous support of Vizcayan industry. The Marqués de Casa Torre, a Conservative, vowed at a meeting in favor of protectionist measures for the local economy that the interests of the province superseded those of the political parties.[97] Conservative E. Aznar had no qualms about forming partnerships with his Fuerist cousin R. Sota, who became a supporter of the Basque Nationalist Party toward the turn of the century.

For these entrepreneurs, parliamentary politics was business. They fervently came to believe that what was good for their industries was good for the country. Of course, their lobbying power did not rest in the small delegation that Vizcaya had in the Cortes; even when increased by the political migrations of some of the businessmen, their number was too small to affect parliamentary votes. Their power, as James Harrington might have said, was based on property, on the strength of an economy that had grown by leaps and bounds during the 1880s. In the following decades, they tried to increase that strength through political measures designed to help their industries.

Industrial Politics

During the 1880s and the 1890s, Vizcayan businessmen succeeded in persuading the national government to promote their region's economy. To sway the authorities, they claimed to have patriotic aims. Although they admitted that they would personally benefit from state measures designed to foster their businesses, the entrepreneurs felt that the alternative to having no national industry was chronic poverty and strategic dependency on foreign countries. In many instances, they based their requests on the well-known argument that infant industries merited protection until they were strong enough to compete in the open market with long-established companies. Often, too, they decried the unfair trade practices of foreign competitors, who dumped industrial products in Spain at prices that were below those charged in their home markets.

The man who articulated these ideas in their clearest form was Pablo Alzola. The son of a wealthy landowner from Guipúzcoa, Alzola probably settled in Bilbao in the early 1870s. He was a man of many talents: an engineer by profession (he drafted plans for a mining railroad for the Ibarras, and also for the Bilbao-Portugalete line); a member of the board of directors of the Altos Hornos de Bilbao steel company after 1895; a politician (head of the Liberal Committee in Bilbao, mayor of the city from 1877 to 1879, president of the Diputación during the 1887–1890 period, and a deputy in the Cortes in 1899); and a high-level national civil administrator (general director of public works in 1900 during a Conservative government). Drawing on his many experiences, he wrote extensively on economic, political, and administrative matters.

Much of what Alzola and a few other businessmen wrote had a polemical character, which sometimes contradicted the economic behavior of Vizcaya's entrepreneurs. For instance, in 1896, Alzola lamented that the government had not set certain barriers to slow down the export of minerals, which barriers would have led businessmen to develop a strong metallurgical sector.[98] Curiously, when the government tried to tax the export of minerals in 1891, steel producers such as La Vizcaya complained about the measure, since the company also sold ore in foreign markets.[99] Although one of La Vizcaya's main stockholders and directors was Víctor Chavarri, he made no attempts to supply the firm with the ore from his mines in Triano, which were cheaper to exploit than the Galdames deposits that provided the mineral for the factory's blast furnaces. Still, Chavarri, too, advocated economic protectionism as a way to promote a budding national steel industry unable to compete with German or British manufacturing giants.

A literary interpretation of Alzola's writings has led historians to posit a confrontation between free-trade mine owners and protectionist metallurgists.[100] Yet, as Chavarri's and other cases presented in chapter 4 show, such a division is artificial. Indeed, the diversified investments of the business elite avoided the potential confrontations that might have pitted ore exporters against iron manufacturers. Thus it was easy for a Chavarri to call for measures to promote the national steel industry and at the same time continue to export most of his mineral production. He could not lose; both policies benefited him. The steel companies required lobbying and heated rhetoric to convince the government of their need for protection; the mining industry prospered quietly on its own, favored by market conditions.

When did Vizcaya's businessmen turn toward protectionism? This question seems to divide historians. Traditionally, Spanish historiography has seen the Basque businessmen as part of a protectionist alliance that included the Catalan textile industrialists and the Castillan wheat growers.[101] According to this view, such an alliance was forged during the early 1840s and lasted throughout the nineteenth century. Recently, however, some historians have questioned the idea of an alliance among the three economic sectors, stressing that the protection eventually obtained by each one resulted from unconnected accommodations between each group and the government. In addition, these scholars have claimed that Vizcaya's iron manufacturers held free-trade ideas until the early 1890s.[102]

Although recent historiographical contributions seem right in denying the existence of a formal protectionist alliance, it is doubtful that Vizcaya's iron manufacturers adhered to a free-trade ideology prior to the 1890s. On the contrary, the few documents available tend to show that protectionist ideas had prevailed among this group since at least 1850. In that year, José Vilallonga, a partner of the Ibarra brothers in their Guriezo iron factory, published a pamphlet in which he countered free-trade tenets with the argument that infant industries needed protection in order to develop. In addition, he believed that the steel industry merited special encouragement from the government because it promoted growth in other economic sectors and was linked to matters of national defense.[103]

The modifications of the import tariffs in 1852 must have encouraged Vilallonga and his partners, since they started the construction of Nuestra Señora del Carmen in Baracaldo near Bilbao in 1855. In 1863, a few years after the factory was finished, the import barriers on several iron products were lowered, over the protest of Vizcaya's manufacturers. In 1866, responding to a questionnaire prepared by a parliamentary commission in charge of studying further reductions to the import tariffs, Vizcaya's iron manufacturers decried the constant manipulation of the duties, noting that the uncertainty fostered by those changes played havoc with their business plans.[104] The industrialists requested long-term tariff protection to improve the competitiveness of their factories. In addition, to overcome the slump affecting the industry during the mid-1860s, the manufacturers proposed to end the exemptions enjoyed by railroad companies, which were permitted to import their material duty free. Finally, they called for the elimination of the tax on the importation of coal and coke. Stretching the internal logic of their protectionist arguments, they claimed that this last measure would not hurt the small mining industry in Asturias because the higher cost of international freight already gave them an advantage over foreign coals.

The same questionnaire of 1866 serves as a window into the beliefs of entrepreneurs involved in other sectors of Vizcaya's economy. The answers given by mine owner Manuel Lezama Leguizamón and by the provincial Board of Trade (Junta de Comercio) help dispel certain myths about the supposed adherence to free-trade ideas. Although ore production increased during the early 1860s as a result of a stronger foreign demand, mining entrepreneurs such as Lezama still saw the local iron factories as their main customers. Thus he wrote that the mining industry did not require direct protection, but could be helped indirectly by promoting iron manufacturing "in a comprehensive, general, and stable way, since the protection given so far is neither certain . . . nor complete because, although some foreign iron products are taxed with a fair duty, there are many others that . . . are assessed only nominal impositions."[105]

The Board of Trade's opinion on the tariff matter resembled that expressed by owners of the local iron factories. The board called for a delay in the reduction of import duties, arguing that it would harm the factories to eliminate them at a time of economic distress. Once the situation improved, the tariff could be reduced gradually to grant the iron companies enough time to prepare to compete in an open market. Finally, as a way to alleviate the factories' problems, the board asked the government to reduce the cost of transportation, to eliminate the tariff on the importation of coal and coke, and to apply duties to imports for the construction of the railroad lines.[106]

What emerges from these answers to the questionnaire is a consensus often expressed by Vizcaya's entrepreneurs when it came to promoting the local economy. These businessmen appear to have perceived the local economy as a unity and not as an aggregate of sectors with irreconcilable, conflicting interests. This view is clear not only in Lezama's opinion, but also in the Board of Trade's comment that, although a reduction in the import tariffs might be beneficial for the external trade of the country, the advantages might be offset by a loss of internal economic traffic caused by the lesser needs of the industrial sector.[107]

Contrary to what is usually asserted, then, Bilbao's merchants did not always advocate free trade.[108] In 1869, for example, the national government enacted legislation designed to end all fiscal protection of Spanish industry within a relatively short time. As the subsidy on freights transported in Spanish bottoms was abolished, several Bilbao merchants, among whom were such well-known names as Abaitua, Epalza, Gurtubay, Ibarra and Co., Olaguivel, Real de Asua, and Uriguen, protested the measure. These merchants complemented their trading activities with shipping, and thus were directly affected by the abolition of the subsidy. That protest led to the creation, in 1870, of the first protectionist association in Bilbao, presided over by E. Real de Asua, a merchant involved in shipping and the wine trade.[109] Unfortunately, not much is known about this association, which must have disappeared during the civil war.

Although there seems to have been a protectionist bent among several sectors of Vizcaya's business community, their pleas were generally ignored during the 1840–1876 period. The region was still underdeveloped, and its businessmen relatively unorganized. In contrast, the more economically advanced Catalonia had seen its textile manufacturers organize protectionist groups since the 1830s. For several decades, the Catalans bore alone the brunt of advocating these ideas before the national government. Their lobbying efforts had mixed results.[110] After an increase in duties during the 1840s, the tariff was reduced in 1869; there were no new major modifications until 1890. The lowered tariffs were met with protests and demonstrations in Barcelona, but in the rest of the country negative reactions, if they occurred at all, were more subdued. During the 1880s, protectionist demands grew stronger when Castilian wheat growers, alarmed by the appearance of cheap American grain in the European markets, joined those requesting higher tariffs.

After the Second Carlist War, the mining boom in Vizcaya quelled overt manifestations in favor of protectionism during the late 1870s and the 1880s. Bilbao's business elite did not join the movement led by Castilian wheat growers and Catalan textile manufacturers in favor of higher tariffs and against the commercial treaties signed by the Spanish government with England and France.[111] Reflecting on this period some twenty years later, Alzola attributed this silence in the Basque region to the lack of major iron and steel factories.[112] In fact, it was precisely during those years that Vizcaya's steel industry was modernized and expanded. Yet, Alzola had to ignore this fact because at the time of the expansion the import tariffs were at a historical low.[113] During the 1890s, when protectionist agitation revived, the development of those factories was presented as self-sacrifice, patriotism, and even foolhardiness on the part of Vizcaya's entrepreneurs.[114] Although the businessmen might have been influenced by patriotism and abnegation, there were also solid economic incentives. Throughout the 1880s, the capacity of the factories to compete in the international market proved that their creation stemmed from a rational expectation of profits for shareholders.

Thus the silence of Vizcaya's businessmen over the signing of commercial treaties during the early 1880s must be understood in a different context from the one provided by Alzola. There was an explicit interest in making Vizcaya's products more competitive abroad. In 1887, supported by the local Chamber of Commerce and the provincial representatives in the Cortes, Vizcaya's three major steel companies obtained a reduction on the export duty levied on iron bars.[115] That same year, when the government solicited the opinion of the Bilbao Chamber of Commerce regarding the signing of bilateral commercial treaties with other European states, Basque businessmen did not express any opposition. On the contrary, they requested that the national government negotiate provisions that would exempt their iron bars from hefty import duties in the signatory countries.[116] In January 1888, the three largest factories in the region—Altos Hornos, La Vizcaya, and San Francisco—renewed this petition as a treaty with Italy was being negotiated. Stressing the importance of this market, the manufacturers wrote, "in 1887, out of a total production of 180,000 tons of bars, the factories we represent have sent to that kingdom approximately 80,000 tons. . . . The day this market closes (and it will close if a significant duty is established), where will our national products be sold?"[117]

To the Basque companies' chagrin, the Italians established a 20 percent duty on Vizcaya's iron bars.[118] Exports to that country dropped immediately from their peak of eighty thousand tons in 1887 to slightly fewer than thirty thousand in 1890. The descent continued during the decade as Italian imports averaged around thirteen thousand tons per annum.[119]

Historians have attributed the protectionist attitude of Vizcaya's steel producers to the loss of their export markets during the 1890s. This explanation is only partially correct, since it also assumes that during the 1880s the desire of the manufacturers to serve foreign customers led them to support free-trade policies in Spain. Yet, their interest in foreign markets does not exclude the possibility that they could also have sought protection at home. The records of Altos Hornos de Bilbao show that in 1886 the company lobbied, albeit unsuccessfully, the ministers of development (Fomento) and finance (Hacienda) to request an end to the railroad companies' permission to import materiel duty free.[120] In 1888, the company obtained better results, since it convinced the national government to alter a public contract that had called for the purchase of foreign-made pipes for a harbor.[121]

Furthermore, during the 1880s, Vizcaya's businessmen sought a more direct kind of government protection for their industry than import tariffs. In 1885, only a few years after the company was established, Altos Hornos de

Bilbao began to look at the state as its best potential customer.[122] Given the strategic needs of modern armies and navies, it is not surprising that some ministries within the state also wanted to promote the steel industry. During the early 1880s, the Spanish navy commissioned one of its officers to study whether the new industries could supply armaments and ships. The officer wrote an optimistic report in which he recommended support for the steel companies, although they still could not supply all the navy's needs.[123]

Altos Hornos was not the only manufacturer in Vizcaya seeking state contracts. In May 1887, Federico Echevarría, a founder of La Vizcaya and La Iberia, wrote a report for the Bilbao Chamber of Commerce in which he advocated the creation of a shipyard on the Nervión banks to build battleships for the navy.[124] Soon after the chamber heard the report, it sent four delegates on a mission to Madrid, seeking to convince the navy to contract the construction of vessels with private national shipyards.

In what became a typical Vizcayan lobbying effort, the provincial representatives in the national parliament and delegates from Altos Hornos and La Vizcaya joined the envoys of the Chamber of Commerce in Madrid in petitioning the national authorities in late May 1887.[125] They had plans to join ranks with a Catalan delegation of businessmen, but the latter canceled their appearance at the last moment. Undaunted, the Vizcayans mounted a lone offensive by publicizing the progress of their industry in the local press and meeting with members of the government, leading politicians, and the queen regent.

The Vizcayan proposal presented four essential points: (1) a guarantee that their industries could provide the navy with all the necessary materials for shipbuilding at competitive international prices; (2) a request to build ships of up to fifteen hundred tons in several companies around the country, such as the Maquinista Terrestre y Marítima of Barcelona; (3) a promise by the Basque companies to establish a major shipyard in Vizcaya capable of constructing large battleships and delivering them to the navy in four years; and (4) if the government did not believe that points 1–3 could be achieved with only national resources, a plan to call for an international competition for the construction of battleships with the condition that the vessels be built in the country. In this last case, the Vizcayans would seek partnerships with foreign manufacturers.

The proposal was cleverly phrased in broad terms and included regions beyond Vizcaya to avoid the impression that the project would benefit only Bilbao or just one industry. The proposal for other regions, however, involved the construction of small ships, while Vizcaya's companies were seeking contracts for large war vessels. The Chamber of Commerce's report narrating the negotiations with the minister of the navy and other important officials portray Vizcaya's delegates as heroic figures trying to enlighten the utterly ignorant politicians about the advances made by industry in Vizcaya. In the end, a happily "surprised" minister of the navy, who could not have been so uninformed about the Basque steel factories as the report claimed, and another influential cabinet member, Segismundo Moret, sponsored the project before the government.[126]

Soon after the delegation left Madrid, their plan came to fruition. Toward the end of the summer of 1887, the queen regent and Prime Minister Sagasta paid an official visit to Bilbao. They toured the mining district and the three largest steel factories in the region to get a direct impression of Vizcaya's industrial might. In December 1887, the government called for a competition to build three cruisers and three torpedo boats. As the Vizcayans had requested, the ships had to be constructed in the country and with national materials. Nine months later, the navy selected the proposal presented by J. Martínez Rivas and his Scottish partner, Charles Palmer. Astilleros del Nervión, as their company came to be known, made Martínez Rivas a hero in Bilbao and won him a seat in the Cortes in the 1891 election.

Unfortunately for the Basque businessmen, the poor condition of state finances precluded the Spanish government's being the major client that they had hoped it would be. In addition, the financial difficulties of Astilleros del Nervión, which led the government to seize the company in 1892, dampened the initial enthusiasm of the authorities to build ships for the navy in the country. These disappointments, together with the tougher international climate, which had reduced the export base of Vizcaya's factories, led to campaigns to seal off the Spanish home market. It was a desperate reaction because the national demand for steel products was extremely low and could be satisfied with only a fraction of the industrial capacity built in Vizcaya during the 1880s. A closed market, however, looked promising because, with price-fixing arrangements, Basque producers could extract extremely high profits.[127]

The fight for protectionism during the 1890s received an unexpected benefit when the national leader of the Conservative Party, Cánovas del Castillo, decided to support the industrialists. In addition, a faction of Cánovas's Liberal rivals had also raised the protectionist banner in an attempt to increase its power within the party.[128] Protectionists were also helped by a climate that was causing tariff wars in Europe and promoting measures to boost sagging national economies. Under these conditions, the new import tariffs approved by the Cortes in 1891 gave ample protection to the two groups that had been clamoring against foreign competition during the 1880s: the Catalan textile manufacturers and the Castilian agriculturalists.

Vizcaya's steel manufacturers, who had joined forces with metallurgical companies from Asturias and Catalonia to form a lobbying group to participate in the hearings prior to the drafting of the new tariff law, also obtained an increase in the import duties to protect their production. According to I. Arana, although the Basques made gains, they were disappointed with the new tariffs for three reasons. First, their request to reduce the import tax on coal and coke was disregarded, and the duties on these raw materials were actually tripled to promote the Asturian mining industry. Second, the tariff protection given to steel rails was slightly decreased. Third, Vizcaya's manufacturers did not obtain the repeal of the privilege of the railway companies to import materiel exempted from customs taxes.[129]

The reduction in the coal tax always remained an elusive goal for the Vizcayans. It was a delicate matter because to oppose protection to the Spanish coal industry undermined their own argument in favor of defending national producers. During the 1890s, the conflict was somewhat defused as several of Vizcaya's principal industrialists bought coal mines in Asturias and León, but it still remained a source of friction because the low quality of Spanish coal forced the Basque factories to continue to import this mineral from Great Britain. When the duties on coal were raised once more in 1895, a compromise was reached, exempting the steel factories from this last increase. In 1897, however, a new tax on freights raised the duties on imported coal even for steel production.[130]

The point concerning steel rails did not seem to worry Vizcaya's industrialists, who were willing to concede some reduction in the duty on this item.[131] Thus they must have been pleased with the actual outcome. The tariff laws included two schedules, one with a maximum duty and another with reductions for trading partners with most-favored-nation status. Arana is only partly right when he claims that the duty on rails was lowered, since the maximum schedule was reduced, but the other was increased. Yet, as a percentage of the value of the rails, both schedules boosted protection.[132]

The problem for Vizcaya's steel producers was that most railroad companies were not taxed according to the two schedules of the tariff law. Thanks to special legislation, the railroads were allowed to import their materiel practically duty free. Thus Vizcaya's companies faced stiff competition from foreign manufacturers, which kept prices low. Not surprisingly, the factories preferred to target their production into areas where they were shielded by the customs tariffs. In 1886, Altos Hornos, for instance, considered lowering its production of rails if it could increase its sales of steel beams for urban construction.[133] Not only the competition for rails was tough, but also the demand was rather narrow, as their consumption decreased during the 1890s.[134]

Nevertheless, as market conditions at home and abroad deteriorated, Vizcaya's industrialists decided to fight for every possible customer within the country. In the summer of 1891, a few months prior to the approval of the new tariff law, they took advantage of the presence of Prime Minister Cánovas in the Basque city of San Sebastián to send a mission led by members of Altos Hornos, La Vizcaya, and La Iberia to express their opposition to the continued exemptions enjoyed by the railroad companies.[135] Cánovas was sympathetic to the Vizcayans' complaint, and he included in the preamble to the 1891 tariff law a promise to review the situation of the steel factories.[136] Although this probably did not fulfill all the expectations of the Basque producers, it was at least an official recognition of their grievance, which had been consistently ignored since at least 1863. As a result, the Basque producers had more reasons to be satisfied than disappointed with the new tariff law.

Despite the higher protection established in the 1891 law, Vizcaya's industrialists had to make sure that the government did not violate the spirit of the tariff by granting special concessions to other countries willing to sign bilateral commercial treaties. Consequently, the Basques began to perceive the treaties as a threat. In contrast to their support for them in the 1880s, they vehemently opposed these pacts during the 1890s. The Liberals, who had returned to power in 1892, wanted to reach such an agreement with Germany. In December 1893, determined to stop the government, businessmen from all regions congregated in Bilbao to protest the signing of commercial treaties. Initially the meeting had been scheduled to take place in Barcelona, but a wave of political disturbances, which led the government to impose a curfew in the city, forced the change of venue.[137] Making Bilbao the meeting place was recognition of its recently developed industrial might.

Although the issues involved in the protest against the commercial treaties affected mostly the metallurgical industry, the Bilbao meeting mobilized practically all sectors of Vizcaya's business and political elites. Among those present were the president of the Diputación, J. M. Arteche, all the local representatives to the national parliament (most of whom had ties to the metallurgical industry), the mayor of Bilbao, the president of the local Chamber of Commerce, the directors of the two local banks, and the leaders of all the major companies in the region.[138]

All the speakers condemned the treaty with Germany as bad for national manufacturers.[139] One of the orators, P. Alzola, also condemned the economic policy that had permitted the construction of railroads since the 1850s without using Vizcayan products. Citing the fact that there were three steel companies where one mill would have been sufficient to cover the national demand, Alzola denied that protectionism would hamper competition among Spanish producers. Alzola cleverly kept silent about the price-fixing arrangements among the local companies, which were doing just what he claimed would not happen.

As a result of the Bilbao meeting, regional business associations were created to lobby against the commercial treaty and to defend national industry. These associations formed a national organization, which came to be known as the Liga Nacional de Productores (National Producers' League) to coordinate their activities. As Arana demonstrates in a thorough study, the Vizcayan group, the Liga Vizcaína de Productores (LVP), became the main pillar of the national organization and surpassed the Catalan business associations, whose internal divisions diminished their effectiveness. The Vizcayans led most of the lobbying efforts, encouraged the formation of protectionist groups in other regions, and tried to smooth the differences that emerged among the different industries represented in the Liga Nacional.

Given the preponderance of the steel industry in Vizcaya, it is not surprising that the main companies in this sector dominated the LVP. Altos Hornos and La Vizcaya, especially, contributed about two-thirds of the total budget of the league. Other industries, such as paper mills, chemical producers, textile manufacturers, and food processors, also became associated, but played a much smaller role in the direction of the league.

The absence of shipping companies among league members has led some historians to believe that there was an ideological confrontation between shippers and metallurgists.[140] The former supposedly wished to maintain free trade to maximize the volume of exports and imports. Therefore, they opposed the tariff protection sought by the metallurgists, which might have dampened international trade. To support this argument, Arana cites a letter written by members of the LVP in which they complain that R. Sota and E. Aznar, partners in a large shipping company, have a plan to demonstrate in favor of the commercial treaties that the steel producers opposed at the end of 1893.[141] This demonstration, however, never took place. Moreover, other facts tend to undermine the idea that there was ideological antagonism between the two groups.

It appears that Sota and Aznar did not object to the tariffs granted to the steel companies, but they did resent the fact that the shipping industry was not protected enough.[142] In 1894, their anger against the steel industry might have come from a perception that the latter promoted a rise in the registration duties for foreign-made ships sailing under the Spanish flag. In 1894, when the Diputación voted to support the LVP in its fight for protection for local manufacturers, Aznar, a provincial deputy at the time, convinced his colleagues in the Diputación that the provincial government should send a petition to the national government supporting both the steel companies and the shipping industry.[143]

Although shipping companies as such never joined the LVP, several important shippers entered the league on an individual basis.[144] In 1896, even Aznar became a member. Furthermore, many steel manufacturers who belonged to the LVP invested in shipping companies, too, defusing the probability of serious confrontations among the different economic groups. For example, the Zubiría family, a branch of the Ibarra clan with interests in mining, metallurgy, and banking, became shareholders in several of Sota and Aznar's shipping companies.[145] When the shippers finally formed their own lobbying group at the turn of the century, they worked closely with the LVP to secure the passage of a law that protected the maritime companies starting in 1909.[146]

Thanks to their interlocking investment pattern, Vizcaya's businessmen were able to present a united front in their lobbying before the national government. None of the protectionist campaigns was ever undermined by regional demonstrations in favor of free trade.[147] Nor did the main business groups—the LVP, the Círculo Minero, the Chamber of Commerce, and the Asociación de Navieros—ever oppose each other's petitions regarding taxes or tariff protection during the 1890s and the early 1900s.[148] Moreover, the same businessmen lobbied for several of these groups. This was particularly true of many of the entrepreneurs who were elected to the Cortes. Indeed, nearly all the national deputies elected between 1893 and 1903 had direct or indirect connections simultaneously with the Círculo Minero, the LVP, and the Chamber of Commerce.[149] Unabashedly, and without concern for potential conflicts of interest, deputies such as Benigno Chavarri or Juan T. Gandarias, major mine owners, did not hesitate to negotiate, on behalf of the Círculo Minero, a reduction in a special tax on the export of minerals in 1896.[150] That year, those same deputies, who were on the board of directors of La Vizcaya, worked closely with the LVP to obtain the repeal of the railroads' import exemptions.[151]

Although several economic policy issues drew the attention of Vizcaya's entrepreneurs, their greatest efforts were concentrated on the rejection of the bilateral commercial treaties in 1894, the repeal of the railroads' import exemptions in 1896, and the revision of the tariff law in 1906. Basque businessmen significantly influenced the outcome of all those issues. In 1894, Chavarri presided over the parliamentary commission that blocked the approval of the commercial treaty with Germany. In a dramatic turn of events, Chavarri, a Liberal at the time, cast the deciding vote by joining the Conservative minority on the commission, thus defeating his own party. In 1896, it was the businessmen's close ties with Prime Minister Cánovas that saved the day, when it seemed that parliamentary maneuvers threatened to ruin once again their hope of seeing the railroad exemptions abolished.[152] And in 1906, Alzola was appointed as the head of a governmental commission in charge of drafting the guidelines for the modification of the tariff laws.

It would be a mistake, however, to believe that the Basque entrepreneurs overran the Spanish state as they sought to promote their industries. Arana has painstakingly demonstrated that the government did not follow every whim of Vizcaya's industrialists. The state not only had to pay attention to their demands but also had to consider other powerful groups whose interests sometimes clashed with those of the Basques. For instance, the 1896 repeal of the railroads' import privileges did not represent a defeat for these firms; it was part of a compromise devised to help the railroads, which were undergoing difficult times. As compensation for giving up their import privileges, their concessions were prolonged and certain subsidies granted. In addition, the compromise established that their imports would be taxed according to the schedule fixed in the 1882 tariff law, which was lower than the one in effect in 1896. Similarly, the 1906 tariff law also required a compromise. Although the Basques played a significant role in drafting the guidelines for the new tariffs, thus assuring that the protection to their industries would not be reduced, they were not able to raise the duties on most of their products because of the opposition of agricultural, mercantile, and industrial groups from other regions.[153]

Moreover, the state itself had its own needs, which sometimes met the opposition of the businessmen. Although government could be sympathetic to requests to have the arsenals buy all war materiel from Spanish factories, the condition of the Spanish treasury delayed this petition several years. In addition, tax matters almost always placed the state in confrontation with the industrialists. For example, in 1897, the LVP protested without success the imposition of export duties, and in 1896 the Círculo Minero could not obtain a reduction on a special ore tax designed to cover part of the cost of the war against Cuba.[154] This last tax was particularly vexing for the Basques because it set a duty on their iron ore that doubled the one fixed for the minerals coming from the southern region of the country.[155] In 1900, the LVP tried to take advantage of the tax agreements between the state and the Vizcayan government to argue that the new taxes on the income of joint stock companies and on the dividends paid out to shareholders created by the fiscal reforms of Finance Minister R. Villaverde should not be collected in the province. The league did not succeed; after a negotiation between the Diputación and the central authorities, the tax was recognized by the Basque authorities as new and therefore not part of the special tax agreement.[156]

Despite some setbacks and compromises, the overall balance of the Vizcayans' lobbying was positive if measured where it counted most: the companies' balance sheets.[157] By this standard, the compromises that the Basque businessmen were forced to accept in the 1891 and 1906 tariff laws and in the 1896 repeal of the railroad import exemptions did not represent major sacrifices. Until the mid-1890s, La Vizcaya and Altos Hornos had lackluster net earnings, averaging approximately 3.2 and 5.3 percent per year, respectively.[158] During the late 1890s, the average annual net earnings shot up to 10 percent for La Vizcaya, and an even more impressive 20 percent for Altos Hornos. During these years, the hefty devaluation of the peseta also helped the Basque steel companies by making foreign products more expensive in the local currency. As the market became increasingly closed, local factories were able to fix prices at high levels through a cartel that curtailed competition within the country. Those high prices and a slight increase in demand after years of stagnation explain the hefty earnings toward the turn of the century. Under those conditions, it was difficult for a company like Altos Hornos, which almost tripled its earnings in 1900 after a very profitable 1899, to justify higher tariff protection.

As the twentieth century progressed, successive Spanish governments became more and more inclined to promote national industry actively and to protect it against foreign competition. Although many of the measures taken to support the Spanish economy were not a direct result of pressure exerted by businessmen, lobbying groups such as the LVP played an important role in changing the general climate that made those measures possible. For the Basque entrepreneurs who had complained in the early 1890s that the state had harmed them by making them bear the cost of the subsidies granted to the railroad companies, the situation was completely reversed. During the 1900–1930 period, the rest of the Spanish economy was saddled with a highly protected steel industry, which supplied 83 percent of national consumption and charged prices that in some cases were twice as high as those paid on the international market.

Conclusion

While natural resources, technological changes, and sociocultural factors played a key role in Vizcaya's economic development, there is no doubt that politics also favored the activities of local businessmen. Just as they ably exploited the opportunities provided by their mines and the invention of the Bessemer converter, they proved adept at lobbying and working within political institutions to further their economic interests. They derived benefits, too, from the fiscal advantages enjoyed by the Basque region during and after the *foral* period.

This does not mean that regional and national institutions became pawns in the hands of Vizcaya's entrepreneurs. Indeed, at the local level, the advantages they obtained came as a result of practically no lobbying activity and only a little participation in provincial government during the 1840–1876 period. After the Second Carlist War, businessmen occupied prominent positions in the new Diputación, but the modified provincial government does not appear to have been an instrument designed to make wealthy entrepreneurs richer at the expense of the rest of the population. It merely continued the same economic and fiscal policies pursued by its *foral* predecessor. These policies do not seem to have gone beyond the broad purpose of encouraging economic growth with the idea of promoting the general prosperity of the region; they did not seek blatantly to favor the business elite.

It was actually at the national level that the businessmen actively sought help from state institutions. Their lobbying became most visible during the 1890s. The idea of asking for government help to develop business was not new. Indeed, it had been consistently expressed since at least 1850, but the message did not carry enough force until the iron industry acquired a certain critical mass during the 1880s. In the 1890s, the request for government help became louder not only because of the bigger size of the factories, but also because of the difficult market conditions that the companies had to face. The businessmen delivered their requests to the authorities in their double capacity as national deputies and members of interest groups such as the LVP or the Bilbao Chamber of Commerce. Their successful lobbying was related to the fact that their diversified investments allowed them to defuse potential conflicts among the different sectors of Vizcaya's economy. Although other issues such as religion, the *fueros,* or forms of government divided the business elite, they were able to put aside those differences to promote their common economic interests. Members of the LVP, for instance, covered a whole gamut of political choices, from Republicans such as F. Goitia, to Liberals such as F. Echevarría, and from Basque Nationalists such as R. Picavea to Conservatives such as Casa Torre.

Of course, not all of their success can be attributed to the businessmen. The political dynamic of the Restoration also helped them by making protectionism an issue between Cánovas's Conservatives and Sagasta's Liberals. Cánovas's conversion to the protectionist side preceded the Vizcayans' lobbying; but once it took place, the Basque businessmen took care to nurture their relationship with the Conservative leader through annual visits to his summer retreats and a stream of messages during the parliamentary season.

The strategic importance of their heavy industry also made it appealing for the state to promote its growth. Such considerations even brought them the support of the Liberals. After all, it was Sagasta's government that agreed to have battleships built in Spain in 1887, permitting the creation of Astilleros del Nervión. In addition, it became easier to convince the Spanish government when all the major European countries, with the exception of England, were raising their import tariffs and trying to promote growth through state intervention.

In the end, lobbying was good business for Vizcaya's entrepreneurs. Protectionist measures increased their profits significantly. Unfortunately, their success came at the expense of the rest of the Spanish economy.

Chapter Seven

Conclusion

The many efforts to design general theories of industrialization attest to the difficulty in explaining not only why but also how economic development takes place in a given society. During the late 1950s and the early 1960s, there was a widespread belief that all countries underwent the same stages in their path toward industrialization. Works such as W. W. Rostow's *The Stages of Economic Growth* (1960) advanced the notion that England's experience as the first industrial nation had set the standard for the rest of the world. Yet, when this idea was tested against the historical record of different countries, economic historians realized that there was more than one way toward industrialization. In a seminal work, A. Gerschenkron designed a typology that tried to explain how the industrialization process differed among countries according to their relative degree of backwardness.[1] Although some aspects of Gerschenkron's typology have come under criticism, his basic framework is still widely used to analyze the history of European economic development. As a way to conclude this study, a brief review of the arguments developed in the previous chapters is presented to analyze how Bilbao's case adapts to or differs from Gerschenkron's and other typologies of economic development.

According to Gerschenkron, the following traits are characteristic of countries that underwent late industrial development: (1) an accelerated growth of factory output; (2) an emphasis on the production of capital goods rather than on consumer goods; (3) stress on large-scale factories and projects; (4) a need for foreign technology; (5) a central role of institutions such as banks or the state in providing capital to the new industries; (6) a virtually negligible contribution of agriculture as a source of demand for industrial production; and (7) a virulent ideology needed to break the inertia that prevented the country from industrializing earlier. Countries displayed these traits not uniformly, but according to existing economic and social conditions on the eve of the industrialization process. For instance, late developers such as Germany or Russia required more capital expenditures to enter the industrial race than had the pioneering countries. Yet, while in Germany banks provided the necessary resources to finance the manufacturing sector, in Russia a greater backwardness required an even more powerful institutional force—the state—to spur economic development.

Do Gerschenkron's seven points characterize Bilbao's industrialization? Certainly, the chronology established for Bilbao's modern economic development qualifies the region as a late industrializer within the European context. At first glance, too, several of Gerschenkron's traits can be discerned in the Basque case. Agriculture's contribution (either at the provincial or the national level) to industrial growth was practically negligible. If one forgets for a moment the modernization attempts of the local iron industry during the 1850s, which increased the productive capacity of the region but were short of a great spurt, the rapid growth experienced after 1875 could surely qualify as a takeoff. The new steel factories were founded during the 1880s using foreign technology and were built on a large scale, with a capacity that surpassed the needs of the Spanish market. In addition, institutions such as the provincial and national governments spurred the industrialization process by building a railroad for the mines or placing orders for the construction of war vessels for the Spanish navy. Finally, it could be argued that in their advocacy of state protection, the Basque industrialists echoed many of the ideas of the economist F. List, who, according to Gerschenkron, provided the ideological underpinnings for Germany's developmental policies.[2]

Yet, despite those similarities, Gerschenkron's model does not fully describe Bilbao's industrialization. The economy's quantum jump after 1875 did not result from acceleration in the production of the manufacturing sector but from increased output of the mining industry. It was primarily export-oriented growth based on the quality and abundance of Basque iron ore, which eventually led to the modernization of the local steel factories and the establishment of other industries. Although the profits generated by the mining industry could not account for total investment in the Bilbao region during the 1880–1900 period, the general prosperity that the extraction of ore generated and the confidence it bestowed on local businessmen must be held responsible for the spillover effect on other sectors of Vizcaya's economy. Mining increased the capital of a local group of entrepreneurs, which reduced the need for the institutional support that seems to characterize other late industrializers. Although the national and provincial governments did play a role in the development of the region, their involvement was rather modest, if one considers that most investment capital came from private sources. Similarly, and despite their later importance in the economy, banks were not so much the midwife of industry as its offspring. There were no true local investment or industrial banks until the late 1890s–early 1900s period, and, to a large extent, Bilbao's industrialists played a major role in their creation. Admittedly, some Madrid financial institutions invested large sums in Altos Hornos de Bilbao, but even this did not seem to overshadow the Ibarras' control of the company.

The export-oriented character of Bilbao's economic growth seems to resemble the development pattern of some Third World countries. Yet, the similarity is only superficial. While Third World economies have undergone cycles of rapid growth and pronounced retrenchments, which reflect transient export booms of raw materials, the growth of the Basque region was not a fleeting phenomenon that went bust after the massive sales of ore to foreign countries subsided. Unlike the colonial or neocolonial situation of some poor nations, Bilbao's businessmen always kept control of their mines and thus reaped handsome profits, even when foreigners participated in the enterprise. In fact, the relationship with foreign companies, which is usually blamed for the persistent poverty of the Third World, proved beneficial for the local entrepreneurs not only in mining but also in manufacturing, since it eased the transfer of technology, which permitted the development of modern steel factories in the Bilbao region.

If Bilbao did not follow the path of some Third World countries or even some late European developers as described by Gerschenkron, it was not only because the local entrepreneurs profited handsomely from the mines but also because Bilbao's economic backwardness was not so pronounced as its belated development seems to imply. Mendels has argued convincingly that, throughout Europe, industrialization tended to occur in regions where nonfactory manufacturing activities were well developed.[3] This manufacturing tradition, or "proto-industry," as Mendels calls it, eased the path toward modern development in both early and late industrializers in two basic ways. First, it permitted the accumulation of capital in the hands of groups of merchants who dominated manufacturing production through the putting-out system. Second, the knowledge that those traders acquired of markets and techniques put them in an advantageous position to judge the need for changes in production methods. Thus, for instance, countries like France delayed the adoption of factories in the early nineteenth century because its large traditional manufacturing sector remained competitive by specializing in high-quality goods. In this regard, the prolonged survival of a protoindustrial sector was not a sign of backwardness but a rational economic response.

Of course, the presence of protoindustries alone could not guarantee that a region would become a modern industrial center. Other factors such as the availability of natural resources, technological changes, new fuels, and the sociopolitical framework were also decisive elements affecting the outcome.

Mendels's concept of protoindustrialization permits us to analyze the development of the Bilbao area as a continuous process, not as a revolutionary departure from its earlier economic experiences. The region had a long history of manufacturing and commercial activity prior to its industrialization during the second half of the nineteenth century. Bridging modern and traditional methods of production was the figure of the entrepreneur. In the 1700s, by controlling capital and commercial circuits, merchants dominated the ore trade and the production of the local ironworks. As the latter grew increasingly obsolete toward the late eighteenth century, local businessmen were not able to respond to the intense competition from abroad and even from other Spanish provinces such as Málaga and Asturias because of the wars that plagued the region until 1840. Once peace was reestablished, merchants like the Ibarras proceeded to modernize the local iron industry along the lines developed in other European countries. At the same time, new property regulations and the possibility of exporting iron ore to France encouraged the Ibarras and other merchants to become directly involved in the extraction of the mineral by claiming the concessions to the richest mines in the Triano-Somorrostro district.

Although these first attempts to modernize the local industries illustrate the willingness of the Bilbao business elite to adapt to new competitive pressures, they still faced a barrier to further progress. With the exception of its rich iron mines, Bilbao could not claim any other bountiful natural resources. Its lack of coal, the main fuel of the Industrial Revolution, especially hindered the growth of the region's iron factories. The new plants replaced the traditional ironworks, but their production trailed behind that of the Asturian manufacturers, who had established their blast furnaces near that region's coal mines. Fortunately, the invention of the Bessemer converter gave a second wind to Vizcaya's industrialization process. It not only spurred the intensive exploitation of Bilbao's mines, but also permitted local factories to surpass their Asturian competitors, since Bessemer's method reduced the consumption of coal needed to produce steel. During the 1880s and the 1890s, Vizcaya's metallurgical industry became so efficient that it was able to export steel products to other European countries. This feat took more than one hundred years after traditional production methods were improved by English manufacturers during the mid-eighteenth century. As in the 1850s, local traders–turned–mine owners such as the Chavarris, the Durañonas, the Gandarias, and the Salazars together with families like the Ibarras seized on the opportunities created by the Bessemer converter. Although these entrepreneurs did not act alone, they certainly played a leading role in the creation of the new metallurgical plants during the 1880s and the 1890s.

While the modern economic growth of the Bilbao region rested on technological developments, its iron mines, and a long tradition of manufacturing and commercial activities, such growth also found fertile ground in Basque culture and society. By European standards, an uncommonly early egalitarian ethos characterized Basque society, beginning at least in the sixteenth century, when universal nobility received legal sanction in the local *fueros.* The lack of rigid hierarchies in the region contributed to the relatively easy acceptance of businessmen as part of the local elite. The idea that industry and trade did not degrade a person's social status was widely held in the Basque region and contrasted greatly with Castillian prejudices against the incompatibility of nobility and commercial and manufacturing activities. Economic and social conditions in the Basque region assured that what could be called a gentry of urban and rural proprietors maintained close links with the local business elite. Sometimes the two groups were hard to distinguish, as many retired merchants became part of the gentry, and often the latter helped the business world by financing commercial and industrial enterprises. Moreover, given the practice of passing on family land to only one heir, many sons of the gentry had to earn a living in the business world.

Among the less fortunate families, emigration was a constant feature in Basque history. In order to ease the hardships of the emigrants, the Basques developed close-knit networks that promoted their careers. At the same time, many returned emigrants and even successful Basque businessmen who had settled in other countries contributed financial resources to the industrial development of Vizcaya. The constant back-and-forth movement of people and capital along these networks instilled a certain cosmopolitanism, which helped keep the Bilbao entrepreneurial elite informed of new developments in the business world.

Vizcaya's social networks were mostly built around family connections. Kinship ties continued to play an important role in the economic growth of the region even after corporate structures began replacing family-owned firms toward the end of the nineteenth century. Many families, like the Ibarras, who maintained a position of economic leadership through several generations, displayed a true Schumpeterian attitude toward business. According to Schumpeter, the economic rise and fall of bourgeois families depends on four factors: (1) an ability to branch out from the business that permitted the initial economic success; (2) a disposition to save; (3) efficiency in the management of enterprises; and (4) boldness and vision to attempt new business ventures that open up new markets.[4] An analysis of the Ibarras' economic ascent reveals that the family understood Schumpeter's insight that "mere husbandry of already existing resources, no matter how painstaking, is always characteristic of a declining position."[5] Indeed, as they diversified the vast capital generated by their mines, the Ibarras showed boldness and vision in pioneering new production methods and businesses. In addition, succeeding generations seem to have displayed an attitude that put a premium on investment and capital growth over consumption. Finally, in order to maintain efficient management of their enterprises, the family encouraged several of its children to study engineering and tried to instill the values of hard work, merit, and personal accomplishment into the younger generations.

In addition to kinship networks, a strong work ethic also appears to have contributed to the economic growth of Vizcaya. Although it has been impossible to determine whether this ethic had a purely lay or religious origin, it clearly received the blessings of the Jesuits, who exerted a profound influence in the region. Despite the acceptance of Castilian titles of nobility and social commingling with the Spanish aristocracy toward the 1890s and early 1900s, it does not appear that Bilbao's businessmen abandoned the values that made them succeed. In this regard, the thesis that claims that the nineteenth-century upper bourgeoisie passively accepted the social hegemony of the aristocracy seems exaggerated.

Finally, political developments also promoted Vizcaya's economy. Although the Basque region suffered economically as a result of the two Carlist Wars (which became interlinked with the *foral* issue), the situation after the modification of the *fueros* in 1841 proved beneficial for the region. Local merchants adapted rapidly to the loss of the right to import goods duty free. Under the protective tariffs of the Spanish customs system, Vizcaya's industries began to be rebuilt. The surviving *foral* institutions and rights granted the region an administrative autonomy without equal among the rest of the Spanish provinces. This autonomy permitted the preservation of a competent local government, which kept taxes low, and complemented private economic initiatives by improving provincial infrastructure through the construction of roads, a railroad for the mining sector, and other public works.

Thus, contrary to the view of some historians, Vizcaya's government aided the economic development of the region. The abolition of the *foral* institutions after the Second Carlist War was deeply lamented by all Vizcayans. Still, through special tax accords (the *conciertos económicos*), the province managed to salvage its administrative autonomy, albeit under a different institutional framework. The claim that Vizcaya's government became the fiefdom of the Bilbao business elite during the 1880–1900 period seems an exaggeration. My analysis of the members of the provincial government during those years has shown that the businessmen did not have direct control of local institutions, and the latter's economic policies simply continued the beneficial tradition established during the *foral* regime, without any new signs of overt favoritism toward local entrepreneurs. At the national level, Bilbao's business elite played a more prominent role, monopolizing the provincial seats in the Spanish parliament for more than twenty years after the late 1880s. As national deputies and as members of lobbying groups such as the LVP, Vizcaya's businessmen became interested in obtaining contracts and tariff protection from the government in Madrid. The Vizcayans' intense lobbying and strategical considerations convinced the national authorities to heed the pleas of the Basque businessmen.

In the long run, these policies, which proved advantageous for Vizcaya's entrepreneurs, saddled the Spanish economy with a heavy burden. While tariffs shielded Vizcaya's industries from foreign competition, cartels and mergers among Spanish producers forced high prices on the national market. Thus, during the twentieth century, the industries of the Bilbao area would continue to grow, but under conditions that differed from their rapid development during the 1880s and the 1890s.

Appendix I

The Mining Industry

TABLE I-1
Mine Owners, Somorrostro-Triano, 1861

Owner	Mine	Location
Ibarra Mier and Co.	Barga	Barga
Ibarra Mier and Co.	Orconera	Orconera
Ibarra Mier and Co.	Primavera	Jarrazuela
Ibarra Mier and Co.	—	Bomba Vieja
Ibarra Hermanos	—	Bomba
Ibarra Hermanos	—	Cadegal
Ibarra Hermanos	Sotera	Zarzal
José Gorostiza	San Martín	La Blanca
José Gorostiza	San Benito	Los Castaños
José Gorostiza	Ceítegui	Ceítigui
José Gorostiza	Santa Bárbara	Cadegal
José Gorostiza	San Bernabé	Bodovalle
José Gorostiza	Magdalena	Espinal
José Gorostiza	Nuestra Señora de Begoña	Bodovalle
Sebastián San Martín	Concha	Cayuela
Ramón Castaño	César	Javilla
Matías Salcedo	Cristina	Los Castaños
Francisco Recalde	Buena Fortuna	Cabrito
Francisco Recalde	Trinidad	Costa Buena
Juan Murrieta	Perseguida	Blanca
Juan Murrieta	Despreciada	Benerillas
Leonardo Zuazo	Alhóndiga	Blanca
Rafael Betanro	Otoño	Jarrazuela
Juan Durañona	San Antonio	Fuente Fría
Juan Durañona	Manuela	Urdinales
Juan Durañona	Infeliz	Yerosa
Juan Durañona	Seiguira	Papelera
Pedro de la Bodega	Adela	Bomba

TABLE I-1
Continued

Owner	Mine	Location
Pedro de la Bodega	Socorro	Campillo
Pedro de la Bodega	Indiana	Cobachos
Pedro de la Bodega	Rosario	Blanca
Carlos Sierra	Marquesa	Fuente Fría
José Arana	Julia	Bomba
José Chavarri	San Miguel	Bodovalle
José Chavarri	Aurora	Cobalón
José Antonio Ustara	Rita	Cobachos
José Antonio Ustara	—	Chacarro
Gregorio Olaso	—	Cadegal
Gregorio Olaso	—	Jata
Gregorio Olaso	—	Matamoros
Vicente Bellido	El Sol	Papelera
Santos Zunzunegui	Rosita	Montijo
Clemente Unzaga	—	Caída
Ambrosio Uribe	—	Caída
Cosme Allende	Justa	Costa Buena
Cosme Allende	Juanita	Jarralta
Miguel Loinaz	—	Fuente Fría
Miguel Humarán	San José	Blanca
Felipe Vizcaya	San Fermín	Blanca
José Murua	San Ceferino	Cabrito
Santos Valle	San Ignacio	Cobachos
Clemente Palacio	Santa Cecilia	Matamoros
Manuel Moya	Parcocha	Parcocha
Ramón Causo	San Ramón	Urdimales
José San Martín	Elevación	Urdimales
Manuel Escobal	Nicanora	Cobachos
Miguel Allende	—	Sarlojo
Romualdo Moya	Furia	Laguna
Ramón Uribe	—	Laguna
Francisco Liona	Milagrosa	Zapato
Francisco Lascue	Litiganta	Urdimales
José Salcedo	—	Cadegal
Ramón Gamboa	Ilena (Elena?)	Cadegal
Cecilia de la Hera	Suceso	Zarzal

TABLE I-1
Continued

Owner	Mine	Location
Manuel Valparda	Salve	Cebillas
Raimundo Butrón	Estrella	Harinera
Lorenzo Ibáñez	Catalina	Bodovalle
José Larrazábal	Diana	Campillo
Francisco Ureta	San Felipe	Cobachos
Cecilio Llano	Pacífica	Campillo

Source: A.M.P., L. 98, no. 97.

TABLE I-2
Ibarra Family Share of Orconera Production

Year	Total Production (tons)	Production from Ibarras' Mines	Royalties (pounds)	Royalties (pesetas)
1877	34,539	17,269	862	21,000
1878	270,432	162,837	8,140	204,000
1879	412,150	262,174	13,108	333,000
1880	596,408	438,874	21,943	550,000
1881	640,674	502,048	25,101	631,000
1882	774,691	649,498	32,473	832,000
1883	815,892	685,973	34,297	879,000
1884	749,707	651,379	32,568	835,000
1885	858,995	736,418	36,820	953,000
1886	888,956	757,898	37,894	969,000
1887	834,261	778,611	38,929	988,000
1888	822,736	724,804	36,240	926,000
1889	904,195	768,496	38,424	1,003,000
1890	883,261	765,483	38,274	1,002,000
1891	923,442	775,092	38,754	1,032,000
1892	878,214	698,563	35,781	1,046,000
1893	1,016,465	815,294	40,764	1,243,000
1894	989,348	825,897	40,948	1,204,000
1895	920,683	756,779	37,837	1,103,000

TABLE I-2
Continued

Year	Total Production (tons)	Production from Ibarras' Mines	Royalties (pounds)	Royalties (pesetas)
1896	970,960	753,861	37,677	1,151,000
1897	1,047,386	947,954	47,397	1,542,000
1898	890,131	871,175	43,558	1,699,000
1899	1,056,000	845,000	42,249	1,373,000
Total	18,179,526	15,191,377	760,038	21,519,000

Sources: I.F.A., annual reports of Orconera's manager in Bilbao 1885, 1886, 1888, 1889; and Orconera's directors' reports to the annual meeting of stockholders 1878, 1882, 1885–1886, 1889–1890, 1892–1893, 1895–1897, 1900.
Note: García Merino has a similar table, but he overcalculates the payments to the Ibarras. He believes that they received an annual rent of 5,000 pounds beyond the one shilling per ton royalty. A careful reading of the lease indicates that those 5,000 pounds were a minimum guaranteed payment, which corresponded to the first 100,000 tons extracted (the rate per ton comes out to be 1 shilling), which had to be paid whether the company worked the mine or not. Any amount of ore above the first 100,000 tons also had to pay a royalty of 1 shilling per ton. Thus, the company did not have to pay an annual rent of 5,000 pounds in addition to the royalty of 1 shilling per ton of ore extracted.

TABLE I-3
Ibarra Family Royalties and Dividends from Orconera (pounds)

Year	Orconera's Profits	Orconera's Dividends	Ibarras' Royalty (1)	Ibarras' Dividend (2)	1+2	1+2 As % of Profits
1881	67,966					
1884	91,702					
1885	96,979	80,000	36,820	20,000	56,820	58.59
1886	93,362					

TABLE I-3
Continued

Year	Orconera's Profits	Orconera's Dividends	Ibarras' Royalty (1)	Ibarras' Dividend (2)	1+2	1+2 As % of Profits
1888	82,398					
1889	100,823	85,000	38,424	21,250	59,674	59.19
1891	121,792	95,000	38,754	23,750	62,504	51.32
1892	104,747	106,000	35,781	26,500	62,281	59.46
1894	132,408					
1895	127,150	115,000	37,837	28,750	66,587	52.37
1896	136,852	110,000	37,677	27,500	65,177	47.63
1898	134,806					
1899	196,241					
1900	239,269					

Source: I.F.A., annual reports of the Orconera's directors to the stockholders.
Note: A blank means data not available.

TABLE I-4
Composition of Orconera's Profits (pounds)

Year	Sales to Partners	Sales to Outsiders	Railroad Service	Total
1884	35,191	40,722	19,429	95,342
1885	38,937	48,583	13,637	101,157
1886	41,700	39,709	15,567	96,976
1888	45,897	28,535	10,671	85,103
1889	47,501	41,549	14,052	103,102
1891	47,620	61,602	14,763	123,985
1892	47,450	43,889	18,297	109,636
1894	47,805	73,319	19,468	140,592
1895	47,500	64,579	23,097	135,176

TABLE I-4
Continued

Year	Sales to Partners	Sales to Outsiders	Railroad Service	Total
1896	47,500	67,034	30,913	145,447
1899	52,150	135,289	21,503	208,942
1900	47,500	194,154	21,417	263,071

Source: I.F.A., annual reports of the Orconera's directors.

TABLE I-5
Ore Sold by Orconera to Partners and Outsiders (tons)

Year	Ore Sold to Partners	Ore Sold to Outsiders
1881	450,000	182,000
1882	450,000*	324,000
1883	450,000*	365,000
1884	450,000	289,000
1885	444,500	382,000
1886	492,000*	388,000
1887	500,000*	284,000
1888	550,000	210,000
1889	580,000	260,000
1890	600,000*	283,000
1891	601,500	290,500
1892	599,400	261,000
1893	600,000*	400,000
1894	604,000	387,000
1895	600,000	306,000
1896	600,000	297,000
1897	600,000*	400,000
1898	600,000*	400,000
1899	636,000	300,000

Source: I.F.A., annual reports (1885, 1886, 1888) of Orconera's manager in Bilbao; annual reports to stockholders (1889, 1891–1892, 1894–1896, 1898–1899), and B.O.V.
Note: * estimated quantity, based on average of preceding or subsequent year, and production figures from B.O.V.

TABLE I-6
Orconera's Profits from Ore Transported for Other Companies (pesetas)

Year	Tons Transported	Profit	Profit/Ton
1881	99,590	202,984	2.04
1884	218,415	497,868	2.28
1885	195,926	352,857	1.80
1888	113,852	272,777	2.40
1889	120,857	367,108	3.04
1895	305,000	673,162	2.20
1898	336,000	724,025	2.15

Source: E. Gruner, "Barcelone-Bilbao," p. 276; Orconera's reports to stockholders.

TABLE I-7
British Ore Imports (1,000 tons)

Year	Total British Ore Imports	British Ore Imports from Bilbao	% of Total Ore Imports
1878	1,174	918	78
1879	1,084	870	80
1880	2,634	1,688	64
1881	2,449	1,713	70
1882	3,282	2,450	75
1883	3,178	2,315	73
1884	2,729	1,991	73
1885	2,818	2,050	73
1886	2,878	2,151	75
1887	3,765	2,855	76
1888	3,562	2,481	70
1889	4,032	2,770	69
1890	4,472	3,040	68
1891	3,180	2,245	71
1892	3,778	2,650	70
1893	4,065	3,000	74
1894	4,413	3,024	69
1895	4,450	3,122	70
1896	5,438	3,375	62

TABLE I-7
Continued

Year	Total British Ore Imports	British Ore Imports from Bilbao	% of Total Ore Imports
1897	5,968	3,207	54
1898	5,468	3,043	56
1899	7,054	3,955	56
1900	6,298	3,101	49

Source: Total British ore imports from Carr and Taplin, *History,* pp. 108, 128, 191. Imports from Bilbao compiled from B.P.P.

TABLE I-8
Principal Buyers of Bilbao's Ore (1,000 tons)

Year	Great Britain	Belgium	Holland	France	Germany	United States	Total
1881	1,713	360	73	335		17	2,498
1882	2,450	703	73	450		15	3,691
1883	2,315	547	50	459		6	3,377
1884	1,991	601	102	456		2	3,152
1885	2,050	654	93	491		7	3,295
1886	2,151	524	98	332		41	3,146
1887	2,855	709	96	359		150	4,169
1888	2,481	644	103	348		15	3,591
1889	2,770	640	93	378		4	3,885
1890	3,040	647	106	388		89	4,270
1891	2,245	631	66	342		31	3,315
1892	2,650	766	75	390		34	3,915
1893	3,000	569	108	329		11	4,017
1894	3,024	691	82	324		1	4,122
1895	3,122	600	148	288		16	4,174
1896	3,375	792	128	324		45	4,664
1897	3,207	812	194	348	60	32	4,653

TABLE I-8
Continued

Year	Great Britain	Belgium	Holland	France	Germany	United States	Total
1898	3,043	803	165	253	45	3	4,312
1899	3,955	903	209	272	44	75	5,458

Source: B.P.P.

Appendix II

The Business Elite

TABLE II-1
The Mercantile Elite during the 1850s

1. Abaitua, Juan
2. Abaitua, Serafín
3. Aguirre, José Pantaleón
4. Aguirre, Eugenio
5. Aguirre, Máximo
6. Aldape, Isidoro
7. Ansótegui, Vicente
8. Arana, José Blas
9. Arana, Vicente
10. Arellano, Romualdo
11. Artiñano, Vicente
12. Ayarragaray, Sebastián
13. Barroeta, Lorenzo H.
14. Barua, Santiago M.
15. Bergareche, Nicolás
16. Briñas, José*
17. Echevarría y La Llana, Juan
18. Errazquin, Pedro
19. Escuza, Benito
20. Escuza, Miguel
21. Escuza, Pedro
22. Epalza, Pablo, Sr.
23. Epalza, Pablo, Jr.
24. Epalza, Tomás J.
25. González de la Mata, R.*
26. Gorocica, Santiago J.
27. Gorbeña, Valentín
28. Jane, Pedro
29. Jane, Manuel
30. Ibarra, Gabriel M.
31. Ibarra, Juan M.
32. Ingunza, Santiago M.
33. Lequerica, José A.
34. MacMahon, Diego*
35. Maruri, Teodoro
36. Mazas, Joaquín
37. Mier, José A.
38. Mowinckel, Gerardo*
39. Olaechea, Ignacio*
40. Olaguivel, Nicolás
41. Olavarria, Pedro F.
42. Olavarri, Pascual
43. Olave, Juan
44. Orbegozo, Gabriel M.
45. Orbegozo, Ambrosio
46. Patrón, Francisco J.
47. Picaza, Joaquín J.
48. Recacoechea, Tiburcio M.
49. Solaún, José
50. Uhagón, Guillermo
51. Uhagón, Francisco
52. Uhagón, Manuel A.
53. Uriguen, José A.
54. Uriguen, Juan A.
55. Uriguen, Luciano
56. Zubiría, Cosme*

Sources: A.C.J.G., Elecciones, Reg. 42; *from F.C.R. or from list of officials of the Commercial Tribunal in *Libro del centenario de la Cámara de Comercio de Bilbao.*

TABLE II-2
Directors of Metallurgical Companies

Name	Company
Mining Background	
Allende, José María	Santa Agueda
Allende, Plácido	Santa Agueda
Chavarri, Benigno	Chavarri-Petrement, La Vizcaya, Santa Agueda
Chavarri, Leonardo	La Vizcaya
Chavarri, Víctor	Chavarri-Petrement, Basconia, La Vizcaya, Santa Agueda
Gandarias, J. T.	La Vizcaya
Gandarias, P. P.	Basconia, Talleres de Deusto, La Vizcaya
Ibarra, Ramón	Basconia, Altos Hornos de Bilbao, Tubos Forjados
Martínez Rivas, F.	Astilleros del Nervión
Martínez Rivas, J.	Astilleros del Nervión
Salazar, Luis	Euskaria
San Martín, José M.	La Vizcaya
Sota, R.	Euskalduna
Urquijo e Ibarra, Adolfo	Aurrera, Astilleros del Nervión
Vilallonga, Mariano	Alambres del Cadagua, Altos Hornos de Bilbao
Vilallonga, José	Altos Hornos de Bilbao
Yandiola, G.	Aurrera, Euskalduna
Yandiola, F.	Aurrera
Zubiría, Tomás	Alambres del Cadagua, Euskaria, Tubos Forjados
Zubiría, Luis	Altos Hornos de Bilbao
Mercantile Background	
Amann, Juan	Astilleros del Nervión
Aznar, Eduardo	Euskalduna
Costa, J. L.	Alambres de Cadagua
Coste y Vildosola, Eduardo	Euskalduna
Echevarría, Federico	Iberia, La Vizcaya
Echevarría, J.	Iberia
Gurtubay, Juan	Altos Hornos de Bilbao
Palacio, C.	Iberia
Rochelt, Ricardo	Basconia

TABLE II-2
Continued

Name	Company
Uriguen, Braulio	Altos Hornos de Bilbao
Outsiders	
Alonso, F. (engineer)	Aurrera
Angoloti, J. (banker)	Alambres del Cadagua, Altos Hornos de Bilbao
Barat, J. (banker)	Altos Hornos de Bilbao
Comyn, A. (?)	Euskaria, Tubos Forjados
Davies, J. B. (mining)	Aurrera
Disdier, Enrique (engineer)	Alambres del Cadagua, Euskaria, Tubos Forjados, Santa Agueda
Girona, J. (banker)	Altos Hornos de Bilbao
Girona, M. (banker)	Altos Hornos de Bilbao
Goitia, F. (industrialist)	Iberia
Larrínaga, R. (shipping)	La Vizcaya
Olano, E. (shipping)	La Vizcaya
Olano, J. A. (shipping)	La Vizcaya
Palmer, Ch. (industrialist)	Astilleros del Nervión
Petrement, J. (industrialist)	Chavarri-Petrement
Rivas, Francisco (banker)	San Francisco de Mudela
Rodríguez San Pedro, F. (?)	Altos Hornos de Bilbao
Tous, N. (engineer)	Talleres Zorroza
Urquijo, Marqués de (banker)	Alambres del Cadagua, Altos Hornos de Bilbao, Euskaria, Tubos Forjados
Others	
Arrótegui, M. (shipping)	La Vizcaya
Aznar, Luis M. (shipping)	Euskalduna
Casa Torre, Marqués de (landowner)	Euskaria
Gana, Enrique (landowner)	Alambres del Cadagua, Basconia, Euskaria, Santa Agueda, Tubos Forjados
Mendialdua, N. (shipping)	Euskalduna
Orueta, José (lawyer)	Talleres Zorroza
Urquijo-Aguirre, A. (shipping)	Euskalduna

TABLE II-2
Continued

Name	Company
Unclassified	
Ajuria, A.	Iberia
Belausteguigoitia, J.	La Vizcaya
Chapa, R.	Santa Agueda
Eguren, R.	Euskaria
Madaleno, J.	Aurrera
Sagarduy, M.	Euskalduna
Zaldembide, T.	Alambres del Cadagua, Euskaria, Tubos Forjados

Sources: Anuario de la minería, metalúrgia y electricidad de España, 1894–1904; Sociedad de Altos Hornos de Bilbao, *Memorias (1883–1899);* Compañía Euskalduna, *Memorias* (1901, 1903); A.B.E., box 42, no. 2663 (partnership contract of Aurrera).

TABLE II-3
Primary Investors in Shipping Companies

Vapores Serra
Manager: J. Serra y Font

Investor's Name	No. of Shares
José Serra y Font (s)	1,200
J. Arano y Alegría (merc)	300
D. Arano y Alegría (merc)	300
E. Real de Asua (merc)	200
R. Real de Asua (merc)	200
D. Madariaga (merc)	200
J. B. Alegría y Zavala (?)	100
Total	2,500
Capital: 6.25 million pesetas	

Source: A.H.P.V., C. Ansuátegui, L. 6324, no. 168, Apr. 25, 1879.

TABLE II-3
Continued

La Flecha
Manager: J. Serra y Font

Investor's Name	No. of Shares
José Serra y Font (s)	500
J. Arano y Alegría (merc)	250
E. Real de Asua (merc)	250
R. Real de Asua (merc)	250
D. Madariaga (merc)	250
J. Rochelt y Amann (merc)	250
A. Rochelt y Amann (merc)	250
Total	2,000
Capital: 5 million pesetas	

Source: A.H.P.V., C. Ansuátegui, L. 6331, no. 526, Dec. 12, 1880.

Bilbaína de Navegación
President: J. B. Longa
Directors: M. Basabe, F. Abaitua, J. M. Olavarri
Managers: E. Aznar, J. B. Astigarraga

Investor's Name	No. of Shares
Juan Bautista Longa (s)	459
J. M. Olavarri (merc)	380
A. Uria y Urresti (l)	309
B. Salazar (min)	290
M. Basabe (merc)	280
E. Aznar (merc)	265
F. Abaitua (merc)	250
J. B. Astigarraga (merc)	230
F. Martínez Rodas (mil)	230
M. Astigarraga (merc)	152
N. Rodríguez Lagunilla	150
E. Uriguen (merc)	150
B. Chavarri (min)	138
J. Echevarría & Sons (merc)	102

TABLE II-3
Continued

P. D. Arana (min)	102
E. Coste y Vildosola (merc)	100
V. Chavarri (min)	78
Total	3,665
Capital: 2 million pesetas	

Source: A.H.P.V., C. Ansuátegui, L. 6348, no. 447/48.
Note: There were 4,000 total shares in the company; the remaining stock was in the hands of small investors, with 50 or fewer shares.

Marítima Vizcaya
President: F. Carranza
Board members: J. A. Uriarte, R. Balparda, E. Sustacha, C. Linares

Investor's Name	No. of Shares
F. Carranza (s)	1,027
J. A. Uriarte (merc)	371
J. R. Uriarte (merc)	266
G. Uribarri (l)	126
C. Linares (merc)	121
R. and F. Carasa (l)	121
Escuza bros. (merc)	95
E. Sustacha (merc)	57
R. Balparda (lp)	49
Total	2,233
Capital: 2.44 million pesetas	

Source: A.H.P.V., C. Ansuátegui, no. 295/96, July 4, 1885.
Note: There were 2,443 total shares in the company; the remaining shares were in the hands of small investors with fewer than 40 shares.

General notes:
(l) = proprietor
(lp) = liberal profession background
(merc) = mercantile background
(mil) = military background
(min) = mining background
(s) = shipping background

TABLE II-4
Boards of Directors of Major Banks

Banco de Bilbao (1882)

J. Aguirre (merc)	J. L. Moyua (merc)
P. Alzola (lp)	A. Obieta (merc)
L. Barroeta (merc)	A. Palacio (merc)
E. Coste y Vildósola (merc)	J. M. Sainz (?)
J. Galíndez San Pedro (l)	M. Zabala (merc)
M. MacMahon (merc)	C. Zubiría (min)
P. Mazas (merc)	

Source: J. Lazúrtegui and V. Larrea, *Merchant's and Shipmaster's Practical Guide to the Port of Bilbao*, pp. 12–13.

Banco de Comercio (1892)

D. Toledo (merc), president	R. Nardiz (l)
E. Arana (?)	R. de la Sota (min)
F. Carranza (s)	J. Sevilla (?)
M. Castellanos (?)	J. T. Uribe (merc)
B. Chavarri (min.)	A. Urioste (?)
S. L. Letona (ind.)	S. Varona (?)
L. Longa (s)	L. Villabaso (l)

Source: El Nervión, Feb. 20, 1892.

Banco de Vizcaya (1901)

P. Allende (min)	P. Orue (?)
D. Aresti (?)	R. Picavea (ind)
J. M. Basterra (merc)	F. Ugalde (merc)
E. Borda (?)	B. Uriguen (merc)
D. Escauriaza (merc)	T. Urquijo (?)
G. M. Ibarra (min)	A. Ustara (min)
P. MacMahon (merc)	M. Vilallonga (min)
P. Maíz (merc)	T. Zubiría (min)

Source: J. Ybarra, *Política nacional en Vizcaya*, p. 213.

TABLE II-4
Continued

Crédito de la Unión Minera (1902)	
T. Allende (min), president	F. Ibáñez Aldecoa (l)
A. López (?), vice-president	O. Kreizner (min)
J. Alonso Allende (?)	V. Larrea (min)
J. Amezola (min)	L. Núñez (min)
C. Aramburu (?)	J. M. San Martín (min)
R. Chapa (?)	J. Santisteban (min)
F. Chavarri (min)	I. Ubieta Ibarra (min)
A. Gandarias (min)	

Source: *Anuario de la minería, metalurgia y electricidad de España,* p. 124.

Banco de España (1898)	
J. Arellano Arrospide (merc)	P. Muñoz Rubio (?)
D. Gil Iturraga (?)	J. Solaún Mugaburu (merc)
T. Irasazabal (?)	F. Soltura Urrutia (lp)
F. Martínez Rodas (mil)	E. Vallejo Arana (lp)
H. Múgica Arenza (?)	R. Vicuña y Lazcano (lp)

Source: Memoria del Banco de España (Bilbao branch), 1898.

General notes:
(ind) = industrialist
(l) = proprietor
(lp) = liberal profession background
(merc) = mercantile background
(mil) = military background
(min) = mining background
(s) = shipping background
Years following the bank names indicate the year in which the men listed served on the board of directors.

Appendix III

Family Connections between Boards of Directors of Banks of Vizcaya and Bilbao (ca. 1914)

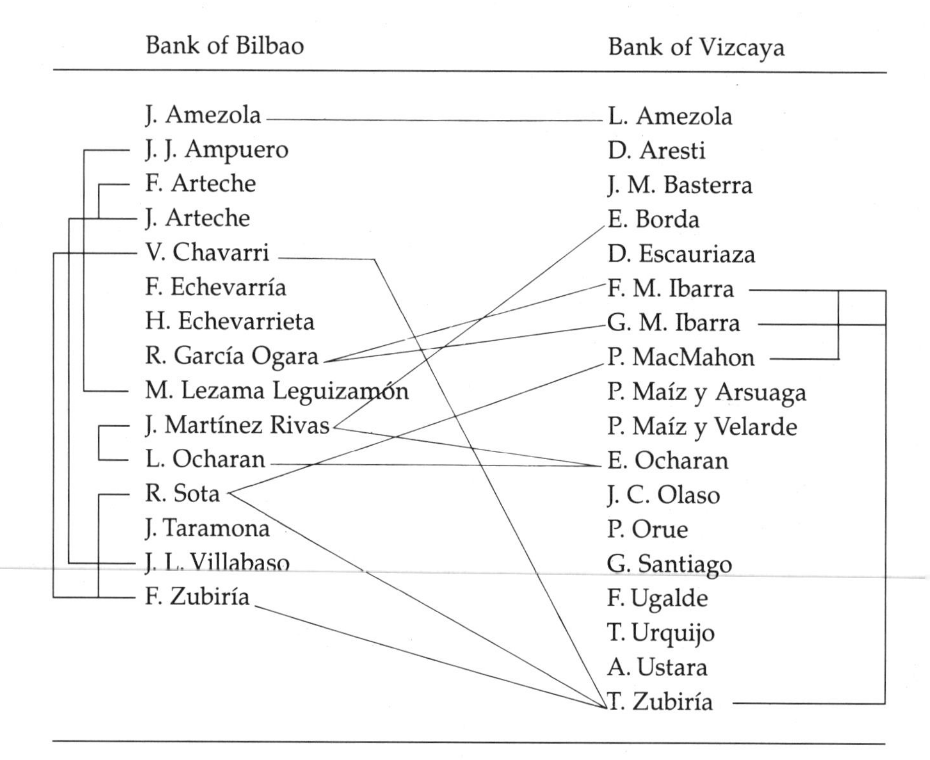

Source: A.B.E., L. 779 for the directorates. For family connections, Pérez Azagra, *Títulos;* idem, *Noticias;* Ybarra, *La Casa.*

Appendix IV

Elected Officials

TABLE IV-1
Members of Diputación, 1850–1870

Deputy	Term	Background
Aguirre, Gregorio	1868–1870	Proprietor
Aldama, Alejandro	1860–1862	?
Allende-Salazar, Castor M.	1850–1852	Proprietor
Antuñano, Alejandro	1856–1858; 1866–1868	Manufacturer
Arrieta Mascarua, Lorenzo	1864–1866; 1866–1868; 1868–1870	Proprietor
Arrieta Mascarua, José Miguel	1854–1856; 1856–1858	Proprietor (B)
Basabe, Julián	1866–1868	Proprietor (B)
Basozábal, Juan José	1858–1860	?
Belarroa, Vicente	1850–1852; 1852–1854	?
Echevarría y La Llana, Juan	1854–1856	Merchant (B)
Echezarreta, Ramón	1858–1860	Manufacturer
Gogeascoechea, Manuel	1858–1860; 1862–1864	?
Guardamino, Rafael	1850–1852	?
Galo Hormaeche, José	1850–1852	?
Jane, Pedro	1850–1852; 1852–1854	Merchant (B)
Jauregui, Juan José	1852–1854; 1862–1864	Manufacturer
Lambarri, José María	1858–1860; 1860–1862	Manufacturer
Larrínaga, Mariano	1856–1858	Proprietor (B)
Loizaga, Timoteo	1856–1858	Manufacturer
López de Calle, Antonio	1858–1860; 1860–1862; 1862–1864	Proprietor
López de Calle, Bruno	1866–1868; 1868–1870	Proprietor
Murúa, José María	1852–1854	Proprietor
Novia Salcedo, Pedro	1852–1854	Proprietor (B)
Olaguivel, Nicolás	1862–1864	Merchant (B)
Santos Orue, José	1854–1856; 1856–1858	Proprietor
Oxangoiti, Cayetano	1866–1868	Proprietor (B)

TABLE IV-1
Continued

Deputy	Term	Background
de la Quadra Salcedo, Andrés	1860–1862	Proprietor
Rotaeche, Ramón Castor	1854–1856; 1858–1860	?
Tellitu y Antuñano, Juan	1854–1856	?
Urcola, Fermín	1862–1864	Proprietor
Urquizu, Fausto	1864–1866; 1868–1870	Proprietor
Urquizu, José Niceto	1852–1854; 1856–1858; 1860–1862; 1864–1866; 1868–1870	Proprietor
Urrecha, Alejandro	1866–1868	? (B)
Urrutia, Manuel	1864–1866	Proprietor (B)
Victoria de Lecea, Eduardo	1868–1870	Proprietor (B)
Victoria de Lecea, Federico	1860–1862	Proprietor (B)
Vidásolo, Juan Francisco	1862–1864	?
Vildósola, José Antonio	1850–1852	? (B)
Yarza, Carlos Adán de	1854–1856	Proprietor (B)
Zabala, Francisco	1864–1866	Proprietor
Zabalburu, José	1864–1866	Proprietor (B)

Sources: The names were taken from Libros de Acuerdos de las Juntas Generales de Guernica. Background was established through the lists of wealthiest proprietors, manufacturers, and merchants published in B.O.V. on Jan. 17, 1860; Oct. 1, 1861; Oct. 8, 1861; Oct. 6, 1863; Oct. 14, 1865; Dec. 3, 1867; Dec. 3, 1872. The lists were complemented with notarial documents.
Note: (B) indicates residence in the Bilbao area.

TABLE IV-2
Municipal Government, Bilbao

Official	Socioeconomic Background
1850	
Orbegozo, Ambrosio, mayor	Merchant
Uriguen, Luciano, deputy mayor	Merchant
Olavarria, José R., deputy mayor	(?)
Arana, José Blas	Merchant
Arana, José Miguel	(?)
Arana, Tomás	Proprietor

TABLE IV-2
Continued

Official	Socioeconomic Background
Barroeta, Lorenzo H.	Merchant
de la Colina, Pedro	Merchant
Echararoeta, Martin	(?)
Elorrieta, Tomás	Proprietor
Garitagoitia, Eugenio	Proprietor
Ibarra y Cortina, José	Merchant
Larrínaga, Eulogio	Proprietor
Mendibil, Ceferino	Proprietor
Vicuña, Millán	(?)

Source: A.H.P.V., S. Urquijo, L. 5841, Apr. 24, 1850. For socioeconomic background, see table II-1 and source note in table IV-1.

1861	
Urrutia y Beltrán, Manuel, mayor	Proprietor
Uhagón, Rafael, deputy mayor	Merchant
de la Colina, Pedro	Merchant
Eguillor, Sebastián	Merchant
Gana, Saturnino	Proprietor
Gortázar, Manuel M.	Proprietor
Larrínaga, Mariano	Proprietor
MacMahon, Pedro	Merchant
Marroquín, Juan	(?)
Maruri, Teodoro	Merchant
Norzagaray, Pedro	Merchant
Palacio, Angel	Merchant
Solaún, Elías	(?)
Somonte, José M.	Merchant
Uriguen, Ezequiel	Merchant
Urrecha, Alejandro	(?)
Violete, Luis	Merchant
Zugasti, Faustino	Merchant

Source: A.H.P.V., S. Urquijo, L. 6216, Jan. 1861.

TABLE IV-2
Continued

Official	Socioeconomic Background
1868	
Victoria de Lecea, Eduardo, mayor	Proprietor
Ansótegui, Luis, deputy mayor	Proprietor
Ansua, Guillermo, deputy mayor	(?)
Amann, Juan	Merchant
Arenaza, Antonio	(?)
Elorriaga, Luis	Merchant
Galíndez, Pablo	Proprietor
Gurtubay, José M.	Merchant
Lazúrtegui, Francisco	Merchant
Llona, Vicento	(?)
Unzurrurzaga, Antonio	(?)
Zarauz, Antonio	Merchant

Source: A.H.P.V., S. Urquijo, L. 6227, July 2, 1868.

TABLE IV-3
Provincial Deputies, 1880–1900

Deputy	Term	Political Affiliation
Abásolo, Joaquín	1882–1884	Liberal
Acha, Tomás (b)	1880–1882	Democrat
Achúcarro, Severiano (lp)	1880–1882	Liberal
Acillona, Esteban (lp)	1886–1890	Carlist
Aguirre y Oxangoiti, Juan	1882–1886	Liberal
Aldama y Hormaza, Carlos	1892–1896	Liberal
Algorta, Claudio	1880–1882	Carlist
Algorta, Pascual	1896–1900	Liberal
Allende, Plácido (B)	1894–1896	Liberal
Allende y García, Antonio	1898–1902	Carlist
Alonso, Rafael (b)	1892–1896	Liberal
Alonso, Natalio (b)	1882–1886	?
Altube y Albiz, Lucas	1886–1900	Carlist
Alzaga y Arrieta, Antonio	1896–1900	Liberal
Alzola, Pablo (B)	1886–1900	Liberal

TABLE IV-3
Continued

Deputy	Term	Political Affiliation
Amézola, José (b)	1894–1898	Liberal
Ampuero, José M.	1880–1886	Carlist
Apoitia, Fernando (lp)	1884–1888; 1888–1892	Liberal
Arana, Sabino (p)	1898–1902	Basque Nationalist
Arana, Pedro D. (B)	1880–1882	Independent
Areitio, Federico	1888–1892	Liberal
Areizaga, Juan C.	1888–1892	?
Aresti, Enrique (B)	1898–1902	Liberal
Arnedo, Fermín (b)	1882–1886	Liberal
Arroita-Jaregui, José N.	1890–1894	Catholic-Fuerist
Arrola, Idelfonos	1894–1898; 1898–1902	Carlist
Arrótegui, Manuel M. (B)	1890–1894	Liberal
Arteche, José M. (B)	1892–1896	Liberal
Astola, Casimiro	1880–1882	Carlist
Aznar, Alberto (B)	1896–1900	Liberal
Aznar, Eduardo (B)	1892–1896	Liberal
Aznar, Luis (B)	1890–1894	Liberal
Barcenas, Antonio	1896–1900	Liberal
Basozábal, Carlos	1894–1898	Carlist
Basterra, José L.	1880–1882	?
Basterra y Esturo, Mario (lp)	1886–1890	?
Bolívar, Bartolomé	1884–1888; 1896–1900	Liberal
Cariaga, Francisco	1880–1882	Independent
Carranza, Fernando (B)	1896–1900	Liberal
Cerrajeria, Cesáreo	1880–1882	Independent
Cobreros, Gustavo (b)	1880–1882; 1884–1886; 1896–1900	Carlist
Cruceño, José	1896–1900	Liberal
Epalza, José F. (lp)	1886–1890; 1898–1902	Carlist
Galarza y Menchaca, Aureliano	1886–1890; 1898–1902	Fuerist
Gáldiz y Aurrecochea, Victoriano	1890–1894; 1894–1898	Liberal
Galíndez y Bermejillo, Manuel	1890–1894	Liberal
García, Eusebio (b)	1880–1882	?
Goldoracena, Benito	1882–1886	Republican

TABLE IV-3
Continued

Deputy	Term	Political Affiliation
González Llaguno, Benito	1884–1888	?
Gordon, Simón	1880–1882; 1894–1898	Indep./ Liberal
Gortázar, José M.	1888–1892	Liberal
Goyarrola, Manuel	1894–1898	Liberal
Goyoaga, Restituto (B)	1884–1888	?
Ibáñez Aldecoa, Fausto (B)	1888–1892	Carlist
Ingunza, Benigno	1892–1896	Carlist
Jáuregui, Juan (p)	1884–1888; 1892–1896	Fuerist
Landecho, Fernando	1886–1890	?
Larrazábal, Pascual	1884–1888	?
Larrea y Echanove, Alejandro	1896–1900	Liberal
Larrínaga, Bernabé (B)	1880–1882	Republican
Larrucea, José	1890–1894	Catholic/ Fuerist
Leicea, Pedro	1890–1894	Liberal
León, Isidoro (b)	1894–1898; 1898–1902	Liberal
López de Calle, Javier (p)	1888–1892	Liberal
de la Mata y Gondra, Perfecto	1894–1898	Carlist
Muñoz, José R. (lp)	1888–1892	?
Nárdiz, Galo (p)	1880–1882	Independent
Olalde y Artázar, Victoriano	1898–1902	Liberal
Olano, Francisco	1880–1882	Carlist
Olascoaga, Fernando	1890–1892; 1892–1896	Catholic/ Fuerist
Orbe, José M. (p)	1882–1884; 1886–1890	Carlist
Oxangoiti, Joaquín (p)	1892–1896	Liberal
Palacio, Cosme (B)	1888–1892; 1892–1896	Liberal
Pértica, Serapio	1882–1886	Carlist
del Rivero y Trevilla, Antonio	1884–1888	?
Sagarmínaga, Pablo (p)	1890–1894	Catholic/ Fuerist
Salazar, Benigno (B)	1880–1882; 1882–1886	Liberal
San Martín, José M. (B)	1892–1896	Liberal
Santo Domingo, Raimundo	1880–1882	Carlist
Saráchaga, Liborio	1882–1884	Liberal

TABLE IV-3
Continued

Deputy	Term	Political Affiliation
Sota y Llano, Ramón (B)	1888–1892	Fuerist
Torres Vildósola, José L. (p)	1884–1888	?
Unamúnzaga, J.	1882–1886	Carlist
Uría, Angel (B)	1882–1884; 1884–1888; 1888–1892	Liberal
Uriarte y Arana, Antonio (B)	1886–1888	?
Uribe, José J.	1882–1886	Carlist
Urizar, Agapito	1884–1888	?
Urizar, Idelfonso	1882–1884	Carlist
Urquizu y Zorrilla, Alfredo	1896–1900	Liberal
Urráburu, José L.	1880–1882; 1882–1886	Carlist
Urrutia, Saturnino	1888–1892	?
Vallejo y Arana, Emilio (B)	1896–1900	Liberal
Velasco, Atanasio	1892–1896	Liberal
Veristain, Nicasio	1896–1900	Carlist
Vicuña, Ramón (lp)	1884–1888	Liberal
Vilallonga, Gabriel (B)	1890–1894	Catholic/ Fuerist
Villabaso, José L. (B)	1882–1886	Liberal
Zabala, Casto	1884–1888	?
Zárraga, Alejandro	1882–1886	Liberal
Zubiaga, Román	1880–1882	Carlist
Zunzunegui, Casimiro (b)	1898–1902	Liberal

Source: A.H.D.V., Actas de Sesiones de la Diputacion. For socioeconomic background, see chap. 4. For the other businessmen and liberal professionals, see R. Fontán y L. Larrañaga, *El libro de Bilbao. Guia artístico-comercial;* and list of members of the Bilbao Chamber of Commerce published in the Nov. 1889 issue of *Boletín oficial de la Cámara de Comercio de Bilbao.* For the proprietors, B.O.V. lists and notarial documents. Political affiliations taken from Ybarra, *Political nacional en Vizcaya,* and Real Cuesta, *El carlismo vasco.*

Notes:
(B) = member of the business elite
(b) = small to mid-size businessman
(lp) = liberal professional
(p) = proprietor

TABLE IV-4

National Deputies, 1879–1910

	Year Elected				
District	1879	1881	1884	1886	1891
Bilbao	M. Zavala	E. Aguirre	R. Mazarredo	E. Aguirre	E. Vitoria Lecea
Durango	F. Sagarmínaga	J. M. Ampuero	J. Ibargoitia	J. Ibargoitia	Marqués Casa Torre (b)
Guernica	B. López Calle	A. Allendesalazar	A. Allendesalazar	L. Landecho	L. Landecho
Valmaseda	G. Vicuña	R. Balparda	G. Vicuña	V. Chavarri (b)	J. Martínez Rivas (b)
	1893	1896	1898	1899	1901
Bilbao	A. Urquijo (b)	J. Martínez Rivas (b)	J. Martínez Rivas (b)	F. Echevarría (b)	T. Zubiría (b)
Baracaldo	*	A. Urquijo-Ibarra (b)	R. Ibarra (b)	R. Ibarra (b)	R. Ibarra (b)
Durango	Marqués Casa Torre (b)	Marqués Casa Torre (b)	**	Marqués Casa Torre (b)	Marqués Casa Torre (b)
Guernica	M. Arrótegui	J. T. Gandarias (b)	J. T. Gandarias (b)	J. T. Gandarias (b)	J. T. Gandarias (b)
Marquina	F. Martínez Rodas (b)	E. Aznar-Tutor (b)	P. Allende (b)	P. Allende (b)	P. Allende (b)
Valmaseda	B. Chavarri (b)	B. Chavarri (b)	B. Chavarri (b)	B. Chavarri (b)	B. Chavarri (b)

TABLE IV-4
Continued

	Year Elected			
District	1903	1905	1907	1910
Bilbao	J. Urquijo-Ibarra (b)	F. Solaegui (b)	F. M. Ibarra (b)	H. Echevarrieta (b)
Baracaldo	T. Zubiría (b)	T. Zubiría (b)	T. Zubiría (b)	F. M. Ibarra (b)
Durango	Marqués Casa Torre (b)	Marqués Casa Torre (b)	Marqués Casa Torre (b)	Marqués Casa Torre (b)
Guernica	J. T. Gandarias (b)	J. T. Gandarias (b)	J. T. Gandarias (b)	J. T. Gandarias (b)
Marquina	J. Acillona (b)	J. Acillona (b)	J. Acillona (b)	J. Acillona (b)
Valmaseda	B. Chavarri (b)	B. Chavarri (b)	B. Chavarri (b)	J. M. Chavarri (b)

Source: Ybarra, *Politica nacional en Vizcaya.*
Notes: (b) indicates membership in business elite.
*Baracaldo district created in 1896.
**Election results canceled.

Notes

ACKNOWLEDGMENTS

1. I use Ibarra instead of Ybarra because the former is the most commonly used spelling of the name in the historical literature.

INTRODUCTION

1. Perhaps the best-known example of this characterization is Nadal *El fracaso de la revolución en España,* an expanded version in Spanish of "The Failure of the Industrial Revolution in Spain 1830–1914."

CHAPTER ONE. HISTORICAL BACKGROUND

1. According to Miguel de Unamuno, the word derives from the Basque word *bizkarr,* meaning "hill." See Miguel de Unamuno, "Cuatro reseñas," vol. VI, pp. 270–271.
2. J. A. García de Cortázar, *Vizcaya en el siglo XV. Aspectos económicos y sociales,* p. 92.
3. See Robert-Henri Bautier, "Notes sur le commerce du fer en Europe occidentale du XIIIe au XIV siècle."
4. W. R. Childs, "England's Iron Trade in the Fifteenth Century," p. 41. Childs calculates the Basque annual output at thirteen thousand tons. E. Fernández Pinedo and L. M. Bilbao estimate that in the sixteenth century annual production hovered between eight thousand and ten thousand tons, which would represent between 10 and 16 percent of all European production. See "Auge y crisis de la siderometalurgia tradicional en el País Vasco (1700–1850)," p. 141, n. 5. Although the figures of these scholars do not match, they all agree on the overall importance of Basque production.
5. The word appears in *Falstaff, Hamlet,* and *The Merry Wives of Windsor.* For this and the *biscaien* reference, see J. Caro Baroja, *Vasconiana,* p. 174, n. 121.
6. For an excellent treatment of this subject, see Fernández Pinedo and Bilbao, "Auge y crisis."
7. Caro Baroja, *Vasconiana,* p. 111.

8. F. Braudel, *The Mediterranean and the Mediterranean World in the Age of Philip II,* v. I, p. 225.

9. J. Caro Baroja, *Los vascos y el mar,* p. 53. In contrast, Abbot Payson Usher, "Spanish Ships and Shipping in the Sixteenth and Seventeenth Centuries," p. 190, argues that, despite their location, the Vizcayans were "closely identified with the traditions of ship-building common to the Mediterranean."

10. García de Cortázar, *Vizcaya en el siglo XV,* p. 75.

11. L. García Merino, *La formación de una ciudad industrial. El despegue urbano de Bilbao,* pp. 199–202, 225–227. In contrast, Ortega y Galindo de Salcedo, *Bilbao y su hinterland,* emphasizes the importance of Bilbao's location.

12. On the rise of Santander and its rivalry with Bilbao, see V. Palacio Atard, *El comercio de Castilla y el puerto de Santander en el siglo XVIII.*

13. T. Guiard, *Historia de la noble villa de Bilbao,* vol. II, pp. 277–285.

14. T. Guiard, *Historia del Consulado y Casa de Contratación de Bilbao,* vol. II, p. 544; also R. Basurto Larrañaga, *Comercio y burguesía mercantil de Bilbao en la segunda mitad del siglo XVIII,* pp. 25–26.

15. Caro Baroja, *Vasconiana,* p. 147. He reiterates the argument in *Los Vascos,* pp. 198–199.

16. In the well-known passage in which Sancho is governor of an island, Don Quixote's squire asks: "Who here is my secretary?" Someone responds: "I, sir, for I can read and write, and I'm a Basque." After he hears this, Sancho remarks ironically: "With that last qualification . . . you could well be secretary to the Emperor himself" (Miguel de Cervantes, *Don Quixote,* p. 767).

17. R. Hernández Ponce, "El aporte vasco a la construcción de Chile. Gobernadores del siglo XVI."

18. A. Otazu y Llana, *El "igualitarismo" vasco y otros mitos,* p. 339, n. 441.

19. D. A. Brading, *Miners and Merchants in Bourbon Mexico, 1763–1810,* pp. 106–107.

20. The figures for the *consulado* of Cádiz are from J. B. Ruiz Rivera, *El consulado de Cádiz. Matrícula de comerciantes 1730–1823.* Guipúzcoa and Vizcaya accounted for the bulk of the registered Basque merchants, with about 5.6 percent each. The remaining 2.8 percent were from Alava. The Basque share of the total Spanish population was calculated from the figures provided by P. Romero de Solís, *La población española en los siglos XVIII y XIX,* pp. 136–137, 170.

21. For Mexico, see Brading, *Miners and Merchants,* p. 108. Caro Baroja shows a similar phenomenon among Navarra's Basques in *La hora navarra del XVIII.*

22. Brading, *Miners and Merchants,* p. 110. Caro Baroja has gone one step farther, challenging Max Weber's thesis of the allegedly inferior capacity of Catholics vis-à-vis Protestants to develop capitalist enterprises. Caro has pointed out the paradox that the most devout Catholics in Spain, the Basques, followed the same business practices as did northern European Protestants. It is perhaps not coincidental that the Jesuits, an order founded by the Basque Saint Ignatius of Loyola, provided much of the theoretical support for their business ethic. See Caro

Baroja, *Los vascos,* pp. 202–203. The role of religion in the lives of nineteenth-century Basque business leaders is treated in chap. 5.

23. R. Herr, *An Historical Essay on Modern Spain,* pp. 40–41.

24. A good brief discussion on the present state of this research is R. Collins, *The Basques,* pp. 9–15.

25. The medieval kingdom of Aragón also included the regions of Catalonia and Valencia.

26. S. Payne, *Basque Nationalism,* p. 17.

27. Collins, *The Basques,* p. 201.

28. E. de Tejada, *El señorío de Vizcaya hasta 1812,* pp. 47–48.

29. R. Barahona, "Basque Regionalism and Centre-Periphery Relations, 1759–1833," p. 273. See also D. J. Greenwood, "Continuity in Change: Spanish Basque Ethnicity As a Historical Process."

30. According to Greenwood, military defense lay behind the granting of collective nobility. In exchange for it, the Basques promised to guard and pay for the defense of their sector of the Spanish frontier with France. Indeed, the grant of nobility coincided with periods of war. Yet, from Greenwood's explanation, it is not clear why the Spanish crown did not grant nobility only to the leaders of such frontier defenses. See "Continuity in Change," pp. 90–96.

31. Otazu y Llana, *El "igualitarismo" vasco.*

32. The argument is made by Greenwood for Guipúzcoa, but it could be made for Vizcaya, too. See Greenwood, "Continuity in Change," p. 91. In fact, the argument was advanced by a late-nineteenth-century supporter of the *fueros,* Aristides Artiñano y Zuricalday, *El señorío de Vizcaya, histórico y foral,* p. 214.

33. For Greenwood, the idea of *pureza de sangre* also supports the Basques' claim to universal nobility. It is strange, however, that, given the success of the Basques, other regions that could have made similar claims did not obtain the privilege. According to Otazu, the idea of keeping other people from settling down in a region with limited resources was key to the adoption of universal nobility. See *El "igualitarismo" vasco,* p. 113. The problem with this argument is that it does not explain why the Castilian crown went along with it.

34. Mauleón Isla, *La población de Bilbao,* p. 35.

35. On the subject of aristocratic prejudices against commerce and industry in Spain, see W. Callahan, *Honor, Commerce, and Industry in Eighteenth-Century Spain,* p. 9.

36. The different aristocratic conceptions held by Vizcayans and Castilians led to strong polemics and resentment against the Basque privileges. At best, Castilians tended to look on the Basques' claim to nobility with deep suspicion; at worst, they denied it outright. This polemic was well reflected in the Spanish literature of the sixteenth and seventeenth centuries. For many examples of this, see Otazu, *El "igualitarismo" vasco,* pp. 121–128; Caro Baroja, *Los vascos,* p. 260, n. 1.

37. Vizcaya's inheritance practices pushed part of the population off the land,

since parents had the right to choose only one child (not necessarily the eldest) to inherit the family property.

38. The *fueros* also prohibited the exportation of iron ore to foreign countries. Principally designed to protect the *ferrerías,* this regulation was finally abolished in 1849, paving the way for the massive exports of the following decades. More details in chap. 2.

39. *Tierra llana* literally means flat land, but the name actually derives from the fact that this territory was not surrounded by walls, as the cities were.

40. M. García Venero, *Historia del nacionalismo vasco,* pp. 87–88.

41. R. Barahona, "The Making of Carlism in Vizcaya (1814–1833)," intro.

42. The interests of these two groups collided in the 1760s. At the time, the proprietors, entrenched in the municipal offices, thwarted a project of the *consulado* to build a warehouse outside the city limits. Fear that such construction might lower their rents seems to have prompted the proprietors' opposition. See Guiard, *Historia de la noble villa de Bilbao,* vol. III, pp. 293–305.

43. A contemporary observer who praised the Vizcayan system was John Francis Bacon, in *Six Years in Vizcaya,* pp. 58–65. An enemy of the *fueros* who nevertheless saw the advantages of the Basque administrative system, was V. de Gaminde, *Intereses de Bilbao. Examen de lo perjudicial que sería la permanencia del sistema foral en el siglo XIX al comercio y a la industria del país y a los liberales de Vizcaya.* Even the politician who eventually abolished the *fueros,* Cánovas del Castillo, praised the Basque administrative institutions; see Payne, *Basque Nationalism,* p. 55.

44. Barahona, "Basque Regionalism," p. 277.

45. For the effects on Bilbao's flour mills, see Palacio Atard, *El comercio de Castilla,* pp. 143–148. On the iron industry, see Fernández Pinedo and Bilbao, "Auge y crisis."

46. Barahona, "Basque Regionalism," pp. 284–285.

47. A good treatment of the tense relationship between Ferdinand and the Basque provinces is found in Barahona, "The Making of Carlism."

48. Caro Baroja, *Los vascos,* p. 202.

49. E. Fernández Pinedo, *Crecimiento económico y transformaciones sociales del País Vasco 1100–1850,* p. 70.

50. Barahona, "The Making of Carlism," p. 216.

51. See, for example, Barahona, "Basque Regionalism," pp. 280–281; and J. H. Elliot, *The Count-Duke of Olivares: The Statesman in an Age of Decline,* pp. 450–451.

52. Barahona, "The Making of Carlism," pp. 482–483.

53. For example, Bacon, *Six Years in Vizcaya,* pp. 103–104; and several contemporary observers cited by Fernández Pinedo in *Crecimiento económico,* p. 473.

54. Barahona, "The Making of Carlism," p. 484; and J. Coverdale, *The Basque Phase of Spain's First Carlist War,* p. 258.

55. Coverdale, *The Basque Phase,* p. 258.

56. Barahona, "The Making of Carlism," p. 477; also Coverdale, *The Basque Phase,* p. 17.

57. Barahona, "The Making of Carlism," pp. 484–485.

58. On the concentration of property from the late eighteenth century to the 1850s, see Fernández Pinedo, *Crecimiento económico,* pp. 263–267.

59. Coverdale, *The Basque Phase,* pp. 17–18.

60. Caro Baroja, *Los vascos,* p. 215.

61. Coverdale, *The Basque Phase,* p. 21.

62. Most of this discussion comes from Coverdale's perceptive analysis of the peasants' motivations for supporting the Carlists. See *The Basque Phase,* esp. pp. 300–306.

63. Ibid., p. 306.

64. Barahona, "The Making of Carlism," p. 485.

65. For example, Fernández Pinedo, *Crecimiento económico,* p. 467.

66. Coverdale, *The Basque Phase,* p. 262, n. 27.

67. Barahona, "The Making of Carlism," p. 487.

68. Coverdale, *The Basque Phase,* p. 260.

69. Bacon, *Six Years in Vizcaya,* pp. 77–78.

70. Coverdale, *The Basque Phase,* p. 262.

71. Ibid., p. 261.

72. This is Barahona's thesis, for example.

73. M. A. Larrea, *Caminos de Vizcaya en la segunda mitad del siglo XVIII,* p. 315.

74. Guiard, *Historia del Consulado,* vol. II, p. 339.

75. This interpretation is favored by Barahona.

76. Cited in Guiard, *Historia del Consulado,* vol. II, p. 21.

77. Ibid., p. 792.

78. Fernández Pinedo and Bilbao, "Auge y crisis," p. 163.

79. E. P. Thompson, *The Making of the English Working Class,* chap. 4.

80. See article by C. Echevarrieta in *Eco Bilbaíno,* May 5, 1865.

CHAPTER TWO. THE DEVELOPMENT OF THE MINING INDUSTRY

1. M. González Portilla mentions that the central government issued a decree ordering the application of the national mining legislation to Vizcaya in 1830. It is doubtful, however, that this decree was enforced. See *La formación de la sociedad capitalista en el País Vasco,* p. 24.

2. See García de Cortázar, *Vizcaya en el siglo XV,* p. 143; Caro Baroja, *Los vascos,* p. 185; and Lucas Aldana, "Descripción de la mina de hierro de Triano," p. 307.

3. García de Cortázar, *Vizcaya en el siglo XV,* p. 426.

4. Aldana, "Descripción," p. 307.

5. I. Goenaga, "El hierro de Vizcaya," p. 297; Aldana, "Descripción," p. 307; F. Elhuyar, "Estado de las minas de Somorrostro," p. 101. The restriction must have been informally enforced (i.e., no written edict or minicipal law backed it), since it flew in the face of the disposition of the *fuero* guaranteeing the right to exploit the mines to *all* Vizcayans.

6. Caro Baroja, *Los vascos,* p. 186.
7. Wilhelm von Humboldt, the German scientist and indefatigable voyager, visited the mines in 1801. See *Guillermo Humboldt y el País Vasco,* p. 249.
8. Elhuyar, "Estado de las minas," p. 101.
9. Ibid., p. 102.
10. Goenaga, "El hierro de Vizcaya," p. 298.
11. Ibid., p. 298; and Humboldt, *Guillermo Humboldt,* p. 249.
12. Humboldt, *Guillermo Humboldt,* p. 249.
13. Elhuyar, "Estado de las minas," pp. 101–102.
14. Fernández Pinedo and Bilbao, "Auge y crisis," pp. 170, 216, 224.
15. Elhuyar, "Estado de las minas," p. 102.
16. Moreover, these calculations do not take into consideration the production of other mining districts within Vizcaya. Yet, it is likely that other areas also supplied ore to the *ferrerías.*
17. Pinedo and Bilbao, "Auge y crisis," p. 224.
18. R. Uriarte Ayo, "El tráfico marítimo del mineral de hierro vizcaíno (1700–1850)," pp. 154–158.
19. Humboldt, *Guillermo Humboldt,* pp. 249–250.
20. Elhuyar, "Estado de las minas," p. 102.
21. M. González Portilla, "La industria siderúrgica en el País Vasco: del verlangssystem [*sic*] al capitalismo industrial," pp. 117–181.
22. Aldana, "Descripción," p. 308.
23. Uriarte, "El tráfico marítimo," p. 137.
24. A copy of the law is in the *Actas de las Juntas Generales del Señorío de Vizcaya* (1818), pp. 17–23, in the Archivo Histórico de la Diputación de Vizcaya (henceforth A.H.D.V.).
25. See the 1825 complaint of José Chavarri, a mine owner, in the Archivo General del Señorío de Vizcaya, in Guernica (henceforth A.C.J.G.), higher section, Mines, Reg. 3, L. 4, no. 2.
26. A.C.J.G., higher section, Mines, Reg. 3, L. 4, no. 1.
27. M. Montero, "La minería de Vizcaya durante el siglo XIX," p. 146.
28. The text of the law is in the *Actas de las Juntas Generales del Señorío de Vizcaya* (1827), pp. 28–47, A.H.D.V.
29. The report entitled "Memoria sobre las minas de hierro de Somorrostro propias del M. N. Señorío de Vizcaya," and dated Mar. 19, 1827, is in A.C.J.G., higher section, Mines, Reg. 1.
30. The partnership contract is found in the Archivo Histórico Provincial de Vizcaya (henceforth A.H.P.V.), among the protocols of the notary Josef Joaquín Barandica, L. 3266, Nov. 22, 1827. By the 1840s, after the departure of Chavarri and Llano, the company became a two-family affair involving the Ibarras, their brother-in-law Cosme Zubiría, and Mier and his son-in-law José Gorostiza. See A.H.P.V., J. P. Castañiza, L. 3356, and V. Uribarri, L. 3719. In 1854, after Mier and José Antonio Ibarra died, the company changed its name to Ibarra Bros. and Co.,

and its partners were Juan María and Gabriel María Ibarra, Zubiría, and Gorostiza. See A.H.P.V., M. Castañiza, L. 6037. Eventually, Gorostiza retired as an active partner, but remained a co-owner of all the company's mines. See A.H.P.V., S. Urquijo, L. 6256, no. 429.

31. Between 1829 and 1842, Ibarra Mier and Co. accounted for almost 75 percent of all the ore exports from Bilbao's harbor. See Uriarte, "El tráfico marítimo," p. 179, table 9.

32. The balance figures are taken from Gabriel María Ibarra's unpublished paper on José Antonio Ibarra's estate, Ibarra Family Archive (henceforth I.F.A.).

33. A.C.J.G., higher section, Mines, Reg. 3, L. 1, no. 6–3.

34. Aldana, "Descripción," p. 309.

35. A.C.J.G., higher section, Mines, Reg. 2, L. 2, no. 2.

36. For instance, the Ibarras came to an agreement with Francisca de los Heros over the Concha mine, which included within its perimeter Heros's Rosita mine: A.H.P.V., F. Uribarri, 1883, no. 73, pp. 312–327. Another example is the lawsuit that Juan Durañona pressed against José Gorostiza and Francisca de los Heros over the César mine. Durañona claimed that the registration of this mine had overlooked his rights to the Castaños mine: A.C.J.G., higher section, Mines, Reg. 2, L. 3, no. 5.

37. Both complaints in A.C.J.G., higher section, Mines, Reg. 3, L. 2, no. 6; and Reg. 2, L. 6, no. 9.

38. The percentage was calculated from a set of receipts for shipments during that year in I.F.A.

39. The list is reproduced as table I-1 in Appendix I.

40. The documents that support this were registered by the notary S. de Urquijo in L. 6256, no. 429, L. 6243, nos. 216–217, L. 6240, no. 4, L. 6239, nos. 352 and 379, L. 6238, no. 197; by the notary M. Castañiza in L. 6041, no. 100; by the notary C. Ansuátegui, L. 6291, no. 66; and by the notary F. Uribarri in his book of protocols for the year 1885, nos. 389–390, pp. 1764–1869, A.H.P.V.

41. The number is actually greater if one considers that the Buena Fortuna, Catalina, Indiana, Perseguida, San Miguel, and Trinidad mines were all included in the 1861 list (table I-1).

42. Size is therefore an imprecise variable to use to figure out when a mine was registered. Not only were mines like the Orconera and the Concha expanded several times, but others, like the Olvido, had areas smaller than the maximum allowed by the law when they were registered. As a result, Manuel Montero's analysis of the chronology for the ownership of the mines is clever but somewhat flawed. See "La minería de Vizcaya," pp. 148–151.

43. See Uriarte, "El tráfico marítimo," pp. 179–181.

44. A.H.P.V., M. Castañiza, L. 6076, no. 181. Arana and Bodega bought the mine in 1855 for 500 pesetas from José Villota.

45. A.H.P.V., R. Vildósola, L. 6661, nos. 6–7; L. 6662, no. 44.

46. A.H.P.V., P. Goicoechea, L. 5986, no. 21; and J. Ansuátegui, L. 5976, no. 380.

47. A.H.P.V., M. Castañiza, L. 6076, no. 181.

48. All of these production figures were extracted from the trimestral reports that appeared in the *Boletín Oficial de Vizcaya* (henceforth B.O.V.) between 1877 and 1900.

49. The mines were bought by the company's representative in Bilbao, José Martínez Rivas. The price for seven-tenths of the Unión was 52,500 pesetas. It is not clear how much he paid for the Amistosa, but it was mortgaged to assure a payment of 7,556 pounds (183,200 pesetas). See A.H.P.V., S. Urquijo, L. 6239, Oct. 14, 1872 and Nov. 5, 1872. If the output of these two mines is added to that of those on table 6, their percentage of total Vizcayan production reached 85 in 1880, 90 in 1885, 78 in 1890, 62 in 1895, and 56 in 1898.

50. A.H.P.V., J. B. Butrón, L. 6643, no. 225; and J. Ansuátegui, L. 5973, no. 2. The example of Vitoria Maruri and Suñol also shows the risks involved in the mining business. The company went bankrupt in 1883. In 1890, the Basque banking house of C. Murrieta and Co., based in London, paid approximately 20 million pesetas for the mine and some other properties related to its exploitation (A.H.D.V., Fondo del Corregimiento, L. 1506, no. 13). Eventually, Murrieta and Co. itself went bankrupt, dragged down by the financial crash of the early 1890s in London. As a result, the mine ended up in the hands of what appeared to be a British partnership, the Parcocha Iron Ore Co.

51. The two British businessmen paid thirty thousand pounds (around 750,000 pesetas) for a 26.8 percent share of the Catalina and Nuestra Señora de Begoña, and two thousand pounds for a 24.6 percent share of the San Miguel (A.H.P.V., F. Uribarri, L. 6393, nos. 48–49).

52. Overall, the company registered twenty-four mines, although not all were actually worked. In addition, the company rented a few concessions like the Juliana, Juliana II, and Aumenta, which were the most productive among all the mines it worked (A.H.P.V., M. Castañiza, L. 6065, no. 139, L. 7081, no. 387, and L. 6081, no. 634; and C. Ansuátegui, L. 6330, no. 221).

53. A.H.P.V., F. Uribarri, L. 6191, no. 57, L. 6192, no. 3, L. 6193, no. 267; S. Urquijo, L. 6246, no. 155. The partners—the mercantile society of José Antonio Uriguen; Luis Torres Vildósola y Urquijo, an engineer; Eduardo Coste Vildósola, a merchant; Carlos Aguirre y Labroche, a merchant; and Simón Ochandategui, an architect—had not been involved in the iron trade until then.

54. Uriguen Coste and Vildósola also obtained concessions in the periphery of Somorrostro, registering the Cerrillo and Moruecos mines in 1871.

55. A.H.P.V., S. Urquijo, L. 6246, no. 155; and M. Castañiza, protocol book corresponding to 1884, no. 357.

56. Aldana, "Descripción," pp. 387–388.

57. A.H.P.V., B. Onzoño, protocols corresponding to Sept.–Oct. 1886, no. 535. The mines listed were Aurora, Felicidad, San Antonio, San Pedro, San Blas, and San Andrés.

58. A.H.P.V., S. Urquijo, L. 6237, no. 173. The mines involved were the San Francisco and the Segunda. In the lease, Ibarra Mier reserved the right to buy up to

six thousand tons of ore from those mines at a fixed price, but with the understanding that they would use it to supply their factory and not resell it.

59. A.H.P.V., F. Uribarri, protocols of 1883, no. 35.

60. A.H.P.V., F. Uribarri, Feb. 7, 1885, nos. 46–47.

61. A.H.P.V., C. Ansuátegui, 2nd trimester 1885, no. 274. Lezama also owned less valuable mines in other districts. Overall, he owned shares in twenty-three mines, valued at 743,000 pesetas.

62. A.H.P.V., F. Basterra, L. 6130, no. 421; F. Uribarri, L. 6067, no. 281; and M. Castañiza, L. 6070, no. 199.

63. Registered in 1869, the mine was sold by its first owners, José Gutiérrez Vallejo and Ignacio Ubieta, to Trevor Berkley and Charles Liddel for 100,000 pesetas. In addition, the buyers agreed to pay a royalty of 0.375 pesetas for forty years for each ton of mineral extracted. Berkley and Liddel transferred the mine to Landore Siemens shortly after signing the contract (A.H.P.V., S. Urquijo, L. 6237, no. 18, and L. 6240, no. 131).

64. A good summary of Bessemer's invention and the old methods is given in D. Landes, *The Unbound Prometheus,* pp. 249–256. A more detailed and technical account is presented in J. C. Carr and W. Taplin, *History of the British Steel Industry,* pp. 1–30.

65. M. W. Flinn, "British Steel and Spanish Ore: 1871–1914," p. 85.

66. Although the trend toward an increase is indisputable, 1873 prices may be somewhat distorted because they probably reflect the fear caused by the outbreak of the Second Carlist War, just as the Bilbao ore fields were beginning to increase their exports. For the prices cited, see ibid.

67. On the yield of the British ore, see L. Bell, *The Iron Trade of the United Kingdom,* pp. 11–13.

68. In North America, the United States had an extensive deposit in the Lake Superior basin. Even with this plentiful supply, a sign of the competitiveness of the Bilbao minerals was the fact that some East Coast steel factories imported ore from Vizcaya.

69. *Journal of the Iron and Steel Institute* (henceforth J.I.S.) 1 (1874): 323. Italy, too, had hematite deposits on Elba. Although much smaller than those of Bilbao, the Italian mines did export some ore during the 1880s. Still, Bilbao's production costs were much lower than those of Elba. While in 1880, the price of mineral f.o.b. (free on board) in Elba was 8.8 French francs, in Bilbao it hovered between 6 and 4 francs. See J.I.S., no. 1, 1880; and E. Bourson, "Les mines de Somorrostro," p. 675.

70. Kilns to refine the *carbonatos* were introduced to the region in the early 1880s by José MacLennan. However, they were not used extensively until the 1890s, when several companies built their own. Washing was another method introduced in the mid-1890s to extract ore that until then had been thrown away as unusable. See Gill, "The Present Position," pp. 49–52; also J. Lazúrtegui, "La industria minera de Vizcaya," pp. 137–141.

71. See Appendix I, tables I-7 and I-8.

72. Archive du Ministère des Affaires Etrangères (hereafter F.C.R.), livre 10, 1875–1892, letter dated Feb. 10, 1882.

73. On Britain's reliance on basic steel (the output of the Bessemer converter or the Siemens-Martin oven), see Landes, *Unbound Prometheus*, p. 263.

74. The idea to build a railroad appeared in González Azaola's report in 1827 (see n. 29).

75. Still, the government had to overcome many difficulties, since the Spanish legislation that regulated the railroads did not recognize the provincial government's right to build lines. For this reason, the line was first entrusted to an agent of the Vizcayan government, Nicolás Urcullu Smith, who acted as a front man. In 1870, the Diputación was recognized as the official owner. See Ignacio Echevarría and Federico Grijelmo, *Las minas de hierro de la provincia de Vizcaya. Progresos realizados en esta región desde 1870 a 1899*, pp. 46–47.

76. See M. González Portilla, "El mineral de hierro español (1870–1914): su contribución al crecimiento económico inglés y a la formación del capitalismo vasco," pp. 78–79.

77. Montero, "La minería de Vizcaya," pp. 149–150.

78. González Portilla, "El mineral de hierro español," p. 79.

79. For a good analysis of transportation costs and their effects on the price of ore, see García Merino, *La formación de una ciudad industrial*, pp. 515–521. A contemporary French visitor also observed that the Triano Railroad charged relatively high tariffs, adding also that "if there had been some sort of agreement among all the mining companies, two railroad lines would have been more than enough to transport all the mineral to the river" (E. Gruner, "Barcelone-Bilbao, notes de voyages (Oct. 1888)," pp. 246–247.

80. In fact, prices averaged approximately 1.9 pesetas during the 1870–1890 period, and 1.4 pesetas during the 1890s. See García Merino, *La formación de una ciudad industrial*, p. 520.

81. Between 1881 and 1883, the average tonnage of the ships dedicated to the ore trade increased 35 percent. See Montero, "La minería de Vizcaya," p. 153.

82. A.H.P.V., S. Urquijo, L. 6244, no. 89. Another document showing their need for credit is the loan for 100,000 pesetas that they secured in 1872 from the merchant house of José Antonio Uriguen (A.H.P.V., F. Uribarri, L. 6194, no. 44).

83. A.H.P.V., M. Castañiza, L. 6087, nos. 265, 300, and L. 6091, no. 902.

84. Gruner, "Barcelone-Bilbao," p. 242.

85. In the 1870s, Bilbao had only two banks: a branch of the Banco de España, and the Banco de Bilbao (founded in 1857). Neither bank was permitted to make loans that used ore or mines as collateral. Both functioned as banks of emission (at least until the mid-1870s, when the privilege was monopolized by the Banco de España), and as clearinghouses for commercial notes.

86. The Mier family received one-fourth of the royalties under the terms of the lease of those mines, but they were not partners in the Orconera Iron Ore Co. For the mine leases, see A.H.P.V., S. Urquijo L. 6241, no. 196; Sept. 20, 1879.

87. A copy of the contract is in A.H.P.V., S. Urquijo, L. 6241, no. 146. The Orco-

nera promised to pay up to three thousand pounds for the plans and the surveys that the Ibarras had already made for the construction of the railway.

88. The Ibarras' stake in the Franco-Belge has sometimes been mistakenly assumed to be the same as their share in the Orconera. See, for example, González Portilla, *La formación de la sociedad capitalista,* vol. I, p. 41. Montero also states that the Ibarras owned 25 percent of the Franco-Belge, as he mentions in "La minería de Vizcaya," p. 165. A. Escudero in "La minería vizcaína durante la primera guerra mundial," p. 368, correctly states that they owned 10 percent, but errs in their share of the Orconera, which he believes was just 14 percent. The original partnership deed of the Franco-Belge found in I.F.A. shows that the Ibarras' stake was indeed only 10 percent. A donation made by Juan María de Ibarra and his wife to their children clearly shows that Juan María de Ibarra, Gabriel María Ibarra, and their brother-in-law Cosme Zubiría owned 25 percent of the Orconera. See A.H.P.V., F. Uribarri, Oct. 21, 1882.

89. A.H.P.V., M. Castañiza, L. 6039, Jan. 1, 1860. In 1866 the Ibarras bought Uribarren's shares: A.H.P.V., S. Urquijo, L. 6222, no. 87, March, 1866.

90. A.H.P.V., S. Urquijo, L. 6241, no. 139.

91. For years after the end of the Second Carlist War, practically all mining leases carried annulment clauses in case a new war broke out.

92. Some historians claim that the arrangement was not beneficial to the Ibarras, since all the ore was sold to the constituent partners below market price, 1 shilling 7 pence over the cost of production, in the case of the Orconera. However, as will be shown, not all the ore was sold to the partners. Moreover, the Ibarras also benefited from the below-market price, since they were guaranteed a quarter of production, which they used at their factory.

93. Historisches Archiv Fried. Krupp (henceforth K.A.), WA IV 1649. According to the balance sheet of the Orconera, the total capital expenditure up to December 1877 was 332,619 pounds. Of those funds, 290,000 pounds were spent in the construction of the railroad and inclined planes for the mines.

94. The sum was also roughly twice the amount of the largest estate left by a Bilbao merchant during the 1850–1875 period.

95. Despite this, the line went bankrupt in the 1860s, causing severe losses.

96. Actually, these companies subcontracted most of the extraction work to a large number of smaller enterprises.

97. Archivo Histórico del Banco de España (henceforth A.B.E.), Section: Branches, Bilbao, Box 39B.

98. González Portilla, "El desarrollo industrial de Vizcaya," pp. 71–72, estimates that the foreigners' share was around 40 percent in the 1876–1890 period, and 48 percent in the 1890s. However, Montero, "La minería de Vizcaya," p. 165, calculates that the foreigners' share during the 1890s was also around 40 percent. My own calculations tend to agree with Montero's.

99. The partnership contract with C. Jacquet, dated July 16, 1863, in A.H.P.V., M. Castañiza, L. 6044, no. 234.

100. In 1883, he formed a partnership with C. Jacquet and Francisco Arana to

exploit the latter's sawmill and to import coal (A.H.P.V., M. Castañiza, book of protocols, Sept.–Oct. 1883, no. 823).

101. MacLennan left an estate valued at 9,850,000 pesetas. Unfortunately, the documents available do not indicate whether he had investments in other countries. This information and the details about his burial are in A.B.E., Section: Branches, Bilbao, Box 49.

102. This position is best illustrated by a classic book on Spanish economic history, Nadal's *El fracaso.* Recently, however, several Spanish historians, like Gabriel Tortella and Leandro Prados, have departed radically from this position.

103. Still, F. Sánchez Ramos's *La economía siderúrgica española,* pp. 270–272, following the works of Basque engineer Pablo Alzola, complains that more than 90 percent of the production of the Somorrostro district was in the hands of English companies. Nadal also believes that foreigners took the lion's share, but that local businessmen still managed to amass vast fortunes from their minority interests. See *El fracaso,* p. 119.

104. Escudero, "La minería vizcaína," p. 368.

105. The original contract signed in 1871 was actually even more favorable to Uriguen Vildósola, since it gave the company a higher royalty on the mines that it had leased to the British company. In 1884, the contract was renegotiated and it was agreed that the royalty would be three pennies both for the ore extracted from the leased mines and the minerals transported from other parties. See A.H.P.V., F. Uribarri, L. 6192, no. 3; and M. Castañiza, book of protocols, Mar.–Apr. 1884, no. 357.

106. According to González Portilla, the British company leased the mines to a local steel manufacturer for a royalty of 0.5 pesetas per ton, guaranteeing a minimum payment of 55,000 pesetas. See *La siderurgía vasca (1880–1901),* p. 114. Interestingly, the royalty of 0.5 pesetas was below the amount Bilbao Iron had agreed to pay Uriguen Vildósola. So actually the British company was losing money on the lease, but presumably it must have recovered part of the loss by assuring a steady flow of mineral for its railway, since La Vizcaya agreed to pay penalty charges if it failed to meet a minimum annual quota.

107. Moreover, it has been assumed that this company was completely owned by foreigners. Yet, the estate of the widow of Juan Antonio Uriguen included twenty-six shares (with a nominal value of fifty pounds each) in the firm. See A.H.P.V., F. Uribarri, L. 6205, no. 37.

108. See C. Harvey and P. Taylor, "Mineral Wealth and Economic Development: Foreign Direct Investment in Spain, 1851–1913," p. 196.

109. A.H.P.V., C. Ansuátegui, protocol book for May–June 1886, no. 360. Edwards leased the three mines to J. MacLennan in 1886 for ten years.

110. See trimestral reports in B.O.V.

111. A.H.P.V., R. Vildosola, L. 6668, no. 154.

112. Kreizner could have been an agent of the company, although this is not verified.

113. The lease contracts are in A.H.P.V., C. Ansuátegui, protocol book for May–June 1886, no. 360, and L. 6329, no. 7; R. Vildósola, L. 6665, no. 3, L. 6664, no. 118, and L. 6661, no. 7; S. Urquijo, L. 6256, no. 377.

114. Escudero, "La minería vizcaína," p. 368.

115. Fernández Pinedo, "La industria siderurgica, la minería y la flota vizcaína a fines del siglo XIX. Unas puntualizaciones," p. 167.

116. Bell, *Iron Trade,* p. 75.

117. In the case of the Franco-Belge, the agreed 1.25 francs per ton were practically the same as the top, one-shilling, royalty cited by Bell, *Iron Trade,* p. 75.

118. The exchange rates between the peseta and the British pound are taken from the *British Parliamentary Papers* (hereafter B.P.P.)

119. A.H.P.V., S. Urquijo, L. 6246, no. 200. Martínez Rivas guaranteed a minimum payment corresponding to twenty thousand tons, which had to be paid whether the mineral was extracted or not. But for any amount exceeding the first forty thousand tons in any given year, Martínez Rivas had to pay only 1 peseta.

120. A.H.P.V., C. Ansuátegui, protocol book for the fourth trimester of 1884, Oct. 9, 1884.

121. A.H.P.V., S. Urquijo, L. 6256, Sept. 29, 1880. According to a chart in I.F.A., the royalties paid during the 1888–1899 period averaged 0.83 pesetas, but during six of those years, the prevailing rate was 0.75 pesetas per ton.

122. A.H.P.V., F. Uribarri, July 30, 1885. The mines leased were San Miguel, Begoña, Indiana, Esperanza, Buena Fortuna, Catalina, Trinidad, Ser, Bilbao, Cristina, Mingolea, Antón, Presentación, Asunción, and San Antonio.

123. The contract and information about the exploitation of this mine are in I.F.A.

124. The 1879 contract is in A.H.P.V., R. Vildósola, L. 6665, no. 3. Although the contract established a 1.25-peseta royalty, this was the rent paid on the first twenty thousand tons. For the second twenty thousand tons, the royalty was set at one peseta; anything exceeding forty thousand tons paid only 0.75 pesetas per ton. The second contract is in I.F.A.

125. As mentioned earlier, the Ibarras' biggest mistake concerned the royalties they agreed to for the transport of ore on the Orconera railway. In effect, they forwent all royalties from the mineral transported from mines not served by the Orconera. Uriguen Vildósola negotiated a much better contract with the Bilbao River and Cantabrian Railway; see pp. 91–92 here.

126. Some historians, like Fernández Pinedo, argue that the Orconera's profits were low because the company's purpose was to provide cheap ore to its partners. While part of the argument is true, the Orconera still sold important quantities of ore at market prices to outsiders. See App. I, table I-4, for a breakdown of the company's profits. In any case, as partners of the Orconera and stockholders of an important steel factory in Bilbao, the Ibarras also benefited from the company's policy of maintaining low prices.

127. González Portilla, "El desarrollo industrial," p. 60.

128. García Merino, *La formación de una ciudad industrial,* pp. 510–514. The tax was assessed on the price of the ore at the mine site. The higher the price, the greater the revenue for the government.

129. Fernández Pinedo, "La industria siderúrgica," p. 160.

130. The point is well illustrated in the case of the San Bernabé mine, worked by the Franco-Belge. As this company informed its stockholders, the transport of mineral from that mine, which could not be served by its own railroad, cost the Franco-Belge 3.12 francs per ton in 1889. In contrast, mineral transported on its own line cost the company only 0.54 francs per ton (I.F.A.).

131. For the Franco-Belge, see Gruner, "Barcelone-Bilbao," p. 251. For the Orconera, see mid-year balance sheet for 1883 and summary of costs for 1897–1902 period in I.F.A.

132. A.H.P.V., F. Uribarri, 1884 protocols, no. 9., ff. 33–37.

133. I calculated this amount in the following manner. According to Goenaga ("El hierro de Vizcaya," p. 329), the transport cost per ton of the aerial ropeway from the mines to the Triano Railroad was 0.73 pesetas. The approximately four kilometers left to the port must have cost Martínez Rivas around 1.40 pesetas (0.345 pesetas was the tariff of the Triano Railroad per ton and per kilometer. See García Merino, *La formación de una ciudad industrial,* p. 516).

134. A.H.P.V., S. Urquijo, L. 6254, no. 102.

135. A.H.P.V., M. Castañiza, L. 6091, no. 902.

136. A.H.P.V., J. Ansuátegui, L. 6355, no. 61.

137. See "Extracto de una memoria remitida al excmo. sr. ministro de negocios extranjeros en Stockolmo por Alfred Kirsebom," in Archivo de la Cámara de comercio de Bilbao (hereafter A.C.C.B.), section: company reports, no. 260.

138. The 35 percent does not include the Unión and Amistosa mines, but does include the other mines worked by Martínez Rivas.

139. For example, the Sociedad Vizcaya agreed to pay a tariff of 3.75 pesetas per ton to transport ore from the Tardia, Cenefa, Escarpada, and Berango mines (all in the Galdames district) to the river. See González Portilla, *La siderurgia vasca,* p. 114.

140. This becomes immediately apparent if one compares García Merino's prices, especially during the 1890s, with González Portilla's. The latter cites prices by type of ore. See "El mineral de hierro español," p. 98.

141. Here my calculation was based on the estimate for 1884. In 1883, production and price were 10 percent higher than in the following year. Thus, the 1883 profit was roughly 11.64 million. In 1882, production and price were about equal to those in 1883. In 1881, the price was the same but production was only three-fourths that of 1882. Hence the yearly gains were around 8.73 million pesetas. The 1880 production was similar to that of 1881, but the prices were 20 percent higher. The profits, then, were 10.5 million. In 1879, production was about half that of 1880 and price was 33 percent lower; profits must have been around 3.5 million pesetas. In 1878, production was nearly the same as in 1879, but prices were 14 percent higher; hence profits were about 4 million. Finally, in 1877, pro-

duction was 23 percent lower than in 1878 and price was about the same. Therefore, profits were around 3 million.

142. García Merino correctly assumes that the royalties should not be considered as a cost but as a profit for the mine owners. The figures for the Orconera and the Franco-Belge, however, include the royalties as a cost. The 30.5 million pesetas was obtained by adding the Orconera's and the Franco-Belge's royalty payments. Table I-2 in App. I shows that the Orconera's total royalty payment was 21.5 million pesetas. I estimated that, since the Franco-Belge's royalty payment approximately equaled that of the Orconera, and since the former company's production was slightly less than half of that of the Orconera, the Ibarras must have received around 9 million pesetas in royalty payments; hence the 30.5 million pesetas.

143. The figure is also lower than González Portilla's estimate, which was 559 million for the period 1876–1900. See *La formación de la sociedad capitalista,* p. 68.

144. Taking the Ibarras as a model, it is not unrealistic to claim that royalty payments could have reduced profits by one-third. Consider the 30 million the family obtained from the Orconera and the Franco-Belge, whose combined profits were around 95 million pesetas.

145. I. Echevarría and F. Grijelmo, *Las minas de hierro de la provincia de Vizcaya,* p. 51.

146. My calculations of service for other mining companies of the Orconera's and the Franco-Belge's railroads are based on the statistics published by Gruner, "Barcelone-Bilbao," p. 276, and the B.P.P. 1901. Knowing the total volume each railway transported, I obtained the percentage carried for other producers by deducting the Orconera's and the Franco-Belge's yearly production.

147. In the Orconera's and the Franco-Belge's estimates, the mineral carried from the respective mines of these two companies was included because it was considered to be a cost and therefore was already computed in the profits obtained from the sale of ore.

CHAPTER THREE. GENERAL ECONOMIC DEVELOPMENT

1. See, for example, comments by novelist Benito Pérez Galdós after a visit at the turn of the nineteenth century, in "Fisonomías sociales, Bilbao," pp. 27–38. For Galdós, the entrepreneurial activities of the city's inhabitants had as much to do with its development as did its iron deposits.

2. See, for example, A. Carrera de Odriozola's index of industrial production in "La producción industrial catalana y vasca 1844–1935. Elementos para una comparación." Although this index deals with the three Basque provinces as a unit, it is still useful as a broad measure of Vizcayan development, since Vizcaya had by far the largest economy among the three provinces. The index clearly shows accelerated growth from the mid-1870s until the 1900s.

3. See F. Mendels, "Proto-industrialization: The First Phase of the Industrialization Process."

4. The economic history of the interwar period remains terra incognita except for a few studies such as M. Basas's collection of articles, *Aspectos de la vida económica de Bilbao 1861–1866, Economía y sociedad bilbaína, El Lloyds bilbaíno hace un siglo,* and a book by J. Aguirreazkuenaga, *Vizcaya en el siglo XIX: las finanzas públicas de un estado emergente.* The iron industry is the best-studied topic of the period thanks to several articles by Fernández Pinedo and Bilbao.

5. For problems concerning Vizcayan statistics and especially the survey planned in 1864, see Aguirreazkuenaga, *Vizcaya en el siglo XIX,* pp. 356–372. In the words of this author, "what might have been the first modern statistics of Vizcayan economic activities ended as a new failure due to the constant boycott of the local authorities and commissions" (p. 371).

6. For the rest of Spain, Nadal has cleverly used fiscal sources (especially, the so-called *contribución industrial*) to assess the industrial capacity of the different regions. See "La industria fabril española en 1900. Una aproximación," pp. 23–61.

7. The mercantile register was established in Spain by the Commercial Code of 1829. In all of Spain, except the Basque region, the office of the register was administered by the central government. Due to their peculiar autonomy, the Basque provinces administered their register through local institutions. In Vizcaya, the Diputación was in charge of this task. I have found references in notarial documents in the A.H.P.V. to the existence of this register, where not only companies were recorded, but also the dowries of merchants' wives. Unfortunately, efforts to locate it in A.H.D.V. were not successful.

8. P. Madoz, *Diccionario geográfico, estadístico e histórico de España y sus posesiones de ultramar,* vol. 15, p. 402.

9. Madoz tended to believe that the higher 1826 estimate, which came from a police census, was the most accurate, while he doubted the early 1830s figures.

10. For the growth of the Spanish population in general, see Nadal, *El fracaso,* p. 12. In contrast, during the eighteenth century, Vizcaya's population doubled while for Spain in general the population grew only by a factor of 1.34 (ibid., p. 19).

11. Madoz, *Diccionario,* vol. XVI, p. 408.

12. Luna's figures are taken from Aguirreazkuenaga, *Vizcaya en el siglo XIX,* pp. 232–233.

13. Ibid., p. 234.

14. See *Anuario Estadístico de España,* 1858.

15. There are also problems with the professional breakdown of this census. While the number of industrial workers for the whole province is put at 684, other sources mention that the two biggest iron factories in the region alone employed 800 workers in the early 1860s. See García Merino, *La formación de una ciudad industrial,* p. 766; and J. Delmas, *Guía histórico descriptiva del viajero en el señorío de Vizcaya en 1864,* p. 374. In addition, according to the French consul, in the 1850s, there were 2,000 miners, a category that is conspicuously absent from census classifications (F.C.R., 1856). It is possible that some factory workers were included under the rubric of artisans, and the miners were considered agricul-

tural workers, since it may still have been a part-time occupation. The census figures appear in B.O.V., Nov. 16, 1861.

16. *Censo de población de Bilbao tomado del padrón que de su vecindario han formado los señores alcaldes de barrio en marzo de 1869.*

17. See Fernández Pinedo, "Estructura de los sectores agropecuarios y pesqueros vascos (1700–1890)," p. 103.

18. See G. Tortella, *Los orígenes del capitalismo en España,* pp. 100–101.

19. See J. Nadal, "Industrialización y desindustrialización del sureste español, 1817–1913."

20. According to Madoz, while the transfer of customs collection to the seacoast opened opportunities for the big merchants, it harmed the small retailers. See *Diccionario,* vol. XVI, p. 386.

21. B.P.P., 1855 report.

22. The analysis of Bilbao's foreign trade is based on the B.P.P. for 1855–1870, the F.C.R. for 1850–1858, and the *Estadística de Comercio Exterior* for 1858, 1861, and 1862.

23. Unfortunately, I could not consult the complete series of Spanish foreign trade (*Estadística de Comercio Exterior*). The other source for these years, F.C.R., does not distinguish between Bilbao's international and national trade. In any case, all these statistics present a high degree of incertitude.

24. B.P.P., 1872 report.

25. This, at least, was the impression of the British consul in 1859, when he analyzed the future of both ports.

26. Figures taken from *Anuario Estadístico de España,* 1858. In contrast, Santander's fleet totaled only 20,200 tons.

27. Archivo Histórico de la Marina (henceforth A.M.), L. 7102.

28. According to documents at the A.M., in 1858 there were seven shipyards on the Nervión banks (L. 62308).

29. A.M., L. 7102.

30. F.C.R., 1859 report.

31. On the predominance of British shipyards in the world market once the use of iron steamers became widespread, see S. Pollard and P. Robertson, *The British Shipbuilding Industry, 1870–1914,* p. 38.

32. Madoz, "Bilbao," *Diccionario,* vol. IV, p. 326.

33. A.M., Estadística General, 1858, L. 62306.

34. A.H.P.V., S. Urquijo, L. 5841, Oct. 26, 1850. Basas mentions that an insurance company by the name of La Unión had existed earlier and was dissolved in 1848. See *El Lloyds.*

35. Basas, *El Lloyds,* p. 68.

36. The inventory after the death of José Briñas, a wealthy merchant, exemplifies this practice (A.H.P.V., S. Urquijo, L. 5848, ff. 214–314). Briñas owned shares in nine ships, ranging from one-half to one-sixteenth of each of those vessels.

37. On the 1855 legislation and its influence on the creation of banks and railroads, see Tortella, *Los orígenes.*

38. It continued to enjoy this privilege until 1875, when the Bank of Spain monopolized the right.

39. The bank lost its right to issue money in 1873, but because of the Second Carlist War, the minting monopoly assigned to the Banco de España was not enforced until 1875 in Bilbao.

40. See B.P.P., 1859 report, "Report by Mr. Young, British Consul at Bilbao, on the relative importance of the Ports of Bilbao and Santander."

41. Actually, the total cost of the project was estimated at 55,250,000 pesetas, but the local and national governments granted subventions for approximately 21 million pesetas; bonds were to provide 9.5 million pesetas. See A.H.P.V., S. Urquijo, L. 5849, ff. 12–30.

42. The list of the first board of directors and first stockholders reads like a who's who of the most important Bilbao merchants. See ibid. The inventories after death of the notarial archive also show the widespread distribution of the railway's stocks among the general population.

43. B.P.P., 1860 report.

44. See chap. 4.

45. See Palacio, *El comercio de Castilla,* pp. 143–44. Even some flour mills moved from Bilbao to Santander to take advantage of the direct trade of the latter port with America.

46. A brief description of these industries is given by Aguirreazkuenaga, *Vizcaya en el siglo XIX,* pp. 147–170.

47. Madoz, "Vizcaya," *Diccionario,* vol. XVI, p. 381.

48. Aguirreazkuenaga, *Vizcaya el siglo XIX,* p. 146.

49. In addition to Máximo Aguirre's mill, cited below in the text, other large factories belonged to Eugenio Aguirre in Arrigorriaga (see A.H.P.V., I. Ingunza, L. 5798, no. 129, and F. Uribarri, L. 6187, no. 8), Romualdo García in Galdacano (A.H.P.V., C. Ansuátegui, L. 6277, no. 216), and the Berge family in the outskirts of Bilbao (A.H.P.V., B. Onzoño, L. 6007, ff. 320–323).

50. Madoz, "Bilbao," *Diccionario,* vol. IV, p. 320.

51. A.H.P.V., F. Basterra, L. 6117, no. 177. Aguirre's mill also included a textile factory that produced jute sacks to package the flour. In 1874, when the business was managed by Aguirre's son-in-law, Eduardo Coste y Vildósola, the two factories were valued at 789,250 pesetas. See A.H.P.V., C. Ansuátegui, L. 6311, no. 234. In 1864, Aguirre had also built another flour mill in the province of Burgos in partnership with Pedro Zaraúz, which was assessed at 390,000 pesetas. See A.H.P.V., S. Urquijo, L. 6219, no. 147.

52. It is possible that the factory rented was that mentioned by Madoz as being located in Abando. The rental usually represented around 5 percent of the value of the factory. See A.H.P.V., S. Urquijo, L. 6229, no. 177; and F. Basterra, L. 6119, no. 179.

53. A.H.P.V., F. Basterra, L. 6117, no. 140, L. 6119, no. 179, L. 6131, no. 614; and S. Urquijo, L. 6229, no. 177.

54. Aguirre appears on a list of the most important merchants, industrialists,

and property owners drafted to select members for a provincial economic board (the Junta de Agricultura, Industria y Comercio). Other flour producers, such as Baliscueta, Berge, and Eugenio Aguirre, also appeared on it. See B.O.V., Oct. 1, 1861. The list is also published in Basas, *Aspectos de la vida económica,* pp. 21–26.

55. For instance, Eduardo Coste and one of Máximo Aguirre's sons obtained the best mining concessions in the Galdames district. See chap. 2.

56. See Mendels, "Proto-industrialization," pp. 245–246.

57. Landes, *Unbound Prometheus,* p. 91.

58. L. Bilbao, "La siderurgia vasca, 1700–1885. Atraso tecnólogico, política arancelaria y eficiencia económica," pp. 81–93; idem, "Renovación tecnológica y estructura del sector siderúrgico en el País Vasco durante la primera etapa de la industrialización (1849–1880). Aproximación comparativa con la industria algodonera de Cataluña," pp. 211–237.

59. See Nadal, "Industrialización y desindustrialización."

60. Bilbao, "La siderurgia vasca," pp. 86–87.

61. The chemical composition of the Ollargan ore adapted better to the charcoal-fueled blast furnaces than did the Somorrostro mineral.

62. Fernández Pinedo, "Nacimiento y consolidación de la moderna siderurgia vasca (1849–1913): el caso de Vizcaya," p. 10.

63. A.H.P.V., M. Castañiza, L. 3358, June 6, 1846. In the original contract, the Ibarras acted as part of the firm Ibarra Mier and Co., which handled their mines and ore trade. The other partners were Mariano Vilallonga, Andrés Gutiérrez de Cabiedes, and Carlos Dupont. The date for the Ibarras' involvement in this enterprise is sometimes mistakenly given as 1827. This last date, however, signals the formation of the ore trading company of Ibarra Mier and Co. There is no indication that the Ibarras were involved in manufacturing prior to 1846.

64. W. Isard, "Some Locational Factors in the Iron and Steel Industry since the Early Nineteenth Century," p. 203.

65. For the development of the Asturian coal fields and iron factories, see G. Ojeda, *Asturias en la industrialización española, 1833–1907.*

66. Fernández Pinedo, "Nacimiento y consolidación," p. 12.

67. The partnership contract and references to the Vasco-Riojana Co. are in A.H.P.V., F. Uribarri, L. 6185, nos. 5, 8, 18. The Ibarras invested 130,000 reales, and their brother-in-law, C. Zubiría, another 60,000 reales, giving them 19 percent of the partnership.

68. Fernández Pinedo, "Nacimiento y consolidación," p. 14.

69. The investment figures for 1866 are taken from the Cortes' inquiry of 1866, reproduced in García Merino, *La formación de una ciudad industrial,* pp. 765, 773. Santa Ana's initial investment capital is taken from "Sociedad Santa Ana de Bolueta. Centenario de su fundacion," a brochure I found in the Diputación library. The figures given by the companies in 1866 do not match those quoted by Tortella in *Los orígenes,* p. 236. Tortella's figures reflect the capital that appeared in the partnership deeds (in the case of Ibarra and Co., 1.5 million pesetas, and in Santa Ana's, 200,000 pesetas, but not actual investment.

70. A.H.P.V., S. Urquijo, L. 6218, no. 192; Delmas, *Guía histórico descriptiva,* p. 374, describes the factory.

71. One of the company's books with Zarraoa's explanations is catalogued under "Actas sobre fábrica con fuerza motriz de vapor para artículos de hierro dulce, 1862–1871," in A.H.D.V., Armario 14, 20. Although Zarraoa's reasons for moving the factory to Bilbao were sound, his timing was bad. The recession of the mid-1860s and the political unrest that started with the Revolution of 1868 and culminated with the Second Carlist War forced the factory into liquidation.

72. Nadal, "La economía española, 1829–1931," p. 367.

73. A. Gómez Mendoza, "Los ferrocarriles y la industria siderúrgica (1855–1913)." There is an English version of this article in P. O'Brien, *Railways and the Economic Development of Western Europe, 1830–1914,* pp. 148–169.

74. See Nadal, *El fracaso,* App. 6. Gómez Mendoza uses these same figures, taken from the Estadística Minera, in his calculations.

75. They needed to produce approximately 205,000 tons: 155,000 tons for the railroad companies and another 50,000 tons to meet the rest of the national demand.

76. The underutilized capacity of the Basque factories is shown in the Cortes' inquiry of 1865. A published copy of the responses of Santa Ana and Nuestra Señora del Carmen is found in García Merino, *La formación de una ciudad industrial,* pp. 765, 769.

77. For the Asturian factories, see Ojeda, *Asturias,* esp. Pt. II, chap. II.

78. Tortella has also argued that a slower rate of railroad construction might have been beneficial for the economy. See *Los orígenes,* and "La economía española, 1830–1900," p. 81.

79. In the United States, the estimate of pig iron used by the railroad as a proportion of total production hovered between 5 and 21 percent before the Civil War. After the war, the railroads had a greater impact on the steel industry, requiring between 50 and 87 percent of the total output. In England, during the period 1835–1869, the proportion of pig iron used by the railroads reached 13 percent, according to the highest estimate. In France, it was between 12 and 18 percent for the period 1845–1885. See O'Brien, *Railways,* table 1.3, p. 16.

80. Santa Ana and Zarraoa and Co. provided some of the iron tubes used for this project. Zarraoa also supplied the tubes for a similar project in Burgos (A.H.P.V., L. 6232, no. 199; A.H.D.V., Armario 14, 20).

81. A.H.P.V., M. Castañiza, L. 6035, no. 138.

82. Ibid., L. 6042, no. 59. In addition to the needs of the private weapons industry, state arsenals also provided an outlet for the iron factories during those years.

83. González Portilla, *La formación de la sociedad capitalista,* vol. II, p. 154.

84. V. Pérez Moreda, "La modernización demográfica," p. 32.

85. For population figures for the Basque provinces, see A. García-Sanz Marcótegui, "La evolución demográfica vasca en el siglo XIX," p. 20.

86. A. García-Sanz Marcótegui, "El origen geográfico de los inmigrantes y los

inicios de las transición demográfica en el País Vasco (1877–1930). Contribución al estudio de sus interinfluencias," pp. 194–195, tables 4–5.

87. González Portilla, "Los orígenes de la sociedad capitalista en el País Vasco," p. 112. Baracaldo, on the banks of the Nervión River, was the site of Nuestra Señora del Carmen.

88. González Portilla, *La formación de la sociedad capitalista,* vol. II, pp. 170–172; García-Sanz Marcótegui, "El origen geográfico," p. 202.

89. See Olabarri, *Relaciones laborales,* pp. 447–448, tables 1 and 2.

90. Ibid.

91. See Pérez Moreda, "La modernización demográfica," p. 57.

92. The figures from the Mercantile Register are taken from Nadal, *El fracaso,* App. 4.

93. See M. Montero, "Modernización económica y desarrollo empresarial en Vizcaya 1890–1905," p. 234, n. 24.

94. Ibid., pp. 225–253.

95. Montero claims that in 1899 the Bilbao stock market was trading shares for a total nominal value of 139 million pesetas. The smaller figure reflects the fact that not all regional companies raised capital in the local stock market. See "Modernización económica," pp. 234–235. The consul's figures are in B.P.P., report for 1900. His figures are similar to those recorded by the Mercantile Register for the same years.

96. One hundred to 150 million pesetas was the total investment estimated in a report from 1887 found in A.C.C.B., L. 128. See also the discussion later in this chapter.

97. The presence of capital from Cuba and Puerto Rico after 1898 seems to have added to the local resources. Joseph Harrison has stressed this factor, but this remains a poorly studied question. See his "Los orígenes del industrialismo moderno en el País Vasco," pp. 212–213, and "La industria pesada, el estado y el desarrollo económico en el País Vasco, 1876–1936," p. 24, n. 15.

98. González Portilla clearly shows this when he talks about the foundation of Altos Hornos de Bilbao and La Vizcaya in 1882. Olano Larrínaga's share of La Vizcaya was approximately 25 percent of the total capital. See *La siderurgia vasca,* pp. 24–25, 41.

99. A good example is the merchant house of Gurtubay, which participated in the creation of Altos Hornos de Bilbao. The Uriguens also had a mercantile history before they invested in mines in the Galdames district and in Altos Hornos.

100. See Fernández Pinedo, "La industria siderúrgica," p. 166. References sustaining the participation of the cited businessmen in mining appear throughout chap. 2. Other stockholders in Bilbaína de Navegación, such as Eduardo Aznar and Federico Echevarría, also invested in mining during the 1880s.

101. Linares sold the land for twenty-eight thousand British pounds (686,000 pesetas), 5,000 to be paid in shares in the company (representing 8.3 percent of total capital), and the rest in cash (A.H.P.V., S. Urquijo, L. 6239, no. 424).

102. Ibid., F. Uribarri, L. 6204, no. 137. The British company either broke even or lost money in this transaction, since to the 686,000 pesetas paid to Linares must be added what had been invested in the construction of the factory until the moment of the sale.

103. González Portilla, *La siderurgia vasca,* p. 21; Nadal, *El fracaso,* p. 178.

104. González Portilla, *La siderurgia vasca,* pp. 22–29, 33. For the transfer of rights from Orconera and Franco-Belge, A.H.P.V., F. Uribarri, L. 6397, nos. 373, 374, Dec. 2, 1882.

105. A.H.P.V., F. Uribarri, L. 6193, Dec. 15, 1871.

106. González Portilla, *La siderurgia vasca,* pp. 41, 50.

107. Altos Hornos took advantage of the Orconera Railroad, La Vizcaya had a contract with the Galdames Railroad, which carried the ore to the factory. Rivas helped finance the extension of the railroad of Alonso Hnos. Uhagón and Co., which carried the ore from the San José and Perseguida mines to the Triano line, which then transported it to the San Francisco factory (A.H.P.V., F. Uribarri, Jan. 10, 1884, ff. 33–37).

108. Nadal, *El fracaso,* pp. 140–142. During the 1890s, English coke would become more expensive than Asturian because of the severe devaluation of the peseta. However, the quality of the English coke still made it preferable to the Asturian mineral. See P. Fraile, "El carbón inglés en Bilbao: una reinterpretación."

109. Nadal, *El fracaso,* p. 181.

110. See Isard, "Locational Factors," pp. 210–212.

111. Tortella, "La economía española," p. 76.

112. *Journal of the Iron and Steel Institute,* vol. 31, no. 2 (1887): 399.

113. Fernández Pinedo, "La industria siderúrgica," p. 156.

114. For Fernández Pinedo's criticism, see ibid., pp. 152–153. The percentage of La Vizcaya's production exported was calculated from González Portilla's *La siderurgia vasca,* p. 168, table 5-5. Comparing Fernández Pinedo's figures from the E.M.M.E. (Estadística Minera y Metalúrgica de España) for the export of iron bars with those of González Portilla taken from A.H.P.V., one can see serious underestimations in the E.M.M.E. For instance, the figure for 1895 is only 21,936 tons, while according to La Vizcaya's documents, this factory alone exported 30,201 tons of iron bars.

115. I calculated these figures from the B.P.P. for the years 1887–1900.

116. For the politics behind the protectionist fight, see chap. 6.

117. In 1888, La Vizcaya's profit margin on the sale of iron bars in Spain was 6.37 pesetas per ton, whereas it was only 2.42 pesetas per ton on sales to Italy. See González Portilla, *La siderurgia vasca,* p. 196, table 7-3.

118. A convincing argument that Vizcaya's excessive concentration on the Spanish market during the early decades of the twentieth century hindered its growth when compared to similar sectors of the Italian and Swedish economies can be found in P. Fraile, "El País Vasco y el mercado mundial, 1900–1930," pp. 226–251.

119. Capital investment in the lines: Bilbao's urban tramway, 0.45 million pesetas (A.H.P.V., J. Ansuátegui, May–June 1884, no. 347); Bilbao-Santurce tramway,

2.5 million pesetas (A.H.P.V., J. Ansuátegui, Mar.–Apr. 1883, no. 251); Bilbao-Durango-Zumárraga line, 2.85 million pesetas (A.H.P.V., S. Urquijo, L. 6255, ff. 1169–1170 and B.O.V., May 30, 1884); Bilbao-Portugalete line, 1.32 million pesetas (A.H.P.V., F. Uribarri, May 10, 1884, nos. 138–139); Bilbao–Las Arenas line, 2.5 million pesetas (A.H.P.V., F. Uribarri, L. 6205, no. 86); Bilbao-Guernica line, 1.25 (Archivo del Banco de Bilbao [hereafter A.B.B.], E-1, copy of the partnership contract).

120. Abando's official incorporation into the Bilbao municipality took place in the late 1860s. Its first construction boom occurred in the early 1860s, when the Bilbao-Tudela line located its terminus within Bilbao's jurisdiction. During the 1880s, the pace of construction accelerated, as many wealthy and middle-class Bilbaínos fled the crowded conditions of the old center of town.

121. V. Shaw, "Exportaciones y despegue económico; el mineral de hierro de Vizcaya, la región de la ría de Bilbao y algunas de sus implicaciones para España," p. 102.

122. Montero argues that businessmen with interests in mining and metallurgy did not finance most of the railroad construction of the 1880s, with the exception of the Bilbao-Portugalete line. See "Modernización económica," p. 229. For the supply of rails, see A. M. Ormaechea, "Los ferrocarriles vascos y su dependencia tecnológica," pp. 135–149. Unfortunately, this article has a confusing chronology, which makes it difficult to determine how much of the rail market was provided by the local factories.

123. See G. Tortella, "La primera gran empresa química española: la Sociedad Española de la Dinamita," pp. 433–434.

124. A.H.P.V., F. Uribarri, Sept. 22, 1884, no. 290.

125. By 1897, there were three more companies producing explosives in Vizcaya: the Vasco-Asturiana, La Manjoya, and Explosivos Burcena. That year, these three joined the Sociedad Española de la Dinamita and other explosives factories in Spain to create a monopoly within the country under the name Unión Española de Explosivos. See Ignacio Villota, *Vizcaya en la política minera española,* p. 273, n. 9.

126. A.H.P.V., S. Urquijo, L. 6249, no. 254, L. 6254, no. 92, and L. 6332, no. 96.

127. I calculated the 1885 figure from the A.M., estadística de navegación y pesca, L. 2369. This source provides a list of all the Spanish merchant ships according to the harbor where they were registered, and also gives the date of registration. The ninety thousand tons includes only those ships that were registered in Bilbao up to 1885.

128. To arrive at the 30.3-million-peseta estimate, I assumed that a steamer built in 1875 with a total tonnage of 1,000 was being sold in 1879 for around 20,000 pounds in Liverpool, as several contracts from the Libro de Actos Públicos del Consulado Español en Liverpool indicate. This record is kept in the Archivo de Protocolos de Madrid (hereafter A.P.M.), L. 35159. Since 1,341 tons was the average tonnage of the approximately sixty-eight steamers of the Bilbao fleet, the 20,000-pound (505,000-peseta) figure is a conservative estimate.

129. Unsuccessful bids came from La Vizcaya, Factoria Naval Gaditana, Thames

Iron Work, F. Gil and Co., Messers. Vila, Oswald Mordaunt and Co., and Construcciones Navales y Armamentos. The Martínez Rivas–Palmer bid was not the lowest, at 15 million pesetas per battleship, but it promised to deliver the vessels faster than the competition. Martínez Rivas also promised to import only those materials that could not be produced in the country. See A.M., L. 4016, Arsenales.

130. The victory of a Bilbao company in the competition might have been related to the strong lobbying of Vizcayan business people to make the Basque harbor the construction site of the battleships. See A.C.C.B., L. 128 (drafts of the record of the sessions of its ruling commission, Mar. 6, 1887), and L. 301.

131. A glimpse of Palmer's business holdings is found in Pollard and Robertson, *The British Shipbuilding Industry,* p. 90.

132. This was the assessment given to the company when it mortgaged its assets to the government to assure fulfillment of the contract.

133. By the turn of the century, after a protracted lawsuit and arduous negotiations, the shipyard was returned to the Martínez Rivas family.

134. A.M., L. 4227. This, however, was not the only contract that Altos Hornos signed with the navy. For several examples, see A.M., L. 5479, Contracts.

135. See "Informe sobre los aceros moldeados Robert que se elaboran en los Talleres de Deustro," in A.M., Sec. Indiferente, Asuntos Particulares, carpeta July 6, 1893.

136. See Montero, "Modernización económica," p. 243; Fernández Pinedo, "Nacimiento y consolidación," p. 18.

137. The Ibarras and their relatives, the Vilallongas, owned 40 percent of Tubos Forjados. A copy of the partnership contract drawn by the notary Blas Onzoño is located in the library of the Diputación in Bilbao. It was not possible to establish their exact share in the other firms, since the original partnership deeds could not be found. The entries for the companies in the Mercantile Register do not provide the percentage of shares held by each owner. However, the Ibarras' shares in those firms must have been great because members of the family served on the company boards. Tomás Zubiría was the vice-president of the Euskaria and he was the president, and his cousin, Mariano Vilallonga, the vice-president of Alambres del Cadagua. See *Anuario de minería, metalurgia y electricidad de España,* I (1894), p. 83, and II (1895), p. 177.

138. For instance, the Chavarris and the Gandarias invested in La Basconia. Pedro Gandarias was also the president of Talleres de Deusto. See A.M., Indiferente, asuntos particulares, L. 4761. The investment in all these smaller factories may have reached 10 million pesetas.

139. Ojeda, *Asturias,* p. 195.

140. A table compares the different national networks in B.P.P., May 1894.

141. Montero, "Modernización económica," p. 248.

142. Ibid., p. 244.

143. Based on ibid., pp. 235–236.

144. See González Portilla, *La siderurgia vasca,* p. 126.

145. B.P.P., 1899 report.

146. Ibid., 1900 report.

147. Ibid. If the consul's figures are correct, there would seem to be an error in the official Spanish statistics. According to the latter, in 1898, the ore transported in Spanish vessels amounted to only 10 percent of the total exported, which, added to the 65 percent transported in British bottoms, leaves 25 percent to other national fleets. However, according to the British consular reports, the traffic from other nations hovered around 10 percent for all types of cargo during that year, which would leave at least 15 percent of the ore exports unaccounted for.

148. The paid-up capital was actually fifty million for La Polar, and four million for Aurora (ibid.).

149. Ibid.

150. See G. Ibáñez, ed., *Anuario Financiero de Bilbao,* pp. 276–279.

151. Montero, "La modernización económica," pp. 244–245, gives several examples of Vizcayan companies dedicated to the extraction of minerals outside the Basque province.

152. See B.P.P., 1903 report.

153. Montero, "La modernización económica," p. 246.

154. B.P.P., 1901 report.

155. See J. Maluquer de Motes, "Cataluña y el País Vasco en la industria eléctrica española 1901–1935," pp. 247–249.

156. B.P.P., 1901 report. Overall, in 1901, twelve hydroelectric companies were formed in Bilbao with a combined nominal capital of 73.5 million pesetas.

157. In 1901, in addition to the large factories such as Altos Hornos and San Francisco, there were sixty-six other foundries and workshops dedicated to the manufacture of all sorts of iron products. Of those, thirteen were founded in 1901 with a combined capital of 21.5 million pesetas. See ibid.

158. With capital of twenty million pesetas, the Unión Resinera produced resins for the manufacture of turpentine and other chemicals. A copy of the 1898 annual report to its stockholders with a brief history of the company can be found in the A.C.C.B.

159. Fraile, "El País Vasco," p. 229. Fraile does not distinguish among the different Basque provinces, but undoubtedly Vizcaya's share was the largest component of those percentages.

CHAPTER FOUR. THE FORMATION OF BILBAO'S MODERN BUSINESS ELITE

1. Basurto used the tax records of the *consulado* to establish the names of the most important merchants. See *Comercio y burguesía.*

2. Between 1850 and 1855, the F.C.R. listed the main Bilbao mercantile houses. The list of members of the commercial tribunal is also a good source to identify top businessmen, since its officials had to be well-established merchants. This list was published in *Libro del centenario de la Cámara de Comercio de Bilbao,* p. 86.

3. A.H.P.V., S. Urquijo, L. 6223, no. 365.

4. Whereas I identified the merchants using the documents previously mentioned, I distinguished proprietors through notarial records and a list of the wealthiest real estate owners of the region published in the B.O.V. in 1860. The list is described in the next section.

5. See Basurto, *Comercio y burguesía,* pp. 72, 79, 81, 143, 146, 148. I have not been able to verify that the Recacoechea mentioned by Basurto and the one who appears on the French consul's list were directly related. The same surname does not always imply kinship. The same applies to Echevarría and Bergareche.

6. Ibid., p. 242.

7. During 1757 and 1759, Martín Tomás Epalza was prior of the *consulado;* in 1766, Antonio de Epalza occupied the same position; and in 1785, it was Pablo Epalza who did so. Again, it has not been established that these Epalzas were directly related to those mentioned by the consul. Yet, Epalza seems to be a less common name than Echevarría, for instance, and thus the probability of kinship increases. For a complete list of the *consulado*'s authorities from 1701 to 1829, see T. Guiard, *Historia del Consulado,* vol. II, pp. 545–556, 889–890.

8. Ibid., p. 831.

9. See F.C.R. 1850–1855.

10. Epalza and Errazquin invested in two iron factories after the First Carlist War, but the commercialization of iron products does not seem to have been their main trading activity.

11. A similar process seems to have taken place in France during the early nineteenth century among the elite businessmen. See A. Daumard, *Les bourgeois et la bourgeoisie en France,* p. 258.

12. See chap. 1.

13. See R. Forster, *Merchants, Landlords, Magistrates: The Depont Family in Eighteenth Century France.*

14. González Portilla, "La industria siderúrgica," esp. p. 147, n. 74.

15. The list, published in B.O.V., is described in the next section.

16. González Portilla has shown cases of merchants who, during the eighteenth century, established entailed estates to pass on to their descendants. See "La industria siderúrgica," p. 140, nn. 53 and 54.

17. For example, in 1775, a Basque political economist, Juan Antonio de los Heros, wrote a work praising commerce as a way of improving the well-being of Spain. He found it to be an appropriate activity even for the highest ranks of the aristocracy. See ibid., p. 148, n. 78. The original reference appears in Fernando de la Quadra Salcedo, *Economistas vascongados,* pp. 39–42. For even earlier references in Spanish literature to the disdain that Castilians expressed for Basque commercial activities and the Basques' claim that trade did not soil the reputation of a nobleman, see Caro Baroja, *La hora navarra,* pp. 31–32.

18. For a short history of the family and its genealogy, see F. de Uhagón, *Los Uhagón, señores de Hodítegui.*

19. The MacMahons came to Bilbao in the early 1700s. For their genealogy,

see A. Pérez Azagra y Aguirre, *Noticias genealógicas sobre los Primo de Rivera y los Salcedo,* pp. 262–264. The Mowinckels were perhaps Norwegians and probably settled in the city during the 1840s.

20. The Zabalburus belonged to a family of Vizcayan landowners, but they were also involved in national finances. One family member became president of the Bank of Spain and eventually married into the Castilian aristocracy. The Violetes were merchants who built up a glass factory that went bankrupt in the mid-1860s.

21. Joining Epalza were Abaitua, Arellano, Ingunza, MacMahon, Olaguivel, Uhagón, and Uriguen. Other directors, such as Luis Violete and Ramón San Pelayo, were also important local merchants and could have been included among the local business elite (A.H.P.V., L. 5849, ff. 12–30).

22. F. Crouzet, *The First Industrialists,* p. 99.

23. J. Kocka, "Entrepreneur and Manager in the Industrial Revolution," pp. 518–519.

24. See González Portilla, "Los orígenes de la sociedad," p. 75. González Portilla also mentions that José Vilallonga had studied engineering in Bordeaux and had become familiar with modern production techniques through his travels in Europe. Eventually, Vilallonga married one of the Ibarras' daughters.

25. See *Sociedad Santa Ana de Bolueta. Centenario de su fundación,* p. 12.

26. The exact dates of publication were Jan. 17, 1860, Oct. 1, 1861, Oct. 6, 1863, Oct. 14, 1865, and Dec. 3, 1867. An addendum to the 1861 list was published on Oct. 8, 1861. Basas published the first 1861 list without the addendum in *Aspectos de la vida económica,* pp. 21–26.

27. The council had only an advisory function. The provincial governor presided over it and other officials of the regional administration served as nonelected members. Fifteen members were elected (five for each economic classification on the electoral lists). See ibid., p. 20.

28. Sebastián Ayarragaray, Santiago Gorocica, and Gerardo Mowinckel are among the important merchants omitted from the list. Yet, nineteen of the thirty-five merchants who did appear on the B.O.V. list in 1860 also appeared on an electoral list published in 1867 as being part of a select group of electors enjoying annual incomes (*rentas estimadas*) of more than fifty thousand reales. Many of those who did not appear on the 1867 list either died or went bankrupt during the mid-1860s. See B.O.V., Jan. 29, 1867.

29. In the case of landowners, the percentage increases from 32 to 56, if one includes the names of those listed for Abando and Begoña during those same years.

30. These are Juan and Serafín Abaitua, Máximo Aguirre, Romualdo Arellano, Vicente Arana, Vicente Artiñano, Lorenzo H. Barroeta, Pablo Epalza, Tomás J. Epalza, Benito and Miguel Escuza, Juan María and Gabriel María Ibarra, Santiago María Ingunza, Manuel Jane, Teodoro Maruri, Pedro F. Olavarria, Pascual Olavarri, Ambrosio and Gabriel María Orbegozo, Nicolás Olaguivel, Francisco J. Patrón, Tiburcio María Recacoechea, Guillermo and Rafael Uhagón, Juan Antonio and Ezequiel Uriguen, Cosme Zubiría.

31. These are Juan Abaitua, Máximo Aguirre, Lorenzo Hipolito Barroeta,

Miguel Escuza, Pascual Olavarri, Justo Violete, Francisco Patrón, Guillermo and Rafael Uhagón, and Ramón San Pelayo.

32. Included in this group are Luis Violete, Gabriel and Benito Orbegozo, and Antonio Bastida. Eventually, although he still appeared on the 1867 list, Manuel Jane would also go bankrupt.

33. These three are Santiago Barroeta, Antonio Ogara, and Joaquín Julián Picaza.

34. Pedro and Francisco MacMahon were the sons of Diego, who in the 1850s was mentioned in the F.C.R. as an important Bilbao merchant.

35. Basilio Gorbeña, a son of Valentín, and José P. Uriarte, a son-in-law and partner of Santiago Gorocica were two merchants whose relatives do not appear on the B.O.V. list, but who are present on my 1850s roster. The Abaitua, Barroeta, and San Pelayo cases involve the substitution of son for father on the 1860–1867 B.O.V. lists.

36. In addition to the Gurtubays, I have established that the brothers Amann—José and Emiliano—and the brothers Rochelt—Adolfo and José—also came from merchant families. The two families were actually related through marriage, and they had emigrated from Germany during the early nineteenth century.

37. The Gurtubays' story is retold in de la Quadra Salcedo, *Calles de Bilbao,* pp. 63–64. Unamuno uses the story with different names to create the character of a Bilbao merchant in *Paz en la guerra,* p. 17.

38. A notarial document explicitly mentions that Simón did not know how to sign, and for that reason, his son José signed in his place (A.H.P.V., S. Urquijo, L. 6230, no. 369).

39. A.H.P.V., S. Urquijo, L. 6228, no. 119, Mar. 29, 1869.

40. Basas thinks that the term "industrialist" may have denoted a lower social status during this period. See *Economía y sociedad bilbaína,* p. 156. I have not found any clear evidence for this. On the contrary, the fact that merchants with industrial investments were closely associated with them would seem to disprove such a notion. A clear example of pride in an industrial enterprise was the Ibarras' celebration of several of their daughters' weddings in their Baracaldo factory's chapel.

41. See, for instance, his will written in 1883, A.H.P.V., F. Uribarri, May–Aug. 1883, ff. 1117–1122.

42. The flour mills were operated by Pedro and Pablo Berge, Máximo Aguirre, Eugenio Aguirre, and José Baliscueta. The iron factories or forges were operated by Gabriel Ibarra, Pedro T. Errazquin, Antonio González Careaga, Pedro Mazas, José Antonio Sagarduy, and Pradera Hnos. (who actually owned a forge that worked copper and not iron).

43. Isidoro José Laraudo was a partner in the Peña paper mill. See A.H.P.V., B. Onzoño, L. 6016, no. 126. Mariano Urrutia and Ramón Sopelana owned tanneries. See Basas, *Economía y sociedad,* p. 52. Nicolás Justo Galíndez owned a food-processing plant in Deusto. See Basas, *Aspectos de la vida económica,* p. 31. The glass factory belonged to the Violetes, who have already been mentioned.

44. For the paper mill's assets in 1869, see A.H.P.V., B. Onzoño, L. 6016, no. 126; for the food-processing company, Basas, *Aspectos de la vida económica,* p. 31.

45. Basas, *Economía y sociedad,* p. 153. Those who appear on both lists are P. Basterra, Luis Abaitua, Errazquin Hnos., V. Artiñano, N. Olaguível, P. Olavarria, G. Iturriaga, D. Epalza, E. Arriaga, Abaitua Hnos., J. Ibarra, T. Miguel, P. Olavarría, A. González Careaga, R. García, S. Arana, E. and J. Uriguen.

46. For instance, although the Uriguens had interests in mining, metallurgy, and banking, their economic rise came through commerce during the interwar period; consequently, they appear in the merchant category.

47. See tables 7 and 8 in chap. 2.

48. For instance, Juan Durañona appears as trading general goods and ore in a list of Portugalete merchants. Similarly, José Chavarri's inventory of assets also suggests that he was involved in the Cuban trade. See Archivo Municipal de Portugalete (hereafter A.M.P.), L. 154, no. 8; and A.H.P.V., B. Zavalla, L. 6631, Nov. 5, 1861.

49. One exception involved the partnership of Vitoria, Maruri, and Suñol, which acquired part of the Parcocha mine in 1876 (see chap. 2). Maruri and Suñol appeared on the 1865 B.O.V. list of wealthy merchants.

50. This firm included four partners who could be linked directly to the members of the mercantile elite for the 1850s and 1860s. Ezequiel and Juan Antonio Uriguen, Eduardo Coste y Vildosola, and Carlos Aguirre y Labroche. See n. 53, chap. 2.

51. I have found no information about Núñez's social background or early career. During the early 1900s, his mining business flourished, leading to investments in other sectors, such as banking and railroads.

52. See A.H.P.V., M. Castañiza, L. 6054, no. 34; and S. Urquijo, L. 6332, no. 23.

53. Aburto's marriage contract to Benita Martínez Rivas is in A.H.P.V., I. Ingunza, L. 6336, no. 153. For the acquisition of the mine, see A.H.P.V., F. Uribarri, Feb. 7, 1885, nos. 46–47.

54. See chap. 2.

55. A José María Martínez appears on the 1865 B.O.V. list of wealthy merchants. Unfortunately, since the matronymic is not listed, I cannot determine whether it is, in fact, Martínez Rivas.

56. The marqués's fortune was not very old, and his title was less so, having been granted in 1867. During the 1830s and the 1840s, Rivas was part of a clique of government contractors and financiers that grew extremely wealthy. Some of these deals are discussed in A. Otazu, *Los Rotschild y sus socios en España.* According to Javier Ybarra, who cites no source, when the marqués died in the 1880s, he left two hundred million reales to his heirs (about two million pounds sterling). See *Política nacional en Vizcaya,* p. 112.

57. For information on La Vizcaya and its operation of the Galdames mines, see González Portilla, *La siderurgia vasca,* p. 50. Unfortunately, I did not find much information on the Compañía Explotadora de la Demasia Ser. From documents I

found in the private archive of the Ibarra family, it is clear that members of the family were partners in this company, but I could not determine who the other owners were.

58. Echevarrieta seems to have been the owner of a small shop in Bilbao prior to his involvement in mining. See Basas, *Economía y sociedad bilbaína,* p. 212. Larrínaga listed his profession in notarial documents as a watchmaker or a merchant. See A.H.P.V., C. Ansuátegui, L. 5931, no. 2, Mar. 11, 1885, no. 102.

59. For Alejandro Sota's early involvement in mining, see A.H.P.V., P. Goicoechea, L. 5986, no. 21.

60. This figure is based on trimestral reports on ore production published in the B.O.V.

61. A.H.P.V., F. Uribarri, Mar. 10, 1886, no. 74. Eduardo Aznar does not seem to have been involved in mining prior to the formation of this company. During the late 1860s, he worked as a ship broker and a commercial agent in Bilbao. Benigno Salazar, a member of Vizcaya's landed elite, seems to have become involved in mining during the years in which he acted as the legal guardian of his nephews, the Chavarris. I could not obtain information on the other two partners.

62. *Revista minera, metalúrgica y de ingeniería,* Jan. 1, 1893, p. 3.

63. Montero, "Modernización económica," p. 226.

64. Chap. 2 gives several examples of loans by important mercantile houses to mining entrepreneurs. Also the *Anuario de la minería, metalurgia y electricidad de España,* 1895, lists a series of bankers involved in mining and metal businesses. These include members of the 1850s and the 1860s mercantile elite, such as the Epalzas, the Isasis, the Gurtubays, and the Ibarras.

65. J. Valdaliso, "Grupos empresariales e inversión de capital en Vizcaya, 1886–1913."

66. According to his table 1 (p. 13), he gathered the names of investors in eighty-five metallurgical companies in business between 1886 and 1900. The average capital per company turned out to be four hundred thousand pesetas. This was one-thirtieth the founding capital of Altos Hornos de Bilbao, for instance.

67. Although companies like Santa Ana de Bolueta, founded during the interwar period, continued to operate, they did not seem to expand their capital or renew their technology until the early twentieth century. For this reason, they were not included in the sample. The complete list is included in table II-2.

68. Those in the Ibarra clan were Ramón Ibarra, Adolfo Urquijo e Ibarra, José and Mariano Vilallonga, and Tomás and Luis Zubiría. Those in the Chavarri group included Víctor, Benigno, and Leonardo Chavarri, and their cousin Luis Salazar.

69. Olano and Larrínaga owned 20 percent of La Vizcaya's shares. Ramón Larrínaga was not related to the Bernabé Larrínaga who was mentioned previously as a leading mining entrepreneur.

70. Disdier was hired by Altos Hornos in the early 1880s; during the 1890s, he participated in the founding of Euskaria, Tubos Forjados, Alambres del Cadagua, and Santa Agueda.

71. They were Juan Amann, Eduardo Coste y Vildósola, Juan Gurtubay, Braulio Uriguen, Valentín Gorbeña, and Ricardo Rochelt.

72. This family also obtained one of the earliest concessions to install a telephone network in Bilbao (during the 1880s). A.C.C.B., L. 360. Federico Echevarría participated in mining ventures during the late 1880s in the Ollargan district (see A.H.P.V., F. Uribarri, Sept. 13, 1883, no. 382); and during the early twentieth century, he became a member of the board of directors of the Bank of Bilbao.

73. Table II-3 lists the 34 investors and their socioeconomic background.

74. A.P.M., Libros de Actos Públicos del Consulado de Liverpool, L. 35159, p. 223.

75. The same contract lists the occupation of his brother Pablo as a mariner (A.H.P.V., C. Ansuátegui, July 4, 1885, nos. 295–296).

76. In addition to Serra, I identified only four other outsiders: Francisco Martínez Rodas, N. Rodríguez Lagunilla, and R. and F. Carasa.

77. Although Coste and Uriguen had mining interests, trading was their main business activity before they registered several mines.

78. These were E. Aznar, the Astigarraga brothers, the Aranos, the Real de Asuas, D. Madariaga, M. Basabe, the Uriartes, and E. Sustacha.

79. A.M., Estadística de Navegación y Pesca, L. 2363 and 2369. In addition to their own vessels, the Ibarras also participated in the formation of another shipping company, Ibarra and Co., whose main stockholder was a branch of the family settled in Seville.

80. Between 1879 and 1881, Arrótegui bought six ships from the Glyns, usually paying about 10 percent of the value of the vessels and mortgaging the rest. Overall, Arrótegui's debt amounted to 149,500 British pounds (about 3.7 million pesetas) (A.P.M., Consulado de Liverpool, L. 35159, pp. 69–73, 113–117, 142–147, 209–210, 262–263.

81. I calculated the percentage from a register of steamers published in I. Echevarría and F. Grijelmo, *Las minas de hierro de la provincia de Vizcaya. Progresos realizados en esta región desde 1870 a 1899.*

82. A.H.P.V., M. Castañiza, L. 6366, Aug. 11, 1881.

83. A.H.P.V., F. Basterra, L. 6130, no. 486. In his testament, Escuza refers to Abasolo as his employee.

84. In the early twentieth century, Félix was made a count.

85. A.B.E., Box 30, Testament of S. Mier.

86. The complete list of board members of the Bank of Bilbao since its founding until 1956 is published in *Libro de actos conmemorativos del centenario del Banco de Bilbao,* pp. 29–35.

87. A.H.P.V., M. Castañiza, L. 6093, no. 163.

88. The complete list of investors appears in B.O.V., May 5, 1891. More than one thousand stockholders participated in the initial subscription. The forty-two largest investors controlled about 35 percent of the capital; the rest of the shares were spread among the other many investors. Urquijo and Co. was the banking

house of the Marqués de Urquijo, who also participated in the formation of many metallurgical companies in Vizcaya. Similarly, the Bank of Castile was among the founders of Altos Hornos de Bilbao.

89. The Ibarra clan accounted for four of the seven directorates held by the three families. The Chavarris and the Gandarias accounted for two and one, respectively. See table II-4.

CHAPTER FIVE. BUSINESS, CULTURE, AND SOCIETY

1. See D. Augustine-Perez, "Very Wealthy Businessmen in Imperial Germany"; and idem, "The Banker in German Society."
2. Biographies and autobiographies of businessmen were apparently a popular genre in England during the nineteenth century. See T. Koditschek, *Class Formation and Urban-Industrial Society: Bradford, 1750–1850,* esp. chap. 7.
3. For Barcelona, McDonogh, *Good Families.*
4. Cited from M. de Unamuno, "Mi raza," p. 187.
5. W. Douglass and J. Bilbao, *Amerikanuak: Basques in the New World,* p. 70.
6. Ibid., p. 104; also Otazu, *El "igualitarismo" vasco,* p. 339.
7. Douglass and Bilbao, *Amerikanuak,* p. 97.
8. On the changing nature of the function of the will in Spain, see J. Fernández Delgado, "Silenciosos, comedidos y esplendidos: La quiebra de la función religiosa del testamento."
9. Among these common formulas were the invocation to God, which headed the testament, and the frequent protestation of testators to being faithful Roman Catholics.
10. A.B.E., Box 33-A, will of R. Ibarra, notarized by Isidro Erquiaga, Aug. 30, 1899 (all references to this archive pertain to the Bilbao branch).
11. I.F.A., inventory of the estate of R. Ibarra.
12. I.F.A., will and inventory of the estate of Juan María de Ibarra y Gutiérrez de Cabiedes.
13. See the hagiography written by P. Camilo María Abad, *Vida de la sierva de Dios, Da. Rafaela Ibarra de Vilallonga.* The book provides some insights into the life of the Ibarra family. Most of it, however, is concerned with Rafaela's spiritual struggle to achieve a pure soul and discusses in detail some of the penances that she imposed on herself to free herself from sin.
14. For the money allocated to the founding of the school, see her will in A.B.E., Box 30, no. 793. For her share of her father's estate, see A.B.E., Box 25, no. 328. It is likely that the 1.3 million pesetas did not include the value of her share of the mines, which probably remained as common property with her other siblings. This, at least, is what the other branch of the family did when Rafaela's uncle, Juan María Ibarra, died.
15. Abad, *Vida,* p. 34.
16. A.B.E., Box 43, Mar. 2, 1905. The will was notarized by Ildefonso de Urizar.

Mariano was among the founders and members of the first board of directors of the Bank of Vizcaya. See table II-4.

17. A.B.E., Box 33-A, no. 1476.

18. A.B.E., Box 30, will notarized by Isidro Erquiaga, Dec. 17, 1900.

19. A.H.P.V., F. Uribarri, Dec. 24, 1885.

20. A.H.P.V., S. Urquijo, L. 6252, no. 189, May 17, 1879.

21. For Olavarria, see A.H.P.V., M. Castañiza, L. 6054, no. 143. In his will Olavarria also requested six hundred masses after his death. For the widow, see A.H.P.V., S. Urquijo, L. 6219, no. 176.

22. See Frances Lannon, "The Socio-Political Role of the Spanish Church—A Case Study," which analyzes the social background of the nuns of the Society of the Sacred Heart in Bilbao from the 1870s to the 1930s.

23. For instance, the inventories of M. Sanginés and S. Abaitua in A.H.P.V., G. Urresti, L. 5871, no. 279, and C. Ansuátegui, L. 6325, no. 311.

24. The connection between the Basques and the Jesuits was not limited to their home territory. In Mexico City, "the ecclesiastical advisors of the Confraternity of Aranzazu . . . were all Jesuits" (Douglass and Bilbao, *Amerikanuak,* p. 100). Most likely, given its name, the Madrid confraternity of Saint Ignatius also had Jesuit chaplains.

25. A.H.P.V., F. Uribarri, Apr. 19, 1883, no. 185.

26. For a brief history of the University of Deusto, the Business School, and some of the Basque leaders included among their graduates, see E. González, "La fragua de la Comercial," and *El Correo Español–El Pueblo Vasco.*

27. A.B.E., Bilbao Branch, Box 42, will of Florentina Urizar, widow of Luis Zubiría. A similar provision reducing the inheritance of two daughters who had become nuns can be found in the testament of T. J. Epalza, dated Oct. 27, 1916, A.B.E. Box 46/47.

28. I.F.A., testaments and inventories of Ramón, Juan Luis, and Juan María Ibarra.

29. Caro Baroja, *Los Vascos,* p. 203.

30. E. González, "La fragua," p. 3.

31. See H. R. Trevor-Roper, "Religion, the Reformation, and Social Change," p. 35.

32. Cited in Otazu, *El "igualitarismo" vasco,* p. 351 n. 475.

33. H. M. Robertson, *The Rise of Economic Individualism.*

34. Ibid., pp. 108–109.

35. See, for example, N. Hansen, "Sources of Economic Rationality," p. 146.

36. E. Ybarra Osborne and E. Ybarra Hidalgo, *Notas sobre José María Ybarra, Primer Conde de Ybarra,* p. 119. José María was the brother involved in mining. He was brought up in Bilbao, but settled in Seville, where he started a shipping company among several other businesses. This branch of the family kept close ties with the Bilbao relatives, and they supported each other's enterprises.

37. A.B.E., Bilbao Branch, Box 43, testament of M. Vilallonga.

38. *Velada necrológica que se celebró en el Círculo Conservador de Bilbao el 20-10-1911*

para honrar la memoria del Excmo. Sr. D. José María de Lizana, Marqués de Casa Torre, esp. pp. 26 and 54.

39. The anecdote is taken from J. Ybarra, *Política nacional en Vizcaya,* pp. 211–212.

40. A. Trueba, "Organización Social de Vizcaya," vol. II, p. 609. The report was written in 1867, and was first published in 1870. It was commissioned by the Vizcayan government as a contribution to a group of studies analyzed by a group led by French sociologist F. LePlay.

41. A.H.P.V., C. Ansuátegui, L. 6323, no. 291. After studying medicine, one of Villabaso's sons, Luis, founded and worked for many years as a managing director of the Banco de Comercio.

42. The distribution of fortunes in cities like Bilbao was done according to Castilian civil law. The dual legislation stemmed from the fact that the cities received their own *fueros* in their charters. These urban *fueros* were based on and evolved according to Castilian principles of civil law. After much fighting between the cities and the rural areas of the province, an agreement was reached in 1630 under which urban dwellers had the option of embracing either of the two legal systems. See J. M. Martín de Retana, "Orígenes, evolución y bibliografía del fuero de Vizcaya," R. Hormaeche, *Leyes civiles de Vizcaya,* pp. 41–67.

43. Trueba, "Organización social," vol. III, p. 353.

44. His story is recounted by a descendant commemorating the seventy-fifth anniversary of the founding of the Deusto Business School. See P. Icaza, "La fundación de la Comercial."

45. Pedro Aguirre became a member of the board of directors of the Bank of Bilbao. See Icaza, "La fundación de la comercial." He also invested in the Machin shipping company (A.H.P.V., F. Uribarri, Nov. 19, 1884, no. 366).

46. The fourteen were José R. Aqueche, Lino Arisqueta, José de las Casas, J. B. Cortina, Sebastián Eguillor, A. González Careaga, E. Mugaburu, N. Olaguível, Joaquín Quintana, José and Mariano Sanginés, Ceferino Urien, L. Uriguen, and A. Ysasi. See Basas, *Economía y sociedad bilbaína,* pp. 142–144.

47. For instance, former émigrés A. González Careaga and J. Quintana invested in an iron factory in 1862 (A.H.P.V., F. Basterra, L. 6115, no. 190), and Sebastián Eguillor owned shares and bonds in Altos Hornos de Bilbao (A.H.P.V., C. Ansuátegui, May–June 1886, no. 382).

48. M. Basas, "Enrique Aresti, política y negocios." Aresti also served on the boards of other Vizcayan companies such as Unión Resinera Española, the Santander-Bilbao railroad line, and Hulleras Sabero.

49. A.H.P.V., M. Castañiza, L. 6039, no. 100. The Uribarrens sold their share of the business to the Ibarras during the 1860s. In contrast, the Murrietas continued their association and participated in the formation of the revamped Altos Hornos de Bilbao during the 1880s.

50. The suppression of the *fueros* in 1877 did not abolish the inheritance regulations established in Vizcayan law.

51. This was the case even if those families owned land in the Vizcayan countryside.

52. The list appears in table II-1.

53. For the diversification of the family fortune, see the comparison between Juan María Ibarra's estate and that of his son Ramón later in this chapter.

54. McDonogh mentions that the business elite in Catalonia described its own historical experience according to that law. Although some of McDonogh's case studies do not fit the model, his work uses the Law of Three Generations as a basic framework. See *Good Families.*

55. In the 1910s, Gabriel's brother Fernando María was a member of the board of Altos Hornos de Vizcaya (E. Ibáñez, ed., *Anuario Financiero,* 1918), and was a founder and director of the Sociedad Española de Construcción Naval shipyard (*Anuario Marítimo,* 1915).

56. See *El País* (Sunday Supplement, Jan. 31, 1988), p. 3.

57. Ramón Sota strongly supported the cause of the Basque Nationalist Party, which sided with the Republican forces during the Civil War.

58. See M. Ugalde, *Hablando con los vascos,* pp. 55–76.

59. Echevarría was among a select group of artists who first used the techniques of the fauve style of painting in Spain. A brief biographical sketch appears in *Diccionario Enciclopédico del País Vasco,* vol. IV, pp. 69–70.

60. Ibid., pp. 68–69.

61. J. de Orueta, *Memorias de un bilbaíno, 1870 a 1900,* pp. 155–156.

62. R. Maeztu, *Autobiografía,* p. 164. In another passage, Maeztu illustrates how the "business bug" infected men of letters such as his friend Fermín Herrán, and even the better-known Basque novelist Pio Baroja; see p. 128.

63. J. F. Lequerica, *La actividad económica de Vicaya en la vida nacional,* p. 29. Interestingly, Lequerica did not think that Vizcayans carried their attitude toward economic success to the extreme of considering it a reward superior to virtue, as had been done, in his view, by "dissident Christians."

64. E. Burges, *Vizcaya or Life in the Land of the Carlists at the Outbreak of the Insurrection,* p. 173.

65. D. Mazas, *La sociedad de Bilbao en 1887,* pp. iii–iv.

66. A.H.P.V., F. Uribarri, L. 6187, no. 20. Urquijo was not poor by any means at the time of his marriage. He had already inherited land from his mother with an estimated income of 3,250 pesetas per year. These assets together with his law degree did not make him rich, but afforded him a cushion, which is what separated the middle from the lower classes.

67. Basabe contributed 67,000 pesetas to the marriage, which almost matched his wife's dowry of 69,750 pesetas. Yet, while the latter's dowry was only a fraction of what she would eventually inherit, Basabe's capital would not increase through inheritances, since his parents had died long before his marriage (A.H.P.V., S. Urquijo, L. 6240, no. 209 [May 30, 1873]).

68. The social and political implications of this bond with landowners are discussed in chapter 6.

69. For instance, the great merchant families of Boston during the eighteenth and nineteenth century used the same kind of marital alliances established by the

Bilbao business elite. See P. D. Hall, "Family Structure and Economic Organization: Massachusetts Merchants, 1750–1850," pp. 42–43.

70. For the marriage alliances of the Vilallonga family, see A. Pérez de Azagra y Aguirre's genealogical study, *Títulos de Castilla e Indias y Extranjeros a Vascos,* p. 353.

71. These campaigns are discussed in chapter 6.

72. Trueba, "Organización social," p. 609.

73. J. E. Delmas, *Diccionario de claros varones de Vizcaya,* p. 90, mentions that it was the custom of wealthy merchants to send their sons to study in England.

74. A.H.D.V., Corregimiento, 386/21.

75. A.H.P.V., S. Urquijo, L. 6218, no. 176.

76. For details on the Ibarras' education, see Juan María Ibarra's testament in I.F.A. Basas mentions that Tomás Zubiría, a cousin and partner of the Ibarras, also studied in Liège, but he does not mention what degree, if any, he received. See "Cosme, Tomás y otros Zubirías." For the Chavarris, see *Anuario de la minería, metalurgia y electricidad de España,* (1896), p. 61, which has a list of engineers who settled in Bilbao.

77. For Arana, see his marriage contract, where his profession is listed as a mining engineer (A.H.P.V., F. Uribarri, L. 6193, no. 283). Other engineers, direct descendants of the mercantile elite of the 1850s and the 1860s, were Ricardo Arellano, Valentín Gorbeña, José María Solaún, Félix and Recaredo Uhagón, Jesús Uriarte Gorocica.

78. As representatives of the factory, José Antonio and his cousin Fernando Luis toured France, Belgium, and England to decide which technology should be used in Bilbao. See A.A.H.B., Records of the Comité de Madrid, Mar. 30, 1883, Apr. 11, 1883, and May 29, 1883.

79. Chastagnaret is also wrong in claiming that until the end of the nineteenth century, the only Ibarra with an engineering degree belonged to the Seville branch of the family. See "Conclusiones," p. 321.

80. See the next section.

81. McDonogh, *Good Families,* p. 126. The author's chronology is somewhat blurred, but he seems to be referring to the early twentieth century.

82. Ibid., p. 48.

83. M. Tuñón de Lara, *Estudios sobre el siglo XIX español,* pp. 156, 178.

84. A.B.E., Bilbao Branch, Box 43. Testament of Mariano Vilallonga recorded by the notary Ildefonso Urizar. The text in the original underlines the importance of the academic degree by playing on the word "título," which in Spanish is used to denote both the university diploma and the noble title.

85. In rough translation: "My father owns the manor / The noblest in Vizcaya / His coming and going to the Indies / Why should that take away his honor?" (Lope de Vega, "El premio del bien hablar," cited in Otazu, *El "igualitarismo" vasco,* p. 127).

86. Amalia Echevarrieta was married to Rafael Echevarría, part of the Marqués de Villagodio's family.

87. Mayer, *Persistence,* p. 100. The quotation obviously refers to Mayer's discussion of the German case, but he uses similar arguments for other European countries.

88. See Maeztu, *Autobiografía,* p. 163, and Unamuno, "Del Bilbao mercantil al industrial," p. 177.

89. According to Castilian law, the assets introduced to a marriage by each spouse continued to be treated as the individual property of the person who brought those assets to the union. The wife, however, lacked direct control over her property, which was managed by her husband. All assets acquired by using the spouses' initial contributions were considered common property. At the death of one of the spouses, the dowries, together with inherited property acquired by one of the partners during the marriage, were returned to the respective owners, and in the case of the deceased spouse, to his or her heirs. The remainder of the family fortune (the so-called gained assets, or "bienes gananciales") were divided equally among the surviving spouse and the legal heirs.

90. A.H.P.V., F. Uribarri, L. 6187, no. 52. 250,000 pesetas of Vilallonga's capital was invested in Ibarra and Co., which ran the Nuestra Señora del Carmen iron factory in Baracaldo.

91. The dowry was indeed conceived as an advance inheritance payment. When parents' estates were executed, a child's dowry was discounted from the inheritance share to ensure that siblings who had not received dowries would get an equal stake of the family fortune.

92. The female counterparts of middle- and lower-middle-class groups actually contributed more than males to their marriages on average. Although it is not altogether clear why this happened, it is possible that the females in these groups were more open to marrying men with fewer resources. Or perhaps families tended to compensate for the lower economic opportunities of females by giving them greater dowries.

93. His sample includes only three randomly chosen years (1838, 1858, and 1878) and 10 percent of Barcelona's notaries. The number of cases recorded is small, and the notaries selected may have inadvertently skewed the results, if, as in Bilbao, some members of the profession had much larger practices than others.

94. McDonogh, *Good Families,* p. 146, table 9. The author does not indicate the currency unit, but it appears to have been pesetas, since other values throughout the book are expressed in this unit. Yet, the peseta did not become the common currency until the mid-1870s, when it replaced the real. Although the latter was the most often used unit during the 1850s and the 1860s, other currency standards such as the peso fuerte, the escudo, and the duro also appear in documents from the period. The equivalency among all these units was the following: 1 peseta= 4 reales; 1 escudo= 10 reales; 1 peso fuerte= 1 duro= 20 reales.

95. In Bilbao's case, I have not been able to compare the business elite's and the local landowners' dowries because the latter involved transfers of land to which no monetary value was assigned, as well as such items as annual pen-

sions, or promises such as free room and board for a number of years. See, for instance, the marriage contract of Celestino Ortiz de la Riva and Sofía Arana in A.H.P.V., S. Urquijo, L. 6228, no. 36, or that between B. Salazar and C. Zubía in A.H.P.V., V. L. Gaminde, L. 5325, no. 249.

96. A.H.P.V., B. Onzoño, L. 6024, no. 161. Arellano is conspicuously absent from Basas's list of Bilbao millionaires. Since the list was compiled for fiscal purposes and on the eve of the Second Carlist War, it omitted several other prominent merchants such as G. Ibarra, J. Gurtubay, and C. Zubiría. Overall, the list included eighty-five individuals whose wealth ranged from 250,000 to 3 million pesetas (*Economía y sociedad bilbaína,* pp. 142–143.

97. A.H.P.V., S. Urquijo, L. 6249, no. 161.

98. A.H.P.V., B. Onzoño, L. 6028, no. 18.

99. A.H.P.V., C. Ansuátegui, L. 6291, no. 114. Another example: in 1855 José Salazar left a landed fortune of about 580,000 pesetas (ibid., L. 6268, no. 154).

100. McDonogh, *Good Families,* p. 87.

101. Otazu, *Los Rotschild,* p. 482.

102. Delmas, *Diccionario,* p. 174.

103. Studies of wealth in France and England show that, during the nineteenth century, fortunes were larger and the number of millionaires was greater in the capitals than in the industrial centers. See W. D. Rubinstein, "The Victorian Middle Classes: Wealth, Occupation and Geography," and A. Daumard, *Les fortunes françaises au XIX siècle.*

104. See list of rents in P. Tedde de Lorca, "Comerciantes y banqueros madrileños al final del Antiguo Régimen," p. 307.

105. G. Tortella proposes such a thesis in several of his works. See, for instance, "Spain."

106. See, for instance, A.H.P.V., S. Urquijo, L. 5849, ff. 489–495, and L. 6231, no. 79.

107. For instance, in 1869, the sale of a large number of properties in Vizcaya belonging to the Marqués del Duero showed that most of the estates yielded annual rents between 2 and 3 percent of the assessed value of the land. Only estates with flour mills were able to collect rents above 3 percent per year: A.H.P.V., F. Basterra, L. 6133, no. 40. In 1863, the inventory of L. H. Barroeta (A.H.P.V., S. Urquijo, L. 6218, no. 100) shows a rural property whose rent yield was 3.14 percent per year.

108. A.H.P.V., B. Onzoño, L. 6383, no. 240. According to his inventory in 1881, Echevarría's fortune was 2.7 million pesetas. His stocks and bonds accounted for 2.05 million of the total, while his real estate holdings were valued at 475,000 pesetas.

109. The connection between real estate ownership and voting rights is explored in chap. 6.

110. Daumard, *Les bourgeois de Paris au XIXe siècle,* p. 262. Daumard's table actually records a 2.7 and 1.8 percent for active and retired businessmen. I averaged both percentages to arrive at the 2.5 percent that appears above in the text.

111. The two inventories appear in A.H.P.V., S. Urquijo, L. 6218, no. 176; and G. Urresti, L. 5871, no. 279.

112. A.H.P.V., V. L. Gaminde, L. 5326, no. 217. This document was a contract between an apprentice and the manager of the Santa Ana de Bolueta iron factory. The worker's wage was set at 2.75 pesetas per day during his first year. A fully skilled iron worker could earn as much as 5 pesetas per day, as protocol no. 218 of the same notary shows.

113. For Salamanca's lifestyle, see R. Carr, *Spain, 1808–1975*, p. 282. On the low number of carriages in Bilbao see J. C. Gortázar, *Bilbao a mediados del siglo XIX según un epistolario de la época*, p. 121.

114. Burges, *Vizcaya*, p. 185.

115. See Abad, *Vida*, pp. 24, 79.

116. Mazas, *La sociedad*, p. 15.

117. Gortázar, *Bilbao*, pp. 28, 38, 84–85; and for the period following the Second Carlist War, Orueta, *Memorias de un bilbaíno.*

118. For instance, at the death of his wife, José Echavarria y Ascoaga's fortune was assessed at 1.3 million pesetas in 1886 (A.H.P.V., C. Ansuategui, Nov. 1886, no. 1085); in 1886, S. Eguillor left an estate of 3.5 million pesetas (A.H.P.V., C. Ansuategui, May-June 1886, no. 382); in 1905, M. Vilallonga's estate was 3.2 million pesetas (A.B.E., box 43); in 1915, J. MacLennan's fortune was 12.5 million pesetas (A.B.E., box 49); in 1899, the inventory of J. L. Ibarra showed a fortune of 6 million pesetas (I.F.A.), while his father J. M. Ibarra left an estate of 10 million pesetas in 1889 (I.F.A.), and his uncle G. M. Ibarra also left behind a 10-million-peseta estate in 1890 (A.B.E., box 25).

119. Chavarri's widow amassed a fortune of 20 million pesetas, and both Benigno Chavarri and Horacio Echevarrieta amassed 15 million pesetas each (A.B.E., box 140/141). Other fortunes mentioned in this 1925 report by the Banco de España were L. Nuñez Arteche (5 million pesetas); V. Llaguno y Durañona (2.5 million pesetas); A. Iza (5 million pesetas); J. M. Gonzalez Ibarra (3 million pesetas); and P. Astigarrage (1 million pesetas).

120. G. Gortázar, *Alfonso XIII, hombre de negocios*, p. 40.

121. Rubinstein, "The Victorian Middle Classes." British wealth peaked at around twenty million pounds in the late nineteenth century.

122. Jean-Marie Moine, *Les barons du fer*, p. 286.

123. Ibid., pp. 286–291.

124. Although they were listed, the mines were not actually valued in the inventory. Using a document found in the Ibarras's private archive, I conservatively estimate the value of Juan María's share of the family mines at around 5.3 million pesetas. The document prepared by P. Celis in February 1895 gives the area, type of iron ore, and value of each mine owned by the family.

125. These houses were mentioned in the testament, but were not listed in the inventory. However, one of the four houses was appraised at 175,000 pesetas in the inventory of his widow in 1899 (testament and inventory of Luz Arregui, widow of Juan María Ibarra, I.F.A.). A property registry of the city of Bilbao

(found uncatalogued in the A.H.D.V.) gives a similar assessment to all the houses that belonged to the Ibarra family in that area. As a result, I estimate the value of the four houses at 700,000 pesetas.

126. Gortázar, *Bilbao,* p. 40.

127. He estimated the cost of educating his sons at approximately eighty-five thousand pesetas.

128. I.F.A., *Libro de actas del consejo de familia de los menores de Ramón Ibarra.*

129. Among the founding members of the Real Sporting Club one can find the names of the most important Vizcayan businessmen: Ibarra, Chavarri, Martínez Rivas, Salazar, Zubiría, Gurtubay, Taramona, and Uriguen, to cite just a few. See the complete list in Conde de Zubiría, *El libro del Real Sporting Club,* pp. 345–346.

130. See J. Ybarra, *Política nacional en Vizcaya,* p. 254.

131. Maeztu, *Autobiografía,* p. 163.

132. For J. P. Morgan as art collector and the effect of his purchases on his business, see R. Chernow, *The House of Morgan,* pp. 116–120. According to Chernow, Morgan's art collection at the time of his death in 1914 was estimated at fifty million dollars, about half of his total fortune.

CHAPTER SIX. BUSINESS AND POLITICS

1. J. J. Solozábal, *El primer nacionalismo vasco,* pp. 264–265; and J. Corcuera, *Orígenes, ideología y organización del nacionalismo vasco, 1876–1914,* p. 60.

2. Carr discusses the political philosophy and constituency of the Progressive and Moderate Parties in *Spain,* pp. 158–169.

3. For a detailed description of the protracted and inconclusive negotiations between the Vizcayans and the central government, see M. Vázquez Prada, *Negociaciones sobre los fueros entre Vizcaya y el poder central, 1839–1877.*

4. Chapter 1 includes a complete discussion of who the Carlists were and why they enjoyed support in Vizcaya.

5. Vázquez Prada, *Negociaciones,* pp. 399–400.

6. M. A. Larrea and R. M. Mieza, "La Diputación General del Señorío de Vizcaya," p. 14.

7. J. M. Ortiz and J. M. Portillo, "La foralidad y el poder provincial."

8. Ibid., p. 116.

9. Vázquez Prada, *Negociaciones,* pp. 173–175, 178–188; E. Fernández Pinedo, "Haciendas forales y revolución burguesa: las haciendas vascas en la primera mitad del siglo XIX."

10. For these inroads, see Vázquez Prada, *Negociaciones,* pp. 231, 234–235.

11. For the rules governing the selection of deputies, A.H.D.V., Acuerdos de la Junta General de Guernica, 1850, pp. 33–46, and 1854, pp. 50–56. The two groups had no significance beyond the ceremonial. It was a distinction that originally re-

flected clan rivalries during the late Middle Ages, but by the nineteenth century neither the rivalry nor the clans existed.

12. The rule did not apply to the alternate deputies, who were not asked to serve as replacements in case of resignation, death, and so on.

13. The threshold for elegibility was set at 12,000 reales per year. Junta members also needed wealth related to land in order to hold their position, but the standard in this case was lower, 550 reales from rent per year.

14. Vázquez Prada, *Negociaciones,* p. 124.

15. Aguirreazkuenaga, *Vizcaya en el siglo XIX,* p. 305.

16. Larrea and Mieza, "La Diputación," p. 16.

17. In this analysis, I followed a method similar to the one used by Larrea and Mieza in their study on the Diputación. My research, however, allows for stronger conclusions, especially regarding the economic background of the deputies.

18. Appendix IV lists the deputies and shows how they were classified.

19. The eight unclassified deputies with the right to vote in national elections were A. Aldama, V. Belarroa, M. Gogeascoechea, R. Guardamino, R. C. Rotaeche, J. Tellitu y Antuñano, J. F. Vidásolo, and J. A. Vildósola. See lists in B.O.V., May 1 and 13, 1851, Oct. 19, 1858.

20. An idea of Jauregui's landholdings can be seen in his marriage contract (A.H.P.V., M. Castañiza, L. 3363, 3/1851). R. Echezarreta, also classified as an industrialist in the B.O.V. lists, seems to have been an important property owner in the Durango region. See Delmas, *Diccionario,* p. 72. Although T. Loizaga also appeared as a manufacturer in the B.O.V., I have not been able to establish his precise occupation. His family owned land in the Guernica-Luno area. See Aguirreazkuenaga, *Vizcaya en el siglo XIX,* p. 263.

21. Larrea and Mieza, "La Diputación," p. 19.

22. For instance, the merchants Olaguivel and Echevarría, who served as deputies, were among the largest landlords in Bilbao. For Echevarría, see B.O.V., Jan. 17, 1860; for Olaguível, see Basas, *Aspectos de la vida económica,* p. 261. Basas also cites other merchants with high real estate rents: P. Errazquin, P. F. Olavarria, and José Briñas. Similarly, the proportion of real estate assets of many of the merchants listed on table 22 must also have met the eligibility requirement.

23. Those with merchant backgrounds were J. B. Arana, S. Ingunza, P. and M. Jane, A. Orbegozo, J. A. Uriguen, and L. Violete. All were part of the Bilbao business elite (see App. II, and for Violete, B.O.V., Jan. 1, 1860). The five proprietors were C. Adán de Yarza, J. B. Barcena, E. and M. Larrínaga, and E. Victoria de Lecea (see B.O.V., Jan. 17, 1860).

24. It is not clear what income or property requirement was used to limit the franchise. During the 1860s, the franchise was broadened, but still remained quite restrictive. For the number of elegible voters, see B.O.V., June 14, 1853, and Dec. 9, 1856.

25. The names of the mayors are readily available in the B.O.V.

26. See A.H.P.V., S. Urquijo, L. 5841, Apr. 24, 1850, L. 6216, Jan. 1861, and L. 6227, July 2, 1868. For the complete list, see App. IV.

27. P. Epalza was elected in 1856, serving for about one year; P. P. Uhagón was elected in Nov. 1858, and served until 1863.

28. For instance, J. M. Arrieta Mascarua, J. Basabe, R. Guardamino, T. Loizaga, F. Victoria de Lecea, and J. A. Vildósola served as *foral* and national deputies.

29. This explanation is based on the 1846 electoral law, which was in effect during most of the 1850s, except for the biennium 1854–1856, when a different, slightly more liberal, electoral law was enacted. In 1857, the restriction imposing 150 electors per district was reestablished and remained in effect until 1865.

30. According to M. Artola, the franchise established by the 1846 electoral law permitted the participation of 1 percent of the Spanish population. See *La burguesía revolucionaria (1808–1874),* p. 213.

31. See a copy of the list in Archivo General del Señorío de Vizcaya, in Guernica (hereafter A.C.J.G.), Elecciones, Reg. 42, L. 1, no. 2. Similar lists were published throughout the 1850s in the B.O.V., for instance, the issues of June 30, 1857, Oct. 19, 1858, and Apr. 1, 1862.

32. L. Stone and J. C. Fawtier Stone, *An Open Elite? England 1540–1880,* p. 420. According to Mayer, in 1868, British landowners still occupied two-thirds of the seats in the lower house of Parliament. See *Persistence,* p. 164. It has never been argued that such a concentration of political power hindered the English industrialization process.

33. A.H.P.V., S. Urquijo, L. 6215, Oct. 17, 1860, ff. 531–544.

34. A.H.P.V., S. Urquijo, L. 5843, Dec. 5, 1852. For Arriaga's participation in the *consulado,* see Guiard, *Historia del Consulado,* vol. II, p. 889.

35. See table II-1.

36. A sister of Juan M. and Gabriel M. Ibarra, Concepción, was married to F. Zabalburu, brother of José and Mariano, who occupied political positions in Vizcaya. A daughter from the Zabalburu-Ibarra family married A. Quadra Salcedo, who was a deputy from 1860 to 1862. For the Ibarras' genealogy, see J. Ybarra y Berge, *La casa de Salcedo de Aranguren.*

37. For Echevarría as merchant and proprietor, see chap. 4.

38. Guiard, *Historia del Consulado,* vol. 2, 572–573. Curiously, to illustrate his point, Guiard mentions that in 1730, after a protest over the election of an official not involved in trade, the process was justified by citing several precedents among which was that of Bartolomé José Novia de Salcedo, an ancestor, perhaps, of Pedro.

39. A.H.P.V., M. Castañiza, L. 3365, Feb. 22, 1853. Uribarren e Hijo was a Basque banking house located in Paris.

40. A.H.P.V., S. Urquijo, L. 5841 (1850) and L. 5843 (1852), ff. 441–457. It is also worth noting that José Pantaleón Aguirre, a well-known Bilbao merchant, was among the four executors of the will.

41. A.H.P.V., S. Urquijo, L. 5841, Nov. 19, 1850.

42. A.H.P.V., S. Urquijo, L. 6231, no. 156.

43. See his marriage contract in A.H.P.V., F. Uribarri, L. 6186, no. 81.

44. See chap. 2 for a discussion of the 1818 and 1827 mining laws enacted by the Vizcayan government. González Portilla is wrong when he claims that these laws contradicted the property provisions included in the National Mining Law of 1825. See *La formación de la sociedad capitalista,* vol. I, pp. 23–24. In fact, the 1827 legislation virtually copied the Law of 1825 except for the principle that concessions had to be granted by the crown.

45. See Aguirreazkuenaga, *Vizcaya en el siglo XIX,* pp. 510–519, 527–532.

46. Vázquez Prada, *Negociaciones,* p. 233.

47. Aguirreazkuenaga, *Vizcaya en el siglo XIX,* p. 582. According to Vázquez Prada, the Diputación's appeal to the crown coincided with the September Revolution of 1868. See *Negociaciones,* p. 248. Neither author mentions whether the tax was ever collected.

48. See J. Aguirreazkuenaga, *Vizcaya en el siglo XIX,* p. 547. It is worth noting that Carlist ideology supposedly resisted changes such as the construction of railroads. Yet, rhetoric and practice did not always seem to match. On the connection between Carlism and capitalism, see V. Garmendia, *La ideología carlista (1868–1876),* pp. 240–247.

49. A.H.D.V., Acuerdos de las Juntas de Guernica (1870), pp. 24–27.

50. It did not help that the new king belonged to the House of Savoy, the ruling family in Italy, which in the eyes of the ultra-Catholic Carlists had betrayed the faithful by removing Rome from papal authority in 1870.

51. Although a few important merchants such as the Abaituas supported the Carlist uprising in Bilbao, they seem to have been the exception. Several of Luis Abaitua's properties were seized by the government after the war. See A.H.P.V., S. Urquijo, L. 6243, no. 313, Oct. 15, 1875.

52. The speech read by Deputy Eduardo Victoria de Lecea before the Junta at Guernica in December 1872 exemplifies this position (A.H.D.V., Acuerdos de las Juntas de Guernica [1872]).

53. See Vázquez Prada, *Negociaciones,* pp. 373, 401–402.

54. The system was put in place in 1882 with only four electoral districts: Bilbao, Valmaseda, Guernica, and Durango. Each district elected four deputies. In 1884, a fifth district, Marquina, was added to the other four. See Corcuera, *Orígenes,* pp. 168–169.

55. See C. Postigo, *Los conciertos económicos.*

56. M. Montero, "Régimen liberal y autonomía vasca," p. 19. The tax on agriculture, commerce, and industry, which had been collected in the other Spanish provinces since the middle of the nineteenth century, started to be levied in Vizcaya only in 1912. In 1900, however, the national government started directly collecting a tax on the income (*utilidades*) of corporations, just as it did in other regions.

57. Ibid., pp. 3–27; also Postigo, *Conciertos,* pp. 41–42.

58. When the Diputación commissioned a group of lawyers to draft a report on the specific prerogatives of the institution, the attorneys had to conclude, after a

lengthy analysis of all pertinent decrees and laws, that there was no legislation that provided a specific base for the exact functions of the provincial government. See A.H.D.V., Sesiones de la Diputación (1897), p. 167.

59. Ibid., pp. 162–163.

60. In 1890, P. Alzola, then president of Vizcaya's Diputación, mentioned that his institution had the largest budget in Spain. See his speech in *Estado y desarrollo de la escuela de artes y oficio de Bilbao* (Bilbao, 1890), part of the collection of reports in A.C.C.B., no. 41.

61. See *Memoria de los actos económico-administrativos de la Diputación provincial interina de Vizcaya.*

62. See *Gaceta de los caminos de hierro,* Aug. 4, 1903, p. 159.

63. See L. Castells's analysis for Guipuzcoa, which could be applied to the Vizcayan case, since both provinces enjoyed similar tax agreements with the central government (*Modernización y dinámica política en la sociedad guipuzcoana de la Restauración,* pp. 248–250).

64. See Diario de Sesiones de Cortes, Nov. 22, 1900, no. 3, app. 4.

65. J. Orueta, "Influencia del régimen autonómico sobre el desenvolvimiento y progreso de la economía guipuzcoana," in *Ante las Cortes Constituyentes. Guipuzcoa y la razón de su autonomía,* cited in Castells, *Modernización,* p. 244. Although Orueta's article seems to deal specifically with Guipuzcoa in 1931, his remark encompasses the other two Basque provinces and their long tradition of autonomy.

66. F. García de Cortázar, "La oligarquia vasca a comienzos del siglo XX," p. 253.

67. Montero, "Régimen liberal," p. 17.

68. For instance, J. Real Cuesta, *El carlismo vasco, 1876–1900,* pp. 218–219; Beltza, *Del carlismo al nacionalismo burgués,* p. 91; Corcuera, *Orígenes,* p. 97, believe the Vizcayan oligarchy monopolized the Diputación.

69. See App. IV. In only two of those eight cases was it possible to establish kinship: Juan Jauregui was the son of the J. J. Jauregui of the *foral* period; and Javier López de Calle's father, Bruno, had been a deputy from 1866 to 1870.

70. Only three of the sixteen deputies elected more than once belonged to the business elite: A. Uría (who served ten years), C. Palacio (eight years), and B. Salazar (six years).

71. See App. IV.

72. There were three merchants and five manufacturers included among the forty-one individuals selected to serve as deputies during the *foral* period.

73. The four were P. Alzola, J. M. Arteche, B. Salazar, and Angel Uría.

74. In addition to the four men mentioned in note 73, Villabaso and Ibáñez Aldecoa could also be classified as proprietors.

75. See a brief description of the electoral laws in effect at the national and provincial levels in Real Cuesta, *El carlismo,* pp. 232–233.

76. Ibid., p. 234, table 13.

77. Ibid., pp. 209–212; and Corcuera, *Orígenes,* pp. 168–180.

78. Toward the end of the century, some Fuerists and some disaffected Carlists

joined forces to form the Basque Nationalist Party (PNV) under the leadership of Sabino Arana. In 1898, Arana won a seat in the Diputación. The party became a major political force only after the first decade of the twentieth century.

79. For this view, see Beltza, *Del carlismo,* p. 119.

80. Ibid.

81. Ibid.

82. A.H.D.V., Actas de las Sesiones de la Diputación (1880), pp. 16–17, and (1898), p. 149. In these sessions, as subsidies for railroads were discussed, two deputies excused themselves from the debate because they had a direct interest in the companies about to receive public funds.

83. A.H.D.V., Actas de las Sesiones de la Diputación (1898), p. 189.

84. *El Nervión,* "Nuestros candidatos," Jan. 26, 1891. It was the construction of the shipyard that convinced the paper to support Martínez Rivas and not his rival, V. Chavarri, another key business figure (Jan. 29, 1891).

85. *La Lucha de clases,* Apr. 7, 1900, cited in F. García de Cortázar, "La oligarquía," p. 235. A statue honoring Chavarri was erected on the pier of his hometown, Portugalete. As far as I know, this is the only monument in the Bilbao area that celebrates the achievements of an industrialist.

86. For a vision of Chavarri as omnipotent politician, see Corcuera, "Orígenes," pp. 250–251, n. 30.

87. I. Arana Pérez, *La Liga Vizcaína de Productores y la política económica de la Restauración, 1894–1914,* pp. 148–150.

88. There seems to be some discrepancy in the historiography about this election. R. Miralles (*Política electoral en Vizcaya,* pp. 96–99) records Solaegui's victory by a very narrow margin, whereas J. Ybarra, in *Política nacional en Vizcaya,* p. 150, writes that the winner was Urquijo. The Cortes's compilation of elected deputies also has Urquijo as the winner, but mentions that he did not take the oath of office. Solaegui's name does not appear among the Vizcayan deputies.

89. For the results of the municipal elections, see Corcuera, *Orígenes,* pp. 312–313. By 1901, Chavarri was dead, but it is doubtful that his presence would have changed the election results.

90. See Real Cuesta, *El carlismo,* pp. 222–223. Unfortunately, Chavarri's intervention in these electoral pacts cannot be fleshed out. By avoiding a contest in the provincial elections, Chavarri's candidates and the Carlists were assured of the three seats corresponding to the majority and the single seat that was given to the minority that received the most votes.

91. To be precise, in 1893, there were only five seats allotted to Vizcaya in the Cortes, one for each of the following districts: Bilbao, Durango, Guernica, Marquina, and Valmaseda. In 1896, a new district, Baracaldo, was added. For a complete list of national deputies, see App. IV.

92. F. García de Cortázar, "La oligarquía," p. 253.

93. Fusi, *Política obrera,* p. 107. See also Real Cuesta, *El carlismo,* p. 221, n. 138, on the difficulties the Carlists faced in finding candidates willing to spend large amounts of money in electoral campaigns.

94. F. García de Cortázar, "La oligarquía," p. 236.

95. J. Varela Ortega, *Los amigos políticos. Partidos, elecciones y caciquismo en la Restauración (1875–1900)*, p. 235.

96. Among the nineteen businessmen elected deputies during the 1893–1910 period, only three did not belong to the Conservative or Liberal Parties: J. Acillona (Integrist), F. Solaegui (Republican), and H. Echevarrieta (Republican).

97. *Meeting protesta contra los tratados de comercio celebrado en Bilbao el dia 9 de diciembre de 1893*, p. 88.

98. P. Alzola, *Memoria relativa al estado de la industria siderúrgica en España*, p. 56.

99. See "Exposición dirigida por los mineros de Vizcaya al ministro de hacienda en 1891," a petition reproduced in Villota, *Vizcaya*, App. V, pp. 312–318. The measure was also protested by a group of steel producers who lobbied for tariff protection in 1891. See J. Angoloti, F. Goitia, and G. Pradera, *La cuestión arancelaria*, p. 52.

100. See González Portilla, *La formación de la sociedad capitalista*, vol. I, pp. 193–194, 202–203. González Portilla was the first to advance the idea of the division; it was then adopted by Corcuera, *Orígenes*, pp. 60–63, and even most recently in Arana, *La Liga*, pp. 182–183.

101. A thorough historiographical discussion of the issue appears in Arana, *La Liga*, pp. 33–45.

102. Ibid., pp. 87, 94–95; and J. Muñoz, S. Roldán, and A. Serrano, "La involución nacionalista y la vertebración del capitalismo español," pp. 21–22.

103. The pamphlet is reproduced in González Portilla, "La industria siderúrgica," pp. 176–181.

104. Tariffs affecting the iron industry were modified in 1841, 1849, 1852, 1862, and 1869. The complaints of Vizcaya's two main iron companies, Santa Ana de Bolueta and Nuestra Señora del Carmen, are in "Respuestas de los industriales de Bilbao a la información sobre el derecho diferencial de bandera, 1866," reproduced in García Merino, *La formación de una ciudad industrial*, pp. 768, 777.

105. Ibid., p. 764.

106. Ibid., p. 780.

107. Ibid.

108. For a view of merchants as free traders, see Corcuera, *Orígenes*, p. 98.

109. See M. Basas, "La vinculación entre el proteccionismo vasco y el catalán," pp. 276–277.

110. For the connection between Catalan businessmen and protectionism, see M. Puges, *Cómo triunfó el proteccionismo en España;* M. Izard, *Manufactureros, industriales y revolucionarios.*

111. See J. Serrano Sanz, *El viraje proteccionista en la Restauración. La política comercial española, 1875–1895*, pp. 130–131; Varela, *Los amigos*, pp. 267–269. The treaties were especially obnoxious for the protectionists because, by granting most-favored-nation status, the Spanish government was authorized to reduce the duties established in the tariff law.

112. P. Alzola, *La política económica mundial y nuestra reforma arancelaria*, p. 226.

113. See Serrano Sanz, *El viraje*, esp. chap. 2, pp. 46–65. For instance, the tariff on imported iron bars was reduced from 130 reales per ton in 1862 (Fernández Pinedo, "Nacimiento y consolidación," p. 10) to 80 reales per ton in 1882 (Serrano, *El viraje*, p. 62).

114. See, for instance, Angoloti, Goitia, and Pradera, *Cuestión arancelaria*, pp. 69–70; P. Alzola's speech in *Meeting*, p. 98.

115. A.C.C.B., L. 83. In addition to the merits of their case against the tax, the three factories must have benefited from the fact that Vizcaya's deputy, M. Allendesalazar, was a member of the parliamentary finance commission that introduced the reduction of the export tax in the 1887–1888 national budget.

116. A.C.C.B., L. 71.

117. A.C.C.B., L. 184.

118. Ibid.

119. The figures for Italian imports are taken from the B.P.P. for the 1886–1900 period.

120. A.A.H.B., Comité de Madrid, Dec. 10, 1886.

121. A.A.H.B., Comité de Madrid, Mar. 13, 1888.

122. A.A.H.B., Comité de Madrid, Sept. 26, 1885.

123. B. Alzola, *Estudio relativo a los recursos que la industria nacional posee para las construcciones y armamentos navales*, p. 284. Coincidentally, the author was Pablo Alzola's brother. At the time the report was written, Pablo was not a director of Altos Hornos de Bilbao.

124. A.C.C.B., L. 301.

125. A report of this mission is in A.C.C.B., L. 128. Three of the four delegates representing the Chamber of Commerce were stockholders in metallurgical companies: Echevarría (La Vizcaya and La Iberia), Alonso (Aurrera), and Uriguen (Altos Hornos). The fourth member (Ugarte) could not be identified.

126. It would be extremely strange if the minister of the navy had been unaware of the progress made by Bilbao's steel industry. In addition to B. Alzola's book, members of the board of directors of Altos Hornos de Bilbao had met several times with high officials of the ministry to discuss possible contracts. See A.A.H.B., Comité de Madrid, Sept. 26, 1885.

127. For the different cartels formed during the 1886–1913 period, see González Portilla, *La siderurgia vasca*, chap. 7. The cartels were the first step toward the merger of three of the main Vizcayan factories in 1901.

128. Varela, *Los amigos*, p. 215.

129. Arana, *La Liga*, pp. 73–74, 82–83.

130. Ibid., pp. 247, 564.

131. Angoloti, Goitia, and Pradera, *Cuestión arancelaria*, p. 74.

132. Serrano Sanz, *El viraje*, p. 205, table 5.1. Whereas in 1882 protection in the first and second schedules was 42 and 28.4 percent, respectively, in 1891 it was increased to 48 and 40 percent.

133. A.A.H.B., Comité de Madrid, Apr. 7, 1886, and Dec. 2, 1886.

134. A. Gómez Mendoza claims that rail consumption in Spain during the 1890s

represented only 4 percent of total pig iron output, and 9.5 percent of total steel production. See "Spain," p. 158.

135. González Portilla, *La formación de la sociedad capitalista,* vol. II, p. 20.

136. Arana, *La Liga,* pp. 84–85.

137. Ibid., p. 99.

138. V. Chavarri, at the time a senator, acted as chairman of the meeting. Also present were B. Chavarri, brother of Víctor and deputy for Valmaseda; the Marqués de Casa Torre, deputy from Durango; M. M. Arrótegui, deputy from Guernica; F. Martínez Rodas, deputy from Marquina and a shipping entrepreneur; J. Martínez Rivas, a senator. For the participation of these figures in local business, see chap. 4.

139. After the meeting, a book with all the speeches and the resolutions adopted was published. See *Meeting.*

140. González Portilla, *La formación de la sociedad capitalista,* vol. I, pp. 202–203; Arana, *La Liga,* pp. 182–183.

141. Arana, *La Liga,* p. 183.

142. In fact, in a letter from the Liga Nacional de Productores to the LVP, the shippers' meeting plan in 1894 was not characterized as in favor of free trade, as Arana relates, but as a protest against the high duties imposed on registering ships under the Spanish flag (A.C.V.E.M., Box 27, letter dated Apr. 24, 1894).

143. A.H.D.V., Sesiones de la Diputación (1894), pp. 38–42. Aznar's motion was opposed by C. Palacio, a deputy who was a member of the LVP and a director of La Iberia. Palacio did not object to the principle of granting protection to the shipping industry but rather to the timing of the request. He thought that the Diputación should concentrate its efforts on denouncing the commercial treaties.

144. For instance, R. Real de Asua, a founder of Vapores Serra and La Flecha; J. M. Olavarri, a director of Bilbaína de Navegación; and F. Martínez Rodas, a founding shareholder of Bilbaína de Navegacion and promoter of Marítima Unión and Marítima Rodas. For a complete list of members of LVP, see Arana, *La Liga,* pp. 613–620.

145. See A.H.P.V., B. Onzoño, Feb. 24, 1893, Aug. 26, 1893, partnership contracts for the steamers *Poveña* and *Algorta.*

146. On the close association between the two groups, see the annual report of the LVP for 1906 in A.C.C.B., collection of reports. Also, a letter from R. Sota thanking the LVP for its support against the parliamentary maneuvers that were delaying the approval of the law protecting the shipping industry in 1909 (A.C.V.E.M., Box 29, letter dated Apr. 20, 1909).

147. In fact, protectionism became so popular that the Socialists, who opposed it, were unable to get the support of the steel workers during the 1890s. See Fusi, *Política obrera,* pp. 146–147.

148. The Círculo Minero was created in 1886. Its main lobbying activities during the last decade of the nineteenth century concerned tax issues. See Villota, *Vizcaya.*

149. The only ones who were not members of all three organizations were Arrótegui, Echevarría, and Martínez Rodas, who were associated only with the LVP and the Chamber of Commerce. Although the Marqués de Casa Torre did not appear as a member of the Círculo, he was a partner in Salazar and Co., which belonged to the association. Aznar was not a member of the Círculo either, but his partner, Sota, was. For the membership lists, see ibid., pp. 321–323; Arana, *La Liga,* pp. 620–621; Monthly Bulletin of the Chamber of Commerce, Nov. 1889.

150. See Villota, *Vizcaya,* pp. 89–90.

151. Arana, *La Liga,* pp. 386–393. Another example: in 1898, R. Ibarra, a deputy for Baracaldo, lobbied to avoid a high tax on the export of iron ore and on some metallurgical products such as bars (A.C.V.E.M., Box 20, letter from Ibarra to president of LVP, dated May 26, 1898).

152. Arana, *La Liga,* pp. 378–393. Arana seems to downplay the lobbying (pp. 390–391) because Cánovas appears to have acted independently at the last moment in a crucial deal with the Liberals that permitted the passage of the law. If Cánovas acted on his own, it cannot be forgotten that he had been in contact with the Basques ever since they had obtained from him the promise to revise the railroad companies' exemptions in 1890.

153. Ibid., pp. 486–492.

154. Ibid., pp. 564–571; Villota, *Vizcaya,* pp. 89–90.

155. Unfortunately, it has been impossible to establish a reason for this discrimination. Villota mentions that it existed and that it bothered the Vizcayans who had to pay the tax, but does not explain how it came to be.

156. The Diputación's decision to accept the tax was severely criticized by the LVP. The league was especially annoyed about the tax on dividends, which it considered a disguised double imposition, since the companies also had to contribute a portion of their income. See A.L.V.P., Box 34, "Impuestos no concertados."

157. Arana's overall conclusion is more pessimistic because he does not focus on the economic consequences that the businessmen's lobbying had on their companies. See *La Liga,* esp. pp. 582–585.

158. The figures represent net earnings as a percentage of the stock capital of the company. The averages were calculated from the data in González Portilla, *La siderurgia vasca,* pp. 180 (table 6-4), and 185 (table 6-9).

CHAPTER SEVEN. CONCLUSION

1. See Gerschenkron, "The Typology of Industrial Development As a Tool of Analysis," pp. 77–97 in *Continuity in History and Other Essays;* idem "Economic Backwardness in Historical Perspective," pp. 5–30; idem, "The Approach to European Industrialization: A Postscript," pp. 353–364 in *Economic Backwardness in Historical Perspective.*

2. Indeed, List was the favorite economic author of Pablo Alzola, a leading figure in Vizcaya's lobbying campaigns for protection. See Arana, *La Liga,* p. 158.

3. Mendels, "Proto-industrialization."

4. See J. Schumpeter, *Imperialism and Social Classes,* pp. 156–158.

5. Ibid., p. 159.

Bibliography

Primary Sources

ARCHIVES

A.A.H.B.—*Archivo Altos Hornos de Vizcaya*
Actas del Comité de Madrid (1882–1896)
Annual Reports (1882–1899)
A.B.B.—*Archivo del Banco de Bilbao*
Partnership contract for Bilbao-Guernica Railroad (E-1)
Fondo Gandarias: documents related to Gandarias family mining ventures
A.B.E.—*Archivo Histórico del Banco de España*
Registro de Testamentarias y Poderes (Sucursal Bilbao), books 114/115, 118/119. The copies of the testaments and powers of attorney held by the bank are filed in several boxes, which can be found by the number assigned in the registry.
Registro de Sociedades Anónimas (Bilbao, 1929), book no. 132
A.C.C.B.—*Archivo de la Cámara de Comercio de Bilbao*
Collection of annual reports of several important Vizcayan companies
Record of Bilbao's ruling commission
A.C.V.E.M.—*Archivo de la Confederación Vizcaína de Empresas del Metal*
Documentation pertaining to the Liga Vizcaína de Productores. Among the records, numerous newspaper clippings of political and economic events affecting Vizcaya
A.C.J.G.—*Archivo General del Señorío de Vizcaya, in Guernica*
Electoral lists
Mining documents [Archivo alto, Reg. 1 (report of González Azaola on the conditions of the mines in 1827), Reg. 2, L. 2, no. 2; L. 3, no. 5; Reg. 3, L. 1, no. 6-3, L. 2, no. 6; L. 4, no. 2]
Comercio. Reg. no. 5, L. 5. (draft of a registry of partnerships formed in 1885)
A.H.D.V.—*Archivo Histórico de la Diputación de Vizcaya*
Sesiones de la Diputación (1850–1900)
Sesiones de las Juntas Generales de Guernica (1850–1876)
Libros de Acuerdos de las Juntas Generales de Guernica (1850–1870)
Fondo del Corregimiento: judicial documentation that in some cases includes testaments, inventories, and business records

Fondo Sota y Aznar: documents related to Sota's mining ventures (especially signatura 2636/6, 2637/10, 2886/10, 2914/1)
Uncatalogued property registry of the city of Bilbao (1895)
Miscellaneous: "Actas sobre fábrica con fuerza motriz de vapor para artículos de hierro dulce, 1862–1871," Armario 14, 20. Partial records of Zarraoa y Cía.
A.H.P.V.—*Archivo Histórico Provincial de Vizcaya*
Protocol books of Bilbao notaries
C. Ansuátegui (1851–1887); J. Ansuátegui (1864–1887); J. J. Barandica (1827); F. Basterra (1842–1870); J. P. Castañiza (1834; 1838); M. Castañiza (1846; 1849–1887); V. L. Gaminde (1850–1870); J. M. Garate (1850–1865); P. Goicoechea (1861–1867); I. Ingunza (1850–1873); B. Onzoño (1855–1887, 1893); F. Uribarri (1850–1887); V. Uribarri (1838; 1844); S. Urquijo (1850–1882); G. Urresti (1850–1869)
Protocol books of Portugalete notaries
J. B. Butrón (1857–1881); R. Vildósola (1860–1882); J. B. Zavalla (1861)
A.M.—*Archivo Histórico de la Marina* (Viso del Marqués)
Statistics on Bilbao's shipyards and merchant fleet (Estadística General L. 62306, 62308; Estadística de Navegación y Pesca, L. 2356, 2361, 2363, 2369
Vizcayan factory contracts with the navy (L. 5479; Sec. Indiferentes, Asuntos Particulares, L. 4761, carpeta 7/6/1893; L. 4016 Arsenales)
A.M.P.—*Archivo Municipal de Portugalete*
List of Triano mine owners in 1861 and other documents related to mining (L. 98, nos. 94, 95, 97, 98, and 100)
List of merchants in the city in 1864 (L. 154, no. 8)
A.P.M.—*Archivo de Protocolos de Madrid*
Notarial documents from the Spanish Consulate in Liverpool
F.C.R.—*Archive du Ministère des Affaires Etrangères (France)*
Correspondance Consulaire Commerciale, vol. 7 (1851–1860); vol. 8 (1861–1866); vol. 9 (1867–1874); vol. 10 (1875–1892); vol. 11 (1893–1901)
I.F.A.—*Ibarra Family Archive*
Wills and inventories of Juan María Ibarra, Luz Arregui, Ramón and Juan Luis Ibarra; copybook of letters related to the mining interests of the estate of Gabriel María Ibarra; annual stockholders' reports of the Orconera; reports of the managing director of the Orconera in Bilbao to headquarters in London; partnership contract of the Franco-Belge Company; P. Celis's reports prepared in 1895 and 1898 on state and value of mines; other documents related to the lease and exploitation of family mines
K.A.—Historisches Archiv Fried. Krupp
Financial statements and annual reports of the Orconera Iron Ore Co. (1878, 1882)

Secondary Sources

NEWSPAPERS, MAGAZINES, JOURNALS

Anuario de la gran industria de España (1918)
Anuario de la minería, metalurgia y electricidad de España (1894–1909)
Anuario marítimo (1915)
Boletín oficial de la Cámara de Comercio de Bilbao (1888–1890)
Boletín oficial de Vizcaya (B.O.V.) (1850–1900)
Eco bilbaíno (1865)
Gaceta de los caminos de hierro (1903)
Irurac Bat (1859–1862)
Journal of the Iron and Steel Institute (J.I.S.) (1877–1900)
El Nervión (1891–1892)
Revista Bilbao (1895–1900)
Revista minera, metalúrgica y de ingeniería (1850–1906)

BOOKS AND ARTICLES

Abad, P. Camilo María. *Vida de la sierva de Dios, Da. Rafaela Ibarra de Vilallonga.* Bilbao: Imp., Lit. y Enc. de Emeterio Verdes, 1919.

Aguirreazkuenaga, Joseba. *Vizcaya en el siglo XIX: las finanzas de un estado emergente.* Bilbao: Editorial Universidad del País Vasco, 1987(?).

Aldana, Lucas. "Descripción de la mina de hierro de Triano." *Revista minera, metalúrgica, y de ingeniería* (1851): 302–388.

Alzola, Benito. *Estudio relativo a los recursos que la industria nacional posee para las construcciones y armamentos navales.* Madrid: Imp. de Infantería de Marina, 1886.

Alzola, Pablo. *Memoria relativa al estado de la industria siderúrgica en España.* Bilbao: Imp. Casa de Misericordia, 1896.

———. *La política económica mundial y nuestra reforma arancelaria.* Bilbao: Imp. Casa de Misericordia, 1906.

Angoloti, J.; F. Goitia; and G. Pradera. *La cuestión arancelaria. Consideraciones acerca del voto particular del Exmo. Sr. D. Segismundo Moret.* Madrid: Establecimiento Tipográfico de Fortanet, 1892.

Anuario Estadístico de España. 1858. Instituto Nacional de Estadística.

Arana Pérez, Ignacio. *La Liga Vizcaína de Productores y la política económica de la Restauración, 1894–1914.* Bilbao: Caja de Ahorros Vizcaína, 1988.

Artiñano y Zuricalday, Aristides. *El señorío de Vizcaya, histórico y foral.* Barcelona: Establecimiento Tipográfico de Mariol y López, 1885.

Artola, Miguel. *La burguesía revolucionaria (1808–1874).* Madrid: Alianza Editorial, 1983.

Augustine-Perez, Dolores L. "The Banker in German Society." In *Finance and Financiers in European History, 1880–1960,* edited by Y. Cassis. Cambridge: Cambridge University Press, 1992.

———. "Very Wealthy Businessmen in Imperial Germany." *Journal of Social History* 22 (Winter 1988): 299–321.
Bacon, John Francis. *Six Years in Vizcaya.* London: Smith, Elder & Co., 1838.
Barahona, Renato. "Basque Regionalism and Centre-Periphery Relations 1759–1833." *European Studies Review* 13 (July 1983): 271–295.
———. "The Making of Carlism in Vizcaya (1814–1833)." Ph.D. dissertation. Princeton University, 1979.
Basas, Manuel. *Aspectos de la vida económica de Bilbao 1861–1866.* Bilbao: Villar, 1967.
———. "Cosme, Tomás y otros Zubirias." *Información* (Oct. 1987): 84–86.
———. *Economía y sociedad bilbaína en torno al sitio de 1874.* Bilbao: Junta de Cultura de Vizcaya, 1978.
———. "Enrique Aresti, política y negocios." *Información* (Mar. 1987): 59–61.
———. *El Lloyds bilbaíno hace un siglo.* Bilbao: Editorial Vizcaína, 1961.
———. "La vinculación entre el proteccionismo vasco y el catalán." In *Industrialización y nacionalismo. Análisis comparativos,* edited by M. González Portilla, J. Maluquer de Motes, and B. de Riquer Permanyer. Barcelona: Universidad Autónoma de Barcelona, 1985.
Basurto Larrañaga, Román. *Comercio y burguesía mercantil de Bilbao en la segunda mitad del siglo XVIII.* Bilbao: Editorial Universidad del País Vasco, 1983.
Bautier, Robert-Henri. "Notes sur le commerce du fer en Europe occidental du XIIIe au XIVe siècle." *Revue d'Histoire de la Siderurgie* 1 (1960): 7–35.
Bell, Lowthian. *The Iron Trade of the United Kingdom.* London: British Iron Trade Association, 1886.
Beltza (pseud. of E. López Adán). *Del carlismo al nacionalismo burgués.* San Sebastián: Ed. Txertoa, 1978.
Bilbao, Luis María. "Renovación tecnológica y estructura del sector siderúrgico en el País Vasco durante la primera etapa de la industrialización (1849–1880). Aproximación comparativa con la industria algodonera de Cataluña." In *Industrialización y nacionalismo. Análisis comparativos,* edited by M. González Portilla, J. Maluquer de Motes, and B. de Riquer Permanyer. Barcelona: Universidad Autónoma de Barcelona, 1985.
———. "La siderurgia vasca, 1700–1885. Atraso tecnológico, política arancelaria y eficiencia económica." *IX Congreso de Estudios Vascos. Antecedentes de la sociedad vasca actual.* San Sebastián: Eusko-Ikaskuntza, 1984.
Bourson, E. "Les mines de Somorrostro." *Revue universelle des mines,* ser. 2, vol. 4 (1878): 648–690.
Brading, D. A. *Miners and Merchants in Bourbon Mexico 1763–1810.* Cambridge: Cambridge University Press, 1971.
Braudel, Fernand. *The Mediterranean and the Mediterranean World in the Age of Philip II.* New York: Harper Colophon Books, 1976.
British Parliamentary Papers. Consular Reports from Bilbao (1850–1905).
Burges, Ellen. *Vizcaya or Life in the Land of the Carlists at the Outbreak of the Insurrection.* London: Henry S. King & Co., 1874.

Callahan, William J. *Honor, Commerce and Industry in Eighteenth Century Spain.* Cambridge: Harvard Graduate School of Business Administration, 1972.
Caro Baroja, Julio. *La hora navarra del XVIII.* Pamplona: Comunidad Foral de Navarra, 1969.
———. *Vasconiana.* Madrid: Editorial Minotauro, 1957.
———. *Los vascos.* Madrid: Ediciones Istmo, 1984.
———. *Los vascos y el mar.* San Sebastián: Editorial Txertoa, 1981.
Carr, J. C., and W. Taplin. *History of the British Steel Industry.* Cambridge: Harvard University Press, 1962.
Carr, Raymond. *Spain, 1808–1975.* Oxford: Oxford University Press, 1982.
Carrera de Odriozola, Albert. "La producción industrial catalana y vasca 1844–1935. Elementos para una comparación." *Industrialismo y nacionalismo. Análisis comparativos,* edited by M. González Portilla, J. Maluquer de Motes, and B. de Riquer Permanyer. Barcelona: Universidad Autónoma de Barcelona, 1985.
Castells, Luis. *Modernización y dinámica política en la sociedad guipuzcoana de la Restauración.* Madrid: Siglo XXI, 1987.
Censo de población de Bilbao tomado del padrón que de su vecindario han formado los señores alcaldes de Barrio en marzo de 1869. Bilbao: Imprenta de Larumbe.
Cervantes, Miguel de. *Don Quixote.* New York: Penguin, 1979.
Chastagnaret, Gérard. "Conclusiones." In *La industrialización del norte de España,* edited by E. Fernández Pinedo and J. L. Hernández Marco. Barcelona: Editorial Crítica, 1988.
Chernow, Ron. *The House of Morgan.* New York: Touchstone, 1990.
Childs, W. R. "England's Iron Trade in the Fifteenth Century." *Economic History Review* 34 (Feb. 1981): 25–47.
Collins, Roger. *The Basques.* New York: Basil Blackwell, 1987.
Corcuera, J. *Orígenes, ideología y organización del nacionalismo vasco, 1876–1914.* Madrid, Siglo XXI, 1979.
Coverdale, John F. *The Basque Phase of Spain's First Carlist War.* Princeton: Princeton University Press, 1984.
Crouzet, François. *The First Industrialists.* Cambridge: Cambridge University Press, 1985.
Daumard, Adeline. *Les bourgeois de Paris au XIXe siècle.* Paris: Flammarion, 1970.
———. *Les bourgeois et la bourgeoisie en France.* Saint-Amand: Aubier-Montagne, 1987.
———. *Les fortunes françaises au XIX siècle.* Paris: École Hautes Études et Sciences Sociales, 1973.
Delmas, Juan E. *Guía histórico descriptivo del viajero en el señorío de Vizcaya en 1864.* Madrid: Diputación de Vizcaya, 1944.
———. *Diccionario de claros varones de Vizcaya.* Bilbao: Edición separada de la Gran Enciclopedia Vasca, 1970.
Diario de Sesiones de Cortes, Nov. 22, 1900, no. 3, app. 4.
Diccionario enciclopédico del País Vasco. San Sebastián: Haranburu, 1985.

Douglass, William, and Jon Bilbao. *Amerikanuak: Basques in the New World.* Reno: University of Nevada Press, 1975.

Echevarría, Ignacio, and Federico Grijelmo. *Las minas de hierro de la provincia de Vizcaya. Progresos realizados en esta región desde 1870 and 1899.* Bilbao: Imprenta y Litografía de Ezequiel Rodríguez, 1900.

Elhuyar, Fausto. "Estado de las minas de Somorrostro." 1783. Reprint *Revista de trabajo* 21 (1968): 99–105.

Elliot, J. H. *The Count-Duke of Olivares: The Statesman in an Age of Decline.* New Haven: Yale University Press, 1986.

Escudero, Antonio. "La minería vizcaína durante la primera guerra mundial." *Revista de Historia Económica* 2 (spring–summer 1986): 365–387.

Estadística de Comercio Exterior de España. 1860–1863. Dirección General de Aduanas.

Fernández Delgado, J. "Silenciosos, comedidos y espléndidos. La quiebra de la función religiosa del testamento." In *Madrid en la sociedad del siglo XIX,* edited by Luis E. Otero Carvajal and Angel Bahamonde. Madrid: Comunidad de Madrid, Consejería de Cultura, 1986.

Fernández Pinedo, Emiliano. *Crecimiento económico y transformaciones sociales del País Vasco 1100–1850.* Madrid: Siglo XXI, 1974.

———. "Estructura de los sectores agropecuarios y pesqueros vascos (1700–1890)." *IX Congreso de Estudios Vascos.* San Sebastián: Eusko-Ikaskuntza, 1984.

———. "Haciendas forales y revolución burguesa: las haciendas vascas en la primera mitad del siglo XIX." *Hacienda Pública Española* 108–109 (1987): 197–220.

———. "La industria siderúrgica, la minería y la flota vizcaína a fines del siglo XIX. Unas puntualizaciones." In *Mineros, sindicalismo y política,* José A. Fernández Villa et al. Fundación José Barreiro, 1985.

———. "Nacimiento y consolidación de la moderna siderurgia vasca (1849–1913): el caso de Vizcaya." *Información Comercial Española* (June 1983): 9–19.

Fernández Pinedo, Emiliano, and Luis María Bilbao. "Auge y crisis de la siderometalurgia tradicional en el País Vasco (1700–1850)." In *La economía española al final del Antiguo Régimen. II. Manufacturas,* edited by Pedro Tedde. Madrid: Alianza Editorial, 1982.

Flinn, M. W. "British Steel and Spanish Ore: 1871–1914." *Economic History Review,* 2nd ser., vol. 9 (1955–1956): 84–90.

Fontán, R., and L. Larrañaga. *El libro de Bilbao. Guia artístico-comercial.* Bilbao: Imprenta de El Porvenir Vascongado, 1893.

Forster, Robert. *Merchants, Landlords, Magistrates: The Depont Family in Eighteenth Century France.* Baltimore: Johns Hopkins University Press, 1980.

Fraile, Pedro. "El carbón inglés en Bilbao: una reinterpretación." *Moneda y Crédito* 160 (1982): 85–96.

———. "El País Vasco y el mercado mundial, 1900–1930." In *La modernización económica de España 1830–1930,* edited by N. Sánchez Albornoz. Madrid: Alianza Editorial, 1985.

Fusi, J. P. *Política obrera en País Vasco 1880–1923.* Madrid: Turner, 1975.

Gaminde, Víctor Luis de. *Intereses de Bilbao. Examen de lo perjudicial que sería la permanencia del sistema foral en el siglo XIX al comercio y a la industria del país y a los liberales de Vizcaya.* Bilbao: Imprenta de Alfonso Depont, 1837.

García de Cortázar, Fernando. "La oligarquía vasca a comienzos del siglo XX." *Historia del pueblo vasco,* vol. III, F. García de Cortázar et al. San Sebastián: Erein.

García de Cortázar, José Antonio. *Vizcaya en el siglo XV. Aspectos económicos y sociales.* Bilbao: Caja de Ahorros Vizcaína, 1966.

García Merino, Luis Vicente. *La formación de una ciudad industrial. El despegue urbano de Bilbao.* Bilbao: Instituto Vasco de Administración Pública, 1987.

García-Sanz Marcotegui, Angel. "La evolución demográfica vasca en el siglo XIX." *Congreso de historia de Euskal Herria,* vol. IV. Vitoria: Servicio de Publicaciones del Gobierno Vasco, 1988.

———. "El origen geográfico de los inmigrantes y los inicios de la transición demográfica en el País Vasco (1877–1930). Contribución al estudio de sus interinfluencias." *Ekonomiaz* 9–10 (winter-spring 1988): 189–223.

García Venero, Maximiano. *Historia del nacionalismo vasco.* Madrid: Editorial Nacional, 1969.

Garmendia, Vicente. *La ideología carlista (1868–1876).* Zarauz: Diputación Foral de Guipúzcoa.

Gerschenkron, A. *Continuity in History and Other Essays.* Cambridge: Harvard University Press, 1968.

———. *Economic Backwardness in Historical Perspective.* Cambridge: Harvard University Press, 1976.

Gill, William. "The Present Position of the Iron Ore Industries of Biscay and Santander." *Journal of Iron and Steel Institute* 50, no. 2 (1896): 36–99.

Goenaga, Ignacio. "El hierro de Vizcaya." *Revista minera* 34 (1883): 296–299; 311–314; 328–329; 339–341; 355–358; 447–451; 459–462.

Gómez Mendoza, Antonio. "Los ferrocarriles y la industria siderúrgica (1855–1913): *Moneda y Crédito* 155 (1980): 3–19.

———. "Spain." In *Railways and the Economic Development of Western Europe 1830–1914,* edited by Patrick O'Brien. New York: St. Martin's Press, 1983.

González, Enric. "La fragua de la Comercial." *El País* (Jan. 31, 1988): 1–3.

González Portilla, Manuel. "El desarrollo industrial de Vizcaya y la acumulación de capital en elúltimo tercio del siglo XIX." *Anales de Economía,* no. 24 (Oct.-Dec. 1974): 43–85.

———. *La formación de la sociedad capitalista en el País Vasco.* 2 vols. San Sebastián: L. Haranburu Editor, 1981.

———. "La industria siderúrgica en el País Vasco: del verlangssystem [*sic*] al capitalismo industrial." In *Crisis del Antiguo Régimen e industrialización en la España del siglo XIX,* edited by M. Tuñón de Lara et al. Madrid: Editorial Cuadernos para el Diálogo, 1977.

———. "El mineral de hierro español (1870–1914): su contribución al crecimiento económico inglés y a la formación del capitalismo vasco." *Estudios de historia social* 1 (1977): 55–112.

———. "Los orígenes de la sociedad capitalista en País Vasco." *Saioak* 1 (1977): 67–127.

———. *La siderurgia vasca (1880–1901).* Bilbao: Servicio Editorial Universidad del País Vasco, 1985.

Gortázar, Guillermo. *Alfonso XIII, hombre de negocios.* Madrid: Alianza Editorial, 1986.

Gortázar, J. C. *Bilbao a mediados del siglo XIX según un epistolario de la época.* Bilbao: Imprenta de la Biblioteca de Amigos del País, 1920.

Greenwood, Davydd J. "Continuity in Change: Spanish Basque Ethnicity As a Historical Process." In *Ethnic Conflict in the Western World,* edited by Milton J. Esman. Ithaca: Cornell University Press, 1977.

Gruner, E. "Barcelone-Bilbao, notes de voyages (Oct. 1888)." *Mémoires et compte rendu des travaux de la Société des Ingénieurs Civils* (1889): 197–277.

Guía de Bilbao y Vizcaya. Bilbao: Imp. Casa de Misericordia, 1903.

Guiard, Teófilo. *Historia de la noble villa de Bilbao.* Bilbao: Editorial La Gran Enciclopedia Vasca, 1971.

———. *Historia del Consulado y Casa de Contratación de Bilbao.* 2 vols. Bilbao: J. de Astuy, 1913.

Hall, P. D. "Family Structure and Economic Organization: Massachusetts Merchants, 1750–1850." In *Family and Kin in Urban Communities, 1700–1930,* edited by T. Haraven. New York: New Viewpoints, 1977.

Hansen, Niles M. "Sources of Economic Rationality." *Protestantism, Capitalism and Social Science,* edited by Robert W. Green. Lexington, Mass.: D. C. Heath, 1973.

Harrison, Joseph. "La industria pesada, el estado y el desarrollo económico en el País Vasco, 1876–1936." *Información Comercial Española* (June 1983): 21–32.

———. "Los orígenes del industrialismo moderno en el País Vasco." *Hacienda Pública Española* 55 (1978): 209–221.

Harvey, Charles, and Peter Taylor. "Mineral Wealth and Economic Development: Foreign Direct Investment in Spain, 1851–1913." *Economic History Review,* 2nd ser., vol. 40 (1987): 185–207.

Hernández Ponce, Roberto. "El aporte vasco a la construcción de Chile. Gobernadores del siglo XVI." *Journal of Basque Studies* 6, no. 1 (summer 1985): 39–44.

Herr, Richard. *An Historical Essay on Modern Spain.* Berkeley & Los Angeles: University of California Press, 1974.

Hormaeche, Ramón. *Leyes civiles de Vizcaya.* Bilbao, 1891.

Humboldt, Wilhelm von. *Guillermo Humboldt y el País Vasco.* San Sebastián: Imprenta de la Diputación de Guipuzcoa, 1925.
Ibáñez, Guillermo (ed.). *Anuario financiero de Bilbao.* Bilbao: Vda. e hijos de Grijelmo, 1918.
Icaza, Pedro. "La fundación de la Comercial." *El Correo Español–El Pueblo Vasco* (Feb. 8, 1992): 48.
Isard, Walter. "Some Locational Factors in the Iron and Steel Industry since the Early Nineteenth Century." *Journal of Political Economy* 56, no. 3 (June 1948): 203–217.
Izard, M. *Manufactureros, industriales y revolucionarios.* Barcelona: Editorial Crítica, 1979.
Kocka, Jürgen. "Entrepreneur and Manager in the Industrial Revolution." In *Cambridge Economic History of Europe.* Vol. VII. Cambridge: Cambridge University Press, 1978.
Koditschek, Theodore. *Class Formation and Urban-Industrial Society: Bradford, 1750–1850.* Cambridge: Cambridge University Press, 1990.
Landes, David S. *The Unbound Prometheus.* New York: Cambridge University Press, 1985.
Lannon, Frances. "The Socio-Political Role of the Spanish Church—A Case Study." *Journal of Contemporary History* 14 (1979): 193–210.
Larrea, María de los Angeles. *Caminos de Vizcaya en la segunda mitad del siglo XVIII.* Bilbao: Editorial La Gran Enciclopedia Vasca, 1974.
Larrea, María de los Angeles, and R. M. Mieza. "La Diputación General del Señorío de Vizcaya." *Journal of Basque Studies* 6, no. 1 (summer 1985): 8–20.
Larrea, Víctor, and J. Lazúrtegui. *Merchant's and Shipmaster's Practical Guide to the Port of Bilbao.* Bilbao, 1882.
Lazúrtegui, Julio. "La industria minera de Vizcaya." In *Geografía general del País Vasco-Navarro.* Vol. V, edited by Carmelo Echegaray. Barcelona: Editorial Alberto Martin.
Lequerica, José F. *La actividad económica de Vizcaya en la vida nacional.* Madrid: Real Academia de Ciencias Morales y Políticas, 1956.
Libro de actos conmemorativos del centenario del Banco de Bilbao. Bilbao, 1956.
Libro del centenario de la Cámara de Comercio de Bilbao. Bilbao, 1987.
McDonogh, Gary. *Good Families of Barcelona.* Princeton: Princeton University Press, 1986.
Madoz, Pascual. *Diccionario geográfico, estadístico e histórico de España y sus posesiones de Ultramar.* 16 vols. Madrid: Establecimiento Tipográfico de P. Madoz y L. Sagasti, 1845–1850.
Maeztu, Ramiro de. *Autobiografía.* Madrid: Editorial Nacional, 1962.
Maluquer de Motes, J. "Cataluña y el País Vasco en la industria eléctrica española 1901–1935." In *Industrialismo y nacionalismo. Análisis comparativos,* edited by M. González Portilla, J. Maluquer de Motes, and B. de Riquer Permanyer. Barcelona: Universidad Autónoma de Barcelona, 1985.

Mauleón Isla, Mercedes. *La población de Bilbao en el siglo XVIII.* Valladolid: Universidad de Valladolid, 1961.

Mayer, Arno. *The Persistence of the Old Regime.* New York: Pantheon Books, 1981.

Mazas, Diego. *La sociedad de Bilbao en 1887.* Bilbao: Imp. Lit. y Enc. de Emeterio Verdes, 1918.

Meeting protesta contra los tratados de comercio celebrado en Bilbao el día 9 de diciembre de 1893. Bilbao: Imp. Casa de Misericordia, 1894.

Memoria de los actos económico-administrativos de la Diputación provincial interina de Vizcaya. Bilbao: Imprenta Juan E. Delmas, 1880.

Memoria del Banco de España (Bilbao Branch). Bilbao, 1898.

Mendels, Franklin F. "Proto-industrialization: The First Phase of the Industrialization Process." *Journal of Economic History* 32, no. 1 (Mar. 1972): 241–261.

Miralles, R. "Política electoral en Vizcaya." Universidad de Deusto, Facultad de Filosofia y Letras, Nov. 1977 (unpublished).

Moine, Jean-Marie. *Les barons du fer.* Nancy: Presses Universitaires de Nancy, 1989.

Montero, Manuel. "La minería de Vizcaya durante el siglo XIX." *Ekonomiaz* 9–10 (winter–spring 1988): 143–169.

———. "Modernización económica y desarrollo empresarial en Vizcaya 1890–1905." *Ekonomiaz* 9–10 (winter–spring 1988): 225–253.

———. "Régimen liberal y autonomía vasca." *Saioak* 5 (1983): 3–27.

Muñoz, J.; S. Roldán; and A. Serrano. "La involución nacionalista y la vertebración del capitalismo español." *Cuadernos Económicos de I.C.E.* 5 (1978).

Nadal, Jordi. "La economía española, 1829–1931." In *El Banco de España. Una historia económica,* edited by J. Nadal et al. Madrid: Banco de España, 1970.

———. "The Failure of the Industrial Revolution in Spain 1830–1914." In *The Fontana Economic History of Europe.* Vol. IV, edited by C. M. Cipolla. London: Fontana Books, 1973.

———. *El fracaso de la revolución industrial en España, 1814–1913.* Barcelona: Editorial Ariel, 1977.

———. "La industria fabril española en 1900. Una aproximación." In *La economía española en el siglo XX,* edited by Jordi Nadal, Albert Carreras, and Carles Sudria. Barcelona: Editorial Ariel, 1987.

———. "Industrialización y desindustrialización del sureste español, 1817–1913." *Moneda y crédito* 120 (1972): 3–80.

Nájera, Maite. "El comercio a través del puerto de Bilbao 1800–1825." In *Historia de la economía marítima del País Vasco,* edited by A. Zabala et al. San Sebastián: Editorial Txertoa, 1983.

O'Brien, Patrick, ed. *Railways and the Economic Development of Western Europe, 1830–1914.* New York: St. Martin's Press, 1983.

Ojeda, Germán. *Asturias en la industrialización española, 1833–1907.* Madrid: Siglo XXI, 1985.

Olabarri, Ignacio. *Relaciones laborales en Vizcaya (1890–1936).* Durango: Leopoldo Zugaza Editor, 1978.

Ormaechea, A. M. "Los ferrocarriles vascos y su dependencia tecnológica." *Congreso de historia de Euskal Herria.* Vol. V. Vitoria: Publicaciones del Gobierno Vasco, 1988.

Ortega y Galindo de Salcedo, Julio. *Bilbao y su hinterland.* Bilbao, 1951.

Ortiz, J. M., and J. M. Portillo. "La foralidad y el poder provincial." *Historia Contemporánea* 4 (1990): 107–121.

Orueta, José de. *Memorias de un bilbaíno, 1870–1900.* San Sebastián: Biblioteca Vascongada de los Amigos del País, 1952.

Otazu y Llana, Alfonso. *El "igualitarismo" vasco y otros mitos.* San Sebastián: Editorial Txertoa, 1986.

———. *Los Rotschild y sus socios en España.* Madrid: O. Hs. Ediciones, 1987.

Palacio Atard, Vicente. *El comercio de Castilla y el puerto de Santander en el siglo XVIII.* Madrid: C.S.I.C., 1960.

Payne, Stanley. *Basque Nationalism.* Reno: University of Nevada Press, 1975.

Pérez Azagra y Aguirre, Antonio. *Noticias genealógicas sobre los Primo de Rivera y los Salcedo.* Bilbao: Imprenta editorial moderna, 1945.

———. *Títulos de Castilla e Indias y extranjeros a vascos.* Vitoria: Editorial Pujol, 1945.

Pérez Galdós, Benito. "Fisonomías sociales, Bilbao." *Obras inéditas.* Vol. I. Madrid: Renacimiento, 1923.

Pérez Moreda, Vicente. "La modernización demográfica." In *La modernización económica de España 1830–1930,* edited by N. Sánchez Albornoz. Madrid: Alianza Editorial, 1985.

Pollard, Sidney, and Paul Robertson. *The British Shipbuilding Industry 1870–1914.* Cambridge: Harvard University Press, 1979.

Postigo, Carmen. *Los conciertos económicos.* San Sebastián: Haranburu, 1979.

Pourcel, Alexandre. "Mines de fer de Bilbao." *Le Génie Civil* (Apr. 6, 1887): 70–74.

Puges, M. *Como triunfó el proteccionismo en España.* Barcelona: Juventud, 1931.

de la Quadra Salcedo, Fernando. *Calles de Bilbao.* Bilbao: Colección el Cofre del Bilbaíno, 1963.

———. *Economistas vascongados.* Bilbao: Editorial El Pueblo Vasco, 1943.

Real Cuesta, J. *El carlismo vasco, 1876–1900.* Madrid: Siglo XXI, 1985.

Retana, J. M. Martín de. "Orígenes, evolución y bibliografía del fuero de Vizcaya." *La Gran Enciclopedia Vasca.* Vol. I. Bilbao.

Robertson, H. M. *The Rise of Economic Individualism.* Cambridge: Cambridge University Press, 1935.

Romero de Solís, Pedro. *La población española en los siglos XVIII y XIX.* Mexico City: Siglo XXI, 1979.

Rostow, W. W. *The Stages of Economic Growth.* Cambridge: Cambridge University Press, 1960.

Rubinstein, W. D. "The Victorian Middle Classes: Wealth, Occupation and Geography." *Economic History Review* 30, no. 4 (1977): 602–623.
Ruiz Rivera, Julián B. *El Consulado de Cádiz. Matrícula de comerciantes 1730–1823.* Cádiz: Diputación Provincial de Cádiz, 1988.
Sánchez Ramos, Francisco. *La economía siderúrgica española.* Madrid: C.S.I.C., 1945.
Schumpeter, J. *Imperialism and Social Classes.* Philadelphia: Orion, 1991.
Serrano Sanz, José M. *El viraje proteccionista en la Restauración. La política comercial española, 1875–1895.* Madrid: Siglo XXI, 1987.
Shaw, Valery. "Exportaciones y despegue económico; el mineral de hierro de Vizcaya, la región de la ría de Bilbao y algunas de sus implicaciones para España." *Moneda y Crédito* 142 (1977): 87–114.
Sociedad Santa Ana de Bolueta. *Sociedad Santa Ana de Bolueta. Centenario de su fundación.* Bilbao, 1951.
Stone, L., and J. C. Fawtier Stone. *An Open Elite? England 1540–1880.* Oxford: Oxford University Press, 1984.
Tedde de Lorca, Pedro. "Comerciantes y banqueros madrileños al final del Antiguo Régimen." In *Historia económica y pensamiento social,* edited by Gonzalo Anes, Luis A. Rojo, and Pedro Tedde. Madrid: Alianza, 1983.
Tejada, Elías de. *El señorío de Vizcaya hasta 1812.* Madrid: Editorial Minotauro, 1963.
Thompson, E. P. *The Making of the English Working Class.* New York: Vintage Books, 1966.
Tortella, Gabriel. "La economía española, 1830–1900." In *Historia de España.* Vol. VIII, edited by Manuel Tuñón de Lara. Barcelona: Editorial Labor, 1981.
———. *Los orígenes del capitalismo en España.* Madrid: Editorial Technos, 1973.
———. "La primera gran empresa química española: la Sociedad Española de la Dinamita." In *Historia económica y pensamiento social,* edited by Gonzalo Anes, Luis A. Rojo, and Pedro Tedde. Madrid: Alianza Editorial, 1983.
———. "Spain." In *Banking and Economic Development,* edited by R. Cameron. Oxford: Oxford University Press, 1972.
Trevor-Roper, H. R. "Religion, the Reformation and Social Change." *The European Witch-Craze of the Sixteenth and Seventeenth Centuries.* New York: Harper Torchbooks, 1969.
Trueba, Antonio. "Organización social de Vizcaya." *La Gran Enciclopedia Vasca.* Vol. II: 602–625; vol. III: 351–366, 463–473. Bilbao.
Tuñón de Lara, Manuel. *Estudios sobre el siglo XIX español.* 1972. Reprint. Madrid: Siglo XXI, 1984.
Ugalde, Martín. *Hablando con los vascos.* Barcelona: Editorial Ariel, 1974.
Uhagón, Francisco R. de. *Los Uhagón, señores de Hodítegui.* Madrid: Establecimiento Tipográfico de Fortanet, 1908.
Unamuno, Miguel de. "Cuatro reseñas." *Obras completas.* Vol. VI. Madrid: Afrodisio Aguado, 1958.

———. "Del Bilbao mercantil al industrial." *Mi Bochito.* Bilbao: Editorial el Cofre Bilbaíno, 1965.

———. "Mi raza." In *Los baskos en la nación Argentina,* edited by José R. Uriarte. Buenos Aires, 1916.

———. *Mivida yotros recuerdos personales,* edited by Manuel García Blanco. Buenos Aires: Editorial Losada, 1959.

———. *Paz en la guerra.* Madrid: Espasa-Calpe, 1980.

Uriarte Ayo, Rafael. "El tráfico marítimo del mineral de hierro vizcaíno (1700–1850)." In *Historia de la economía marítima del País Vasco,* by A. Zabala et al. San Sebastián: Editorial Txertoa, 1983.

Usher, Abbot Payson. "Spanish Ships and Shipping in the Sixteenth and Seventeenth Centuries." *Facts and Factors in Economic History.* New York: Russel & Russel, 1932.

Valdaliso, Jesús. "Grupos empresariales e inversión de capital en Vizcaya, 1886–1913." *Revista de Historia Económica* 1 (1988): 11–40.

Varela Ortega, J. *Los amigos políticos. Partidos, elecciones y caciquismo en la Restauración (1875–1900).* Madrid: Alianza Editorial, 1977.

Vázquez Prada, M. *Negociaciones sobre los fueros entre Vizcaya y el poder central, 1839–1877.* Bilbao: Caja de Ahorros Vizcaína, 1984.

Velada necrológica que se celebró en el Circulo Conservador de Bilbao el 20-10-1911 para honrar la memoria del Excmo. Sr. D. José María de Lizana, Marqués de Casa Torre. Bilbao: Imprenta Casa Misericordia, 1911.

Villota, Ignacio. *Vizcaya en la política minera española.* Zamudio: Publicaciones de la Diputación Foral de Vizcaya, 1984.

Ybarra Osborne, Eduardo, and Eduardo Ybarra Hidalgo. *Notas sobre José María Ybarra. Primer Conde de Ybarra.* Seville: Hijos de Ybarra, 1984.

Ybarra y Berge, Javier. *La casa de Salcedo de Aranguren.* Bilbao: Editorial El Pueblo Vasco, 1944.

———. *Política nacional en Vizcaya.* Madrid: Instituto de Estudios Políticos, 1948.

Zubiria, Conde de. *El libro del Real Sporting Club.* Las Arenas: Club Marítimo del Abra y Real Sporting Club, 1980.

Index

Page numbers with *t* refer to table information.

The Basque Series

A Book of the Basques
Rodney Gallop

In a Hundred Graves:
A Basque Portrait
Robert Laxalt

Basque Nationalism
Stanley G. Payne

Amerikanuak: Basques in the
New World
William A. Douglass and Jon Bilbao

Beltran: Basque Sheepman of the
American West
Beltran Paris, as told to
William A. Douglass

The Basques: The Franco Years
and Beyond
Robert P. Clark

The Witches' Advocate: Basque
Witchcraft and the Spanish Inquisition
(1609–1614)
Gustav Henningsen

Navarra: The Durable Kingdom
Rachel Bard

The Guernica Generation: Basque
Refugee Children of the Spanish
Civil War
Dorothy Legarreta

Basque Sheepherders of the
American West
photographs by Richard H. Lane
text by William A. Douglass

A Cup of Tea in Pamplona
Robert Laxalt

Sweet Promised Land
Robert Laxalt

Traditional Basque Cooking:
History and Preparation
José María Busca Isusi

Basque Violence:
Metaphor and Sacrament
Joseba Zulaika

Basque-English Dictionary
Gorka Aulestia

The Basque Hotel
Robert Laxalt

Vizcaya on the Eve of Carlism:
Politics and Society, 1800–1833
Renato Barahona

Contemporary Basque Fiction:
An Anthology
Jesús María Lasagabaster

A Time We Knew: Images of
Yesterday in the Basque Homeland
photographs by William Albert Allard
text by Robert Laxalt

English-Basque Dictionary
Gorka Aulestia and Linda White

Negotiating with ETA: Obstacles
to Peace in the Basque Country,
1975–1988
Robert P. Clark

Escape via Berlin: Eluding Franco in Hitler's Europe
José Antonio de Aguirre

A Rebellious People: Basques, Protests, and Politics
Cyrus Ernesto Zirakzadeh

A View from the Witch's Cave: Folktales of the Pyrenees
edited by Luis de Barandiarán Irizar

Child of the Holy Ghost
Robert Laxalt

Basque-English English-Basque Dictionary
Gorka Aulestia and Linda White

The Deep Blue Memory
Monique Urza

Solitude: Art and Symbolism in the National Basque Monument
Carmelo Urza

Korrika: Basque Ritual for Ethnic Identity
Teresa del Valle

The Circle of Mountains: A Basque Shepherding Community
Sandra Ott

The Basque Language: A Practical Introduction
Alan R. King

Hills of Conflict: Basque Nationalism in France
James E. Jacob

The Governor's Mansion
Robert Laxalt

Improvisational Poetry from the Basque Country
Gorka Aulestia

Karmentxu and the Little Ghost
Mariasun Landa

The Dancing Flea
Mariasun Landa

Bilbao's Modern Business Elite
Eduardo Jorge Glas